PENGUIN CLASSICS

THE DAWN OF MODERN COSMOLOGY

Aviva Rothman is Associat_____n
University. She is a histo_____es
on the early modern peri_____ *uit
of Harmony: Kepler on Cosmos, Confession and C_____.ity*
(University of Chicago Press, 2017).

The Dawn of
Modern Cosmology

From Copernicus to Newton

Translated and Edited by
AVIVA ROTHMAN

PENGUIN BOOKS

PENGUIN CLASSICS

UK | USA | Canada | Ireland | Australia
India | New Zealand | South Africa

Penguin Books is part of the Penguin Random House group of companies
whose addresses can be found at global.penguinrandomhouse.com

First published 2023
002

Translation copyright © Aviva Rothman

The moral rights of the author and translator have been asserted

Set in 10.25/12.25pt Sabon LT Std
Typeset by Jouve (UK), Milton Keynes
Printed and bound in Great Britain by Clays Ltd, Elcograf S.p.A.

The authorized representative in the EEA is Penguin Random House Ireland,
Morrison Chambers, 32 Nassau Street, Dublin D02 YH68

A CIP catalogue record for this book is available from the British Library

ISBN: 978-0-241-36063-7

Contents

Introduction

At the start of our story, in the late fifteenth century, the earth stood motionless at the centre of a relatively small, hierarchically ordered cosmos. Around it revolved the moon, sun and planets, each perfect and unchanging and embedded in crystalline spheres that moved in perfect, eternally unchanging circles. The earth, by contrast, was the realm of change, of growth and decay, generation and corruption. By the end of our story, in the early eighteenth century, the sun had replaced the earth at the centre, rotating on its axis as spots moved across its surface. The earth was a planet, just like the others, some of which had moons too, and, despite our commonsense perceptions to the contrary, it hurtled through an indescribably large cosmos at incredible speeds. The crystalline spheres had been shattered, the perfect circles swapped for ellipses, and the contrast between heavens and earth abolished in favour of a set of laws that united both.

These changes are commonly described as the Copernican Revolution, a story that conventionally begins with Copernicus himself and ends with Isaac Newton – though where one chooses to start and end the story, as well as whether it was a 'revolution' at all, is a matter of some scholarly debate. I will not use this Introduction to offer a specific argument about what cause lay behind the movement from geocentrism (where the earth was regarded as the fixed centre of the universe) to heliocentrism (where the sun is at the centre). In part that is because this story, like most periods of transformation in human history, is much more complicated than any single account might suggest. There were many distinct and diverse causes that lay behind the shifts in thought you will read about in the following pages. More importantly, though, the purpose of this book is to allow you to see things for yourself, as you follow the story from Copernicus to Newton in their own words, and in the words and images of their contemporaries. I will highlight some central themes and developments in this Introduction, to help orientate you as you begin, but as the story unfolds in these excerpts you will be able to see what kinds of arguments are made (both to support or to oppose the centrality of the sun and the motions of the earth), on what basis, with what motivation and to what effect.

At the same time, you will see a transformation not just in astronomical theories and scientific ideas, but in the ways in which science itself is discussed, practised and organized. At the start of our story, science was largely located in monasteries and universities. The development of printing in the mid-fifteenth century spurred important changes, and we will see how Copernicus and others took advantage of those changes to spread their ideas, through both the printed word and detailed diagrams and images. Alongside this, we will witness conversations and debates in letters, as a vast scholarly network developed that anticipated the scientific journals of the modern era and allowed scholars from across social and political boundaries to engage with new ideas and their consequences. We will likewise begin to see the blending of various kinds of knowledge, as scholars slowly incorporated ideas from other disciplines and embraced forms of learning traditionally located outside the university. We will see science come to be written in languages other than Latin, the traditional language of scholarship, and to include popular dialogues and works of fiction alongside the traditional treatise.

Much, though not all, of the actual story is still highly technical and mathematical, and in that sense the excerpts in this book are not fully representative. Though Copernicus, echoing Plato before him, opened his own book by warning the reader that technical expertise was a prerequisite – 'let no one untrained in geometry enter here', he wrote – this book does not presume such knowledge. Instead, I have chosen excerpts that should be accessible to any reader with some patience and interest and offer short introductory comments for each excerpt to help you get your bearings before you begin. Further reading and more technical detail are contained in the Appendix at the end of the book, where I list full translations of each text (when available), along with other works of interest.

Before Copernicus

To begin our story, we need to briefly consider the ancient conception of the cosmos and its behaviour, which would remain popular and influential well into the sixteenth century, albeit with some modifications.

Most ancient Greek theories of the world, including the especially influential cosmology of Aristotle in the fourth century BC, begin with a spherical, stationary earth at the very centre of the cosmos. Despite modern misconceptions to the contrary, the ancients did not believe in a flat earth, and pointed to a wide variety of evidence to support the earth's spherical shape, including the round shadow cast by the earth on the moon during a lunar eclipse. The cosmos, it was supposed, was also spherical and finite, and the motion of everything in the heavens was centred on the unmoving earth. All celestial bodies were embedded in three-dimensional spheres that carried them in their motions, and those spheres were ordered concentrically. First came the Primum Mobile, the outermost sphere that bounded the cosmos and spun every twenty-four hours, imparting motion to the spheres beneath it. The outermost visible sphere was that of the fixed stars (whose positions with respect to each other never changed), which rotated from east to west once every twenty-four hours, as did the sphere of the sun (positioned in the middle of the planetary spheres); this accounted for the daily motions of the stars and for day and night. The sun also had an annual motion, in which it moved much more slowly and eastwardly across the heavens through the ecliptic, a line that marked the sun's path along the familiar constellations of the zodiac. The planets, in separate spheres of their own, some above and some below the sun, shared the twenty-four-hour rotational motion of the other heavenly objects. They also wandered through the heavens, as their Greek name *planētēs*, or 'wanderer', indicates, though more irregularly than the sun and with varying periods of revolution. Next came the moon, which also moved around the earth as it shared the daily motion of the other heavenly bodies.

The moon was, in fact, the dividing line between two separate realms – the celestial and terrestrial – according to Aristotelian physics. (Physics was also referred to as natural philosophy, as the science of nature was considered one branch of philosophy.) Aristotle highlighted the periodic, regular and unchanging motions of the heavens to argue that the celestial and terrestrial realms were utterly different. The celestial realm was perfect, regular and unchanging; the terrestrial realm was changeable, corruptible and imperfect. He argued that the terrestrial realm was composed of four elements: fire, air, water, and earth. These naturally occupied concentric spheres below the moon, starting with fire and ending with earth, though most terrestrial things were some combination of these elements. Each of these elements had a natural motion in a straight line, either up for fire and air, or down for water and earth. This meant that objects composed of these elements naturally tried to move in that direction; objects made of earth naturally fell down, toward the very centre of the cosmos. The celestial realm was wholly composed of only one element – ether, or quintessence (the fifth element) – a perfect, incorruptible element whose natural motion was a constant speed in a perfect circle, something referred to as uniform circular motion. The spheres in which celestial bodies were embedded were thus made of ether, as were those bodies themselves.

Although Aristotelian physics insisted on perfectly uniform, circular motions for the celestial realm, ancient astronomers were aware that celestial objects did not, in fact, always appear to move in perfectly uniform, circular ways. From the perspective of the earth, they seemed, at times, to speed up and slow down. They did other strange things, too, like the planetary retrograde motions: the five planets (Mercury, Venus, Mars, Jupiter and Saturn) would sometimes seem to stop their motion from west to east and then move backward, from east to west, before stopping again and resuming their normal course.

Greek astronomers developed models to account for these observations. They did so with the general assumption that their models were mathematical and instrumental, but not physically real. Just as not all computer simulations today are understood to represent an actual underlying physical reality, astronomical models in antiquity were typically understood as computational devices that allowed astronomers to account for observed motions and predict future ones, but that did not reflect on or challenge the reality of Aristotelian physics. There were, after all, many different ways that one might mathematically model the same observed phenomena. If a particular model was accurate and useful for prediction, then it served its purpose, but this didn't mean that it represented the truth of things in the world. This perspective was bolstered by the fact that Aristotelian physics itself was not mathematical, for Aristotle did not believe that nature could be properly described quantitatively. When the disciplines were formalized and taught in the universities that arose in the eleventh and twelfth centuries, these divisions between natural philosophy and mathematics were enshrined in disciplinary distinctions, with astronomy generally understood as a mathematical discipline that offered computational models, and natural philosophy generally understood as a qualitative and causal discipline that explained why real things behaved the way they did.

The mathematical techniques that accounted for the complexities of planetary motion were described by Claudius Ptolemy in the second century AD (who himself *did* seem to believe in the reality of his models), though some of them had been advanced by his predecessors in earlier centuries. Ptolemy's *Mathematical Syntaxis* was written in Greek but came to be known to medieval Europeans in translation from Arabic, so it came to be known by a version of its hybrid Greek and Arabic title, the *Almagest* ('the greatest' treatise). In it, Ptolemy described several models that would be utilized by astronomers for centuries to come: the eccentric model, in which planets moved in perfect circles around a point not centred on the earth, called the eccentric; the epicycle-on-deferent model, in which planets were carried on small circles (epicycles) that moved with uniform circular motion around a larger circle (deferent); and the equant model, in which a planet moved uniformly (equal angles in equal time) with respect to a point not centred on the earth, called the equant. While these models, often in quite complex combinations, could account for planetary motions with great accuracy, to some they seemed to contradict basic principles of Aristotelian physics. However, the disciplinary divisions that separated natural philosophy and astronomy, and the instrumental approach to astronomy as a computational discipline that did not reflect an underlying physical reality, often meant that these contradictions were not viewed as problematic.

Often is not the same as always, of course. In fact, there were those, especially in the thriving astronomical centres of medieval Islam, who began to question some of the tenets of Ptolemaic astronomy and offer suggestions for improvement, arguing that astronomy needed to be construed in physical terms. The thirteenth-century astronomers Nasir al-Din al-Tusi and Muayyad al-Din al-Urdi and the fourteenth-century astronomer Ibn al-Shatir all developed mathematical devices that helped bring the physics of Aristotle, which insisted on the perfectly uniform and circular motion of planets, and the astronomy of Ptolemy, which offered models that departed from perfectly uniform and circular motion, closer together. They did so in part by developing ways to bypass models that relied on the problematic equant point, which described planets moving on spheres whose centre of uniform motion was not the physical centre of the sphere, something that seemed to them impossible. These would be important for Copernicus's own later models (see 'Tusi Couple', page 64). Ptolemaic astronomy, along with many of these later developments, was made available in modern, synthesized form to European astronomers in the fifteenth century with the publication of Peurbach and Regiomontanus's *Epitome of the Almagest* in 1496. This became the standard textbook from which astronomers would learn about Ptolemaic astronomy, including Copernicus himself.

Before moving on to Copernicus and his world, we need to note a few final things.

First, philosophers of the medieval Christian west updated the Aristotelian tradition by explicitly Christianizing it. They did so by rejecting certain aspects of Aristotelian philosophy, like the eternity of the world, and by associating some of the elements of Aristotle's world view with their own theological conceptions. Thus, they placed Hell, the most imperfect part of this imperfect world, at its very centre, and the realm of God and the angels at the outermost edges of the cosmos (see the *Nuremberg Chronicle*, page 2). They also developed a vision of nature which positioned it as one of God's two 'books' – the other being Scripture – which philosophers of nature read to better understand and appreciate God, who had created the cosmos and imparted motion to it.

Second, astronomy and astrology went hand in hand from antiquity throughout the medieval period, and well past the time of Copernicus. Often even the terms for astronomy and astrology were interchangeable. Almost all practising astronomers were also astrologers; Johannes Kepler, for example, while insisting that astronomy needed a physical foundation and developing his three laws, was also issuing horoscopes for his political patrons. In addition to its significance for navigation, astronomical data was particularly important for astrological predictions (which were also an essential part of medical practice), and astronomers relied on their observations and models to create books of tables that could be used to predict the future positions of celestial bodies and develop accurate calendars. Copernicus, in fact, highlighted the production of these astronomical tables as a central reason to adopt his theories, even for those who didn't believe they were true (page 29).

Third, the development of moveable-type printing in the west in the second half of the fifteenth century is a crucial part of our story. When Copernicus developed his theories, manuscript culture was still important, and ideas were often shared without being printed, like his early *Little Commentary* (page 4). But not long after the printing of Gutenberg's first copies of the Bible, works of astronomy were being printed, disseminated and read far more widely than would have been possible earlier. This was true not merely for newer works, like Peurbach and Regiomontanus's *Epitome of the Almagest* or Copernicus's *On the Revolutions*, but also for older works from the ancient world that had been previously unknown.

The body of knowledge available to readers after 1450 was therefore more expansive, more stable and more easily shared than that of their predecessors.

Copernicus and the Early Reception of Heliocentric Theory

As early as 1514, Nicolaus Copernicus, a church canon in Royal Prussia (part of the Kingdom of Poland) who devoted himself to astronomy on the side, suggested an alternative to the traditional picture of the cosmos. This alternative had not been seriously considered since antiquity – and was only a minority position even then. Copernicus argued that the celestial appearances could be better explained by moving the sun to the centre of the world system and putting the earth in motion around it, while giving the earth a diurnal (daily) rotation as well. He did so primarily on the basis of the existing record of astronomical data, which he reinterpreted, with few new observations of his own.

Copernicus did not describe the process by which he arrived at his theory (and there are multiple scholarly reconstructions of what it might have been – see the reading list in the Appendix for some of them). However, in his *Little Commentary* and in his *On the Revolutions of the Heavenly Spheres*, he did offer some arguments for why his theory was preferable to the traditional picture. The first related to the Ptolemaic equant model, in which planets only moved with constant apparent speed around a point not at the centre of the sphere that carried them. Like some medieval Islamic astronomers before him, this troubled Copernicus, who strongly believed that planets could only move with uniform circular motion, and that models using equants were physically problematic. This objection suggests that, unlike many other astronomers, Copernicus understood his system to be not just mathematically useful, but also physically real: he saw himself as offering a true picture of the cosmos and its structure.

Copernicus also noted that his new system offered a kind of simplicity that the older one could not: putting the earth in motion explained a host of celestial appearances, particularly the retrograde motion of the planets, that required separate and more elaborate explanations otherwise. Finally, Copernicus emphasized that in his system the order of the planets necessarily followed from the time it took them to make their journeys around the sun. (In the Ptolemaic system, the ordering of the sun, Mercury and Venus were unclear and a matter of some debate.)

While stressing his system's advantages, Copernicus tried to anticipate and respond to possible objections to his system; this gives us a sense of some of its potential shortcomings. First, there were physical problems. Recall that Aristotelian physics linked the central, immobile earth to the fact that objects made of earth fall downward. If the earth were no longer at the centre of the cosmos, then an entirely new physics would be necessary to explain the behaviour of falling objects, along with everything else that Aristotelian physics neatly explained in one comprehensive system. Copernicus gestured somewhat vaguely toward a physics in which multiple centres of heaviness were possible, but could not supply a fully-fledged counterpart to Aristotelian physics. Further, there was the problem of why the supposed motion of the earth was undetectable to the senses, or even how the earth could carry its atmosphere along with it as it spun. His response to the first was to suggest that the earth's motion was 'natural', rather than violent (which would require an external force). To the second, Copernicus argued that things adjacent to the earth 'conform to the same nature as the earth' and were carried along with it, though he could not explain how.

There were also serious observational problems with Copernicus's theory. If the earth was in motion, then this should be apparent in observations of the fixed stars, which should appear to change position relative to one another as the earth moved (this is known as stellar parallax). But stellar parallax had never been observed. Copernicus attempted to bypass this problem by suggesting that the cosmos was much larger than had ever been supposed, so large that stellar parallax would be undetectable. Yet the vastness required for this to be true might easily seem unbelievable (and required a great deal of unexplained empty space), and like a too-convenient sleight of hand (in other words: when lacking the required observational proof, simply make the universe bigger).

Finally, while it is true that the Ptolemaic system relied on a complex array of circles to accurately model the observed planetary motions, the Copernican system did as well. This was in part because unlike Ptolemy, who accounted for the seemingly non-uniform motion of the planets with equant points that effectively allowed them to move at unequal speeds, Copernicus insisted on strictly uniform, circular motion, an idea so entrenched in the traditional physical world view that he saw this as one of the central advantages of his theory. Though he was able to eliminate some of the complexities of Ptolemaic theory by moving the sun to the centre of the system, he also needed to add additional circles of his own to account for the missing equant points in Ptolemy's models. In sum, the models of Copernicus and Ptolemy were both quite complex mathematically, though Copernicus's was simpler insofar as the motion of the earth sufficed to explain many observed irregularities in one fell swoop.

In the end, then, Copernicus's model offered some real advantages, but also some very real disadvantages, particularly when it came to questions of physics, sense perception and observation. And there was simply no evidence that could settle the question in 1543. Both systems were broadly as good as one another at modelling planetary motions and predicting future behaviours.

Copernicus drafted his *On the Revolutions*, the book that outlined his theory in great mathematical detail, well before 1543, but seems to have been reluctant to publicize it, possibly because of the problems mentioned above. Georg Joachim Rheticus, a young mathematics professor from Germany, convinced Copernicus to publish the work – and indeed pre-emptively published a *First Account* of his own, from which many early readers first learned of Copernicus's theories (page 10). Rheticus delivered *On the Revolutions* to the printer but left the management of the publication in the hands of Andreas Osiander, a Lutheran theologian in Nuremberg, who saw the printing through to completion.

Osiander added to the start of the book an anonymous address to the reader (page 26), in which he situated Copernicus's ideas within the traditional disciplinary division which, as we have seen, confined astronomy to computation only, not the description of physical reality. 'These hypotheses need not be true nor even probable,' he insisted. 'On the contrary, if they provide a calculus consistent with the observations, that alone is enough.' While some early readers were quite clear that the author of this 'To the Reader' was not Copernicus himself, many did read the book exactly as Osiander suggested – as a mathematical model that might make calculations easier, but that need not change their conceptions about the reality of an earth-centred cosmos. Erasmus Reinhold, for example, built new astronomical tables for prediction based on Copernicus's mathematical models while remaining opposed to heliocentric theory (see page 66).

Though the later stages of this story – particularly the famous Galileo trial – might make it seem that religious objections would be a prominent part of the early reception of Copernicus's work, this was simply not the case. Copernicus's *Little Commentary*, the early manuscript in which he had briefly outlined his theory, was read by celebrated religious figures like Cardinal Nicholas Schönberg, archbishop of Capua, who urged him to publish his *On the Revolutions*, a book which Copernicus ultimately dedicated to Pope Paul III (see page 29). Copernicus himself only briefly discussed possible scriptural objections to his theory, and he did so by dismissing them out of hand, calling those who opposed his work on religious grounds alone 'babblers'.

Further, though the Catholic Church's response to the Reformation (in particular, its opposition to biblical reinterpretation) would become an important part of the story in later years, the religious atmosphere at the time of the Galileo affair was quite different from when Copernicus first published his work. And while Catholics and Protestants already viewed each other with suspicion and hostility, this did not prevent scholarly collaboration between the two camps, as evidenced by the two-and-a-half-year collaboration of the Catholic Copernicus with the Lutheran Rheticus; nor would it in later years, when the Catholic Galileo and the Lutheran Kepler joined forces to support the new telescopic observations.

Copernicus's *On the Revolutions* was, then, read fairly unproblematically by early readers, most of whom were interested in the computational uses of the book, while rejecting its physical claims. Nearly all these early readers were themselves serious practising astronomers. The book was long, complex and mathematically challenging: it was a book for experts, as Copernicus himself made clear at the start. The impact it made during the years immediately after its publication was largely technical; new tables were produced that relied on its models and mathematics, and when it was taught in universities, it was taught as one mathematical model alongside the other more traditional possibilities.

Post-Copernican Developments

The invention of the telescope in the first decade of the seventeenth century dramatically changed the astronomical landscape, but even before then some significant developments had already altered the way people viewed the cosmos. First, there were some startling and highly visible new objects that had appeared in the sky: a new star (in today's terminology, a supernova) in the constellation Cassiopeia in 1572, and a comet in 1577. Comets were already objects of highly questionable natures before this; though Aristotle had argued that they were terrestrial phenomena – fires in the atmosphere of earth – others were not so sure. Tycho Brahe, a Danish nobleman, was one astronomer who carefully observed both phenomena. His passion was highly accurate astronomical observation, made with instruments he perfected, and recorded over decades, first at his famous observatory, Uraniborg (see page 138), and later in Prague, where he served as mathematician to the Holy Roman Emperor. Tycho's instruments allowed him to measure the positions of the supernova and the comet, and he found that they were clearly celestial. Yet they were also distinctly new and changeable, far from the perfect, unchanging heavens that Aristotelian theory demanded. Tycho's own observations, along with those of others who observed these phenomena, thus served to slowly chip away at the Aristotelian cosmological world view, and they did so in parallel with Copernicus's theories (which positioned the earth itself as a celestial object).

In fact, Tycho considered Copernicus's heliocentric theory, which he read with interest but ultimately rejected. He was bothered by some of the factors noted above, and in particular the vast size required for a cosmos in which the earth moved but no parallax could be observed, even with Tycho's own highly accurate instruments. Yet Tycho also recognized some of the advantages of the Copernican system, like the way it linked the motions of the planets to the position of the sun. He responded by combining both the geocentric and heliocentric systems into a geo-heliocentric system, which would come to be called the Tychonic system. In this system, the earth remained unmoving at the centre of the cosmos, and the sun still revolved around it. However, the five planets revolved not around the earth, but around the sun (see page 108).

This system was mathematically equivalent to the heliocentric system and could produce the same results (this was true of the geocentric system as well, as we saw earlier). Yet in retaining some of the Copernican system's advantages while rejecting some of its disadvantages, it offered a highly attractive alternative to many in the late sixteenth and early seventeenth centuries. It did have a rather revolutionary consequence of its own, however. For the planets to move around the sun while the sun moved around the earth, the solid celestial sphere that carried Mars would have to intersect with the sphere that carried the sun. (The same was true of Tycho's analysis of the path of the comet of 1577 – it too crossed some planetary spheres.) This led Tycho to reject the reality of the solid, crystalline spheres in which, according to the traditional view, the planets were embedded. And so, although in broad disagreement, Copernicus and Tycho were both part of a process in which the Ptolemaic and Aristotelian systems of astronomy and physics were slowly called into question.

If the planets were not embedded in celestial spheres, how did they move? And if Aristotelian physics was no longer sufficient, what might replace it? William Gilbert, an English physician, speculated in the year 1600 that the earth was a giant magnet, and that this might explain the behaviours we see around us. He also linked the earth's magnetism to its diurnal rotation, which he supported, though he did not take a stand on its annual revolution (see page 140). Johannes Kepler was influenced by Gilbert's ideas at the very same time that he moved to Prague to collaborate with Tycho Brahe. Kepler was already a convinced Copernican in part through the influence of his teacher, Michael Maestlin, and in part because of his own Christianized Platonic beliefs about God, whom he thought to be a geometer and who had structured the cosmos to signify the Trinity, with the sun at the centre representing God the Father. He was interested in Tycho's data because he wanted to refine a theory he developed in his first book, which argued that the distances between the planets was linked to the five Platonic solids (see page 119). Tycho was interested in a mathematician who could use his astronomical findings to build models of the planets that might be used to construct more accurate planetary tables, and so he gave Kepler the data from Mars to work with.

After wrestling with that data for years, Kepler published his *New Astronomy* (page 151). It was subtitled a 'celestial physics' because Kepler insisted that, contrary to the traditional disciplinary divisions, astronomy needed to be physical if it was to be reliable. Inspired by Gilbert's idea of the earth as a magnet, Kepler suggested that the sun might be the source of a force akin to magnetism responsible for moving the planets. This motion was not aided by celestial spheres or orbs, but was simply along a path in the heavens – an orbit. Kepler argued that heliocentric theory was clearly superior to the two alternatives, geocentric and geo-heliocentric, because the idea of a solar force supplied a physical explanation for the motions which the other theories lacked. In working out the specific path of Mars with respect to the sun, he also broke with another traditional orthodoxy, one that even Copernicus had clung to – that celestial objects must move in combinations of perfect, uniform circles. Instead, Kepler argued that planets did not move in circles, but rather in ellipses, with the sun at one focus; this would come to be known as his first law. His second law, also developed in that book, was that planets swept out equal areas in equal times as they moved around the sun, sometimes moving more quickly and sometimes more slowly depending on their distance. This reintroduced the Ptolemaic equant point, but transformed it from a mere calculational technique, as Ptolemy had used it, to a real physical effect centred in the sun.

Kepler published his *New Astronomy* in 1609, one year after a Dutch lens maker announced that he had invented a device that made distant objects appear closer. Galileo Galilei was among the first to devise an improved version of the Dutch invention and point the 'telescope' at the heavens. His discoveries with it were astonishing, and he announced them in his 1610 *Starry Messenger* (page 176). He first revealed that the surface of the moon was much like the surface of the earth: filled with mountains and valleys, and far from the perfect, incorruptible space it should have been according to Aristotelian theory. He also noted that there were far more stars in the cosmos than were apparent to the naked eye, and that they were further away than had been supposed. Finally, he revealed four previously unseen moons orbiting Jupiter, which he named the 'Medicean stars' after the powerful Medici family, who gave him a position as court philosopher and mathematician in return.

One powerful objection to Copernican theory had been the following: why was it that everything revolved around the sun except the moon, which instead revolved around the earth? And how was it even possible to have two centres of revolution in the cosmos? Galileo proved that the earth was not alone in having a moon, and that it was possible to have two centres of revolution, since Jupiter clearly revolved around something, whether the earth or the sun. Galileo himself was already a convinced Copernican, in part because his studies on terrestrial motion had led him to oppose Aristotelian philosophy, and in part for physical reasons (for instance, he believed, incorrectly, that the tides were proof of the earth's motion). And while none of the discoveries revealed in the *Starry Messenger* proved that heliocentric theory was correct, they did remove some obstacles in its path, while casting serious doubt on the Aristotelian view of the cosmos.

After the publication of the *Starry Messenger*, Galileo made two further important discoveries. First, his telescope revealed sunspots that moved irregularly across the surface of the sun (see page 226). These, to Galileo, proved that the sun, like the moon, was not the perfect body Aristotelian theory supposed. He also used the sunspots to argue that the sun rotated, thus providing further support for the idea of the earth's own diurnal rotation. Still more significant were his observations of the phases of Venus (the sizes and shapes of Venus that we see illuminated from earth, much like the phases of the moon), which he revealed in 1613 (see pages 219 and 336). Venus should have phases according to both geocentric and heliocentric theories, but these would differ in appearance depending on whether Venus circled the earth or the sun. Galileo's observations proved clearly that Venus circled the sun, and not the earth. These observations were confirmed by others, and this was a conclusive and devastating blow to Ptolemaic theory.

By 1613, then, traditional Ptolemaic theory was no longer an option for serious astronomers. This did not mean, however, that Galileo's observations proved Copernican theory. In the geo-heliocentric theory of Tycho Brahe, Venus did orbit the sun, which then orbited the earth. This meant that Galileo's observations were consistent with both the Copernican and the Tychonic systems. As most astronomers remained sceptical about Copernican theory because of the problems we've already mentioned, the Tychonic world system became the one to which the majority in the field adhered.

Aside from the larger contest between the geo-heliocentric (Tychonic) or heliocentric (Copernican) systems, there were a range of ideas that serious thinkers still debated after 1613. It was possible to deny the earth's annual revolution but still accept its diurnal rotation, like William Gilbert. It was possible to accept a sun-centred system but believe it was finite and bounded by the fixed stars, like Copernicus and Kepler, or to insist that the universe was infinite and unbounded, like Thomas Digges (page 84) and Giordano Bruno (page 89). It was possible to still debate the nature of the celestial spheres; while Tycho Brahe seriously challenged their solidity, there was no clear alternative answer to how

the planets moved, though Kepler, via Gilbert, offered a magneti-
cally related solar force as one option. It was possible to believe
in a heliocentric cosmos but reject Kepler's ellipses. There were
also variations on the Tychonic theory, in which only some, but
not all, of the planets circled the sun.

Crucially, there was little proof available to settle these questions. The phases of Venus constituted proof against a strict
Ptolemaic system, but that was all. Proponents of the various
systems made arguments of probability, or likelihood, on the
basis of differing values: systematicity or simplicity, sometimes,
for the Copernican system; commonsense evidence and biblical
precedent, sometimes, for the Tychonic. And it is not possible
to understand the subsequent Galileo affair without keeping
this distinction between proof and probability in mind.

The Galileo Affair

Galileo's famous clash with the Catholic Church is much more
complex than many of the traditional accounts would suggest.
While we do not have the space to fully unravel all the factors
at play here, there are a few worth highlighting in the context
of the present story. Galileo's initial discoveries were confirmed
and even celebrated by the Jesuit astronomers of the Catholic
Church, though they interpreted them to favour the Tychonic
world system rather than the Copernican. Galileo's particular
troubles with respect to his advocacy for the Copernican system began with his letter to Benedetto Castelli (page 228). There,
he argued that certain biblical passages that seemed to support
a stationary earth or a moving sun (such as Joshua's asking the
sun to stop its motion in order to lengthen the day) could be
reinterpreted to fit within a heliocentric framework.

Galileo's support for Copernican theory was less problematic than his reinterpretation of the Bible, given the reactionary
post-Reformation world in which he lived. With his doctrine of
sola scriptura, Martin Luther, the critical figure of the Protest-
ant Reformation, had argued that the Bible itself was the sole
and unmediated authority for believers. During the Council of

Trent, Catholic theologians responded to Luther by insisting that only the Church's traditional interpretations of the Bible were authoritative, and that no lay person was allowed to use their own judgement to reinterpret Scripture. Galileo's interpretive efforts therefore raised alarm bells for the Church authorities, as did the letter of Paolo Antonio Foscarini (page 244) which likewise focused on biblical reinterpretation in support of Copernican theory.

In response to all this, Cardinal Robert Bellarmine wrote a letter to Foscarini, but also directed at Galileo, in which he referenced the Council of Trent's decision and warned against overly strong support of Copernican theory (page 262). The traditional interpretation of the Bible, he explained, supported a moving sun and a stationary earth. As such, it was not a problem to consider heliocentric theory as an instrumental hypothesis that might help explain the planetary motions (much as Osiander had suggested in his 'To the Reader' at the start of Copernicus's own book). But it was a problem to insist that the theory was a true and real description of the cosmos. Bellarmine noted that if there were proof that the cosmos was truly heliocentric, then biblical passages would need to be reinterpreted accordingly. But since there was not, the traditional interpretations should be embraced.

At the same time, the Inquisition convened a committee of eleven expert consultants to review the Copernican doctrines. In February 1616, that committee unanimously declared that the doctrine of the sun's centrality was 'foolish and absurd in philosophy and formally heretical', and the doctrine of the earth's motion 'at least erroneous in faith' (see page 276). Galileo was subsequently warned by Bellarmine to avoid promulgating the theory. The next month, *On the Revolutions* and Diego de Zúñiga's commentary on the Book of Job (page 278) were placed on the Index of Forbidden Books until certain problematic passages were removed, and Foscarini's letter was banned entirely (see page 244).

The next stage of the Galileo affair centred on the publication of his *Dialogue on the Great World Systems* in 1632 (page 344). Galileo seems to have received permission from Pope Urban

VIII, a friend and supporter of Galileo's from the days before he became pontiff, to publish a book that considered Copernican theory. This was so long as he did so only hypothetically, and if he demonstrated that there were other options that were also possible. The fact that Galileo chose to do so by contrasting only 'two' great world systems, the Ptolemaic and the Copernican, while leaving out the popular Tychonic theory, is telling. It was fairly clear to readers that Galileo's discussion of geocentric theory was perfunctory at best, or even dismissive, and he structured the discussion to clearly support the heliocentric system (see Campanella's letter, page 388). Further, Galileo placed the pope's words about how both systems were equally possible at the very end of the book, in the mouth of a character who acted rather like a simpleton and whose name, Simplicio, made that fact very clear.

The trial revolved, in part, around whether Galileo had followed instructions and truly considered Copernicanism only hypothetically. It also raised the question of just what such a 'hypothetical' approach meant. For the pope, as for Bellarmine before him, it seems to have meant an instrumental approach, rooted in the old idea of astronomy as a discipline devoted to computation but not truth. For Galileo, by contrast, it suggested probability, rather than certainty. And though Galileo himself believed that the proof of Copernican theory could be found in the tides (he mocked Kepler for correctly suggesting that they were caused by the moon), he spent much of the book amassing a variety of evidence for the heliocentric system in order to argue that, at the very least, it was far more probable than the Ptolemaic alternative.

New Physics and the Newtonian Synthesis

Galileo was found guilty and sentenced to house arrest for the rest of his life. This sentence had some significant consequences relevant to the rest of our story, for Galileo resumed the work he had put aside when he had earlier turned to astronomy: a study of terrestrial physics.

Moving the earth away from the centre of things shook the foundations of Aristotelian terrestrial physics, as we've seen. The shattering of the celestial spheres likewise shook Aristotelian understandings of celestial physics, which explained the motion of heavenly objects by embedding them in those spheres. While observational discoveries increasingly challenged the Aristotelian world view and seemed, to an increasing number of scholars, to point toward a Copernican one, a new physics was needed to sustain it.

Aristotelian physics had been challenged before the time of Copernicus, of course. The nature and reality of the celestial spheres had been a longstanding question, and Aristotle's ideas about terrestrial motion had also long been a subject of dispute and had undergone revision already by medieval philosophers. But the debate over world systems lent new urgency to these inquiries. Galileo himself had turned to these questions much earlier, and it was the problem of terrestrial motion that first turned him against Aristotelian physics. To figure out how to understand the problem of falling objects, he had rolled balls down a channel carved into a wooden beam that he placed at an incline and measured their fall with the aid of a water clock. Based on these experiments, he argued that the distance travelled in naturally accelerated motion was proportional to the square of the time. He also investigated projectile motion and argued that it was a combination of straight and circular motion, which he specified was parabolic. Further, Galileo began to develop an idea of inertia, in which it was change of state, rather than either rest or motion, that required explanation. All this was significant not merely for its specific challenges to Aristotelian theory (which had insisted, for example, that objects could only move against their natural motions while being subjected to an external force), but also for its mathematization of terrestrial physics, which had previously been discussed in largely qualitative terms. While under house arrest, Galileo finally published all these ideas in his *Discourses and Demonstrations on Two New Sciences* (page 430).

Galileo's guilty sentence sent shock waves throughout Europe; the young Descartes, hearing about it, decided to pull his newly written *The World* from publication because it too supported a heliocentric viewpoint (see page 426). But Descartes did continue to work on his world system, while developing a physics of his own that would support it. Unlike Aristotle's physics of natural motions, Descartes's 'mechanical philosophy' specified a cosmos completely filled with particles in constant states of motion. All change, in his view, required direct contact of one object with another. Descartes's cosmos was filled with fluid-like, particle-filled vortices whose motion and collision explained the behaviours we observe in both the terrestrial and celestial worlds, from objects on earth to planets in the heavens (see page 419). In the course of his work, he articulated three rules, or laws, of motion, relating to inertia, natural rectilinear motion, and the behaviour of collisions, that he understood to apply everywhere.

William Gilbert had earlier suggested another way of thinking about the physics of the world, based on magnetism rather than mechanical contact. And Kepler had applied that idea to the force that moved celestial objects, which he located in the sun. Kepler had also articulated three laws of planetary motion that he derived empirically, though they had not yet been proved mathematically. Isaac Newton, born on Christmas day in the year of Galileo's death (although calculated by two different calendars), would synthesize the work of his predecessors in his *Mathematical Principles of Natural Philosophy*, or, as it is now more widely known, the *Principia* (page 524). Newton positioned the book directly against Descartes's own *Principles of Philosophy*, not only by centring mathematics as the cornerstone of natural philosophy but also by arguing against Descartes's mechanical model of direct contact in a vortex-filled cosmos. Though he adopted variants of the principles Descartes had articulated in his own laws of motion, Newton argued that the force behind natural motion was not direct contact, but rather gravity, which acted between two bodies at a distance.

That gravitational force, Newton showed, was a universal one, which retained the planets in their orbits around the sun and caused objects to fall on earth. Newton's *Principia* therefore not only explained, in mathematical terms, the physical causes for the heliocentric theory that Copernicus had first articulated, and Kepler had later developed and refined, but also united that celestial explanation to the terrestrial physics developed by Descartes, Galileo and others. Newton also provided mathematical proofs for Kepler's laws of planetary motion, showing why planets moved in ellipses and how their orbits obeyed an inverse square law. He applied his theory to the motions of the planets as well as their satellites and the paths of comets, showing that all obeyed the very same law of universal gravity. Newton, then, is often where this story ends, because his groundbreaking work was able to synthesize that of those who came before him, offered a mathematical grounding of Copernican theory, and united the physics of the heavens and earth into one large system that obeyed the same laws everywhere.

Newton's picture of the world was not immediately accepted, however. He could not explain what the force of gravity actually was or how it physically operated – only that it did and that he could describe it mathematically and demonstrate it empirically. In responding to criticism in later editions of the *Principia*, Newton argued that this was not a problem (see page 540): 'I have not as yet been able to deduce from phenomena the reason for these properties of gravity, and I do not feign hypotheses,' he wrote. 'It is enough that gravity really exists and acts according to the laws that we have set forth and is sufficient to explain all the motions of the heavenly bodies and of our sea.' Others were not so sure, particularly those who followed Descartes's version of mechanical philosophy and believed that forces that mysteriously operated at a distance were 'occult': hidden and non-scientific.

Yet the vast, universal nature of Newton's theory, along with its mounting successes in agreeing with known phenomena and predicting new ones (like the return of Halley's Comet, and the existence of the planet Neptune) led to the gradual acceptance of the Newtonian world view. This was further aided by parallel (and often intertwined) changes in ideas about what science was for and how it was supposed to operate. By the time of Newton, science had come to be seen as an enterprise that offered probable, rather than certain, explanations; it was a mathematical and experimental pursuit, rather than something qualitative and contemplative; it explained how things behaved, rather than why or for what purpose. And though Christian theology and science still often went hand in hand (as is clear in the General Scholium to Newton's *Principia* (page 540), where Newton insists that 'to treat of God from phenomena is certainly a part of natural philosophy'), interpretations of the Book of Nature had grown apart from those of the Book of Scripture. Rational interpretation of nature might point to God, but it was mathematics and experiment, and not Scripture itself, that would reveal what that nature was like. And there was now widespread agreement that it was Copernican.

There was still, of course, no proof of either the annual revolution of the earth around the sun or its diurnal rotation on its axis. Nobody could yet detect any evidence for stellar parallax, though Robert Hooke tried, unsuccessfully, to prove that he had (page 486). Successful detection of stellar parallax was not achieved until 150 years after the publication of Newton's *Principia*. In 1838, Friedrich Bessel used a heliometer to measure the motion of the star named 61 Cygni against two comparison background stars, and – after taking hundreds of position measurements from thousands of observations – determined that the star seemed to move slightly in the opposite direction of the earth's motion. It was only then that astronomers obtained definitive proof that the earth moved. Finally, in 1851, Léon Foucault suspended a pendulum from the dome of the Pantheon in Paris, over a wooden platform filled with sand. As it swung back and forth, the pendulum traced lines in the sand whose angle changed over time as the earth rotated beneath it. And so, three hundred years after Copernicus published his *On the Revolutions*, the earth's diurnal rotation was finally proved.

Genre and Language

So far, I've charted the basic contours of the ideas behind our story. But this story is not only about ideas, but also about the forms that those ideas take. The selections included here span a wide variety of genres, many of which are quite different from the technical, impersonal scientific journal article you would expect to read today. Even Copernicus's *On the Revolutions*, a fairly traditional astronomical treatise in its form, if not its content, was prefaced by elaborate, dedicatory letters that set it apart from the scientific writing of our own time. That book then spurred works in a profusion of other genres in which the new astronomy was developed, argued and spread.

Letters, first, functioned as a crucial scientific genre in this period. Before the first journals were established in 1665, letters were one way to disseminate new scientific findings quickly, easily and at minimal expense. Though addressed to a single individual, they were often written with the expectation that they would be shared and read by a much broader audience than their stated recipient. Sometimes, letters were even published without changing their form, if speed was a primary concern (see, for example, Kepler's letter in support of Galileo's telescopic discoveries on page 189). Letters were also a way for astronomers or natural philosophers to show support for one another (page 135), ask for advice (page 426), debate ideas (page 226), share data, compare interpretations, and work collaboratively, even when separated by great distances, national borders, religious affiliations, or social positions. They could be used to assert one's priority of discovery, like the anagrams in Galileo's letters to Kepler, published by Kepler in the *Dioptrics* (see page 203). They could also be used as a tool of persuasion, addressing a particular individual in order to show them why their intellectual reservations were unwarranted, like Galileo's letter to Grand Duchess Christina (page 266).

Dialogues, too, were an important genre for those writing about the new astronomical ideas. Dialogues had ancient origins in the works of Plato and of Cicero, and thinkers like Bruno and Galileo embraced the form and put it into the service of the new cosmologies they supported (see pages 89 and 344). As fictitious conversations between several interlocutors who hold different points of view, they could serve a number of purposes. If a theory was controversial, an author could put a challenging idea in the mouth of one of the characters, have another character disagree with it, and so shield his own belief.

But dialogues could also allow authors to show the many sides of an idea, or the many ways it might be interpreted; they could model the process of scientific exchange and agreement, rather than just highlighting the conclusion. Finally, they allowed for the use of rhetorical techniques that might persuade readers in ways that a treatise or commentary might not. Though dialogues were structured, in some ways, like plays, and were often enjoyable to read, they could contain complex and novel ideas and were not necessarily directed at a popular audience.

At the same time, though, the new astronomy did slowly make an appearance in works of popular culture. Here, it's worth stepping back for a moment and recalling that only fifty years before Copernicus published *On the Revolutions*, travellers from Europe discovered what they called a 'new world' filled with lands, peoples, plants and animals that stunned all those who read about them back home. Once undreamed-of, these discoveries encouraged bold new ideas of what might be waiting around the next bend. When Galileo's telescope revealed new worlds in the heavens, the imaginations of those who read his *Starry Messenger* immediately caught fire, often in the form of fictional journeys that imagined the kinds of places these strange new worlds might be, and what sorts of inhabitants might live there. Francis Godwin's *The Man in the Moon* (page 444) is a representative of this kind of astronomical fiction, and his book was an important way that the ideas of Galileo, Kepler and others spread from the scholarly world to the larger public. Those ideas also showed up in poems (e.g., that of John Donne on page 216) and Milton's great epic *Paradise Lost* (page 478), in which Adam and the angel Raphael discuss the merits of geocentric versus heliocentric theory.

Another way that the new ideas spread pertains not to genre, but to language. At the start of our story, science was written in Latin, the language of the medieval universities, of the Church, and of scholarship in general. This universal language of learning enabled communication across national and religious boundaries. Something written in Latin could be easily understood by men of learning anywhere, who could then correspond in Latin to discuss what they had read, and even converse in Latin if they happened to meet. Yet despite the use-fulness of this universal language, over the course of our story we see authors increasingly choose to rely on their native ver-naculars, although still often choosing to write in Latin too. Galileo wrote his *Starry Messenger* in Latin but his *Dialogue on the Two Great World Systems* in Italian. He wrote his *Dis-courses on the Two New Sciences* in Italian, but only the conversations between the three interlocutors: when they quote the 'book' they are reading by a famous author (Galileo him-self), the text is in Latin. Descartes wrote some of his books in French and some in Latin, and Newton some in English and some, like his *Principia*, in Latin.

The move to the vernacular was paralleled, and sometimes spurred, by a number of factors. While science had been studied and produced primarily in universities during the medieval era, during our story we see a move from universities to other sites, princely courts among them. Galileo named the moons of Jupi-ter the Medicean stars to seek support from the powerful Medici, from whom he obtained a court appointment. Kepler replaced Tycho Brahe as Imperial Mathematician to the Holy Roman Emperor in Prague; Descartes corresponded with Prin-cess Elisabeth of Bohemia and served at the court of Queen Christina of Sweden; Newton served as Warden and Master of the Mint in England. These courtly connections indicate that knowledge, particularly of something new and exciting, could give patrons valuable political prestige. And while it wasn't always accompanied by a move to the vernacular, it did indicate a tightening of the links between nation and knowledge: once universal, knowledge came increasingly to be seen as relevant to the state. The new scientific societies, both in England and in France, were likewise sponsored by the reigning monarchs, and their new journals were published in their native vernaculars.

Science written in the vernacular sometimes needed to be translated back into Latin to make it broadly accessible across national boundaries (see, for example, Kepler's translation of Galileo's letters from Italian to Latin in the preface to his *Diop-trics*, page 203). Yet if Latin was broadly accessible to men of learning who did not share a native language, this did not mean it was accessible to everyone. Books written in vernaculars opened science up to new audiences not as proficient in the language of learning: to people of lower classes who didn't always have a robust education, as well as to women. As Copernican astronomy was translated into vernacular languages, it spread to a different and broader range of people than those university scholars who first read and debated *On the Revolutions*. The first partial English translation of Copernicus, by Thomas Digges in 1576, helped introduce the idea of heliocentrism to the larger public sphere (see page 73). Indeed, one reason that the Catholic Church worried about the ideas in Foscarini's Letter and Galileo's *Dialogue* is that both were written in Italian, and so accessible to an audience the Church felt was unschooled and more susceptible to dangerous ideas. The books of John Wilkins and of Godwin, discussed above, were particularly popular and influential because they were written in English. The same was true of Bernard Le Bovier de Fontenelle's popular vernacular dialogue on heliocentric theory, the *Conversations on the Plurality of Worlds* of 1686 (see page 496). A work of popular science in dialogue form, the book was written in French and widely admired. It featured Fontenelle in conversation with an intelligent woman, 'the Marquise', and in the preface Fontenelle addressed female readers specifically. In fact, his book was translated two years later into English by the female playwright Aphra Behn, illustrating yet again both the spread of these ideas and the slow inclusion of women in the Copernican conversation.

Another important genre in this story is the image, which could serve a variety of purposes in astronomical texts. Diagrams could have pedagogical value, illustrating in a simple and memorable way an author's primary argument (see Copernicus's image on page 62), or simplifying and explaining mathematical concepts that were hard to grasp without visual aids (see page 369). They could also be a way to include the reader in the process of discovery, as when Galileo shared images of his daily observations of the objects he had observed near Jupiter (page 186), using sketch after sketch to help the reader understand how he came to interpret them as moons.

Galileo's other images in the *Starry Messenger* also reveal that science and art could be closely interconnected in this period. The illustrations of the moon, in particular (see page 181) – which Galileo himself prepared as wash drawings and were then copied by the engraver – show the ways in which Galileo relied on artistic training in perspective and shading to highlight the light and dark fields on the moon, and the craters he had seen there. These images and others that illustrated a cosmos that extended infinitely (see page 84) or one filled with vortices with a sun or star at each centre (see page 464), taught viewers to see the heavens differently than they had before. Finally, frontispieces – the images that readers would encounter when they first opened a book – are also an important part of this story. They often functioned as highly detailed arguments, whether demonstrating the approach of an author in the particular book at hand (see the frontispiece to Galileo's *Dialogue*, page 342), to the discipline as a whole (see, for example, Riccioli's frontispiece to the *New Almagest*, page 472), or to the author's self-conception as part of a larger tradition (see Kepler's frontispiece to the *Rudolphine Tables*, page 338). And like vernacular languages, images could often reach a different, broader audience than the text itself.

Coda: The 'Copernican Revolution' as Metaphor

The Copernican Revolution was not just Copernican, as we've seen; it was also Keplerian, and Galilean, and Newtonian, and the work of many others besides, some of whom contributed to it long before Copernicus was born, and some long after Newton died. Further, some aspects of Copernicus's theory were highly traditional, like his emphasis on perfect circles or his belief in the solid celestial spheres. Yet there has been something about Copernicus and his idea of a moving earth, against all beliefs and perceptions to the contrary, that has struck a chord in the popular imagination, one that has lasted and remained linked to his name in the centuries after he published *On the Revolutions*. Others, who later saw themselves as responsible

for similarly world-shattering shifts, referred back to Copernicus to describe or explain their own achievements.

One such figure was Immanuel Kant (1724–1804). Before
Kant, philosophers had viewed the human observer as passive
and irrelevant to the object being observed. Kant, by contrast,
made the observer the active and central condition of knowledge. This reversal of observer and observed – one that seemed
counterintuitive and contrary to sensory evidence, for many –
was one that Kant explicitly paralleled to Copernicus's reversal
of the stationary and moving earth. As Kant explained in the
preface to the second edition of his *Critique of Pure Reason*
(1787), 'We here propose to do just what Copernicus did in
attempting to explain the celestial movements. When he found
that he could make no progress by assuming that all the heavenly bodies revolved round the spectator, he reversed the
process, and tried the experiment of assuming that the spectator revolved, while the stars remained at rest. We may make the
same experiment with regard to the intuition of objects.' Kant's
philosophical approach has thereafter been referred to by philosophers as his 'Copernican revolution'.

More than a century later, Sigmund Freud (1856–1939)
argued that Copernicus had started a revolution that Freud himself had completed. In Freud's view, the work of science, while
revelatory, had dealt some deep blows to humans' sense of their
own privileged status. In his *Introductory Lectures* of 1916–17,
Freud explained that in the past, the 'naive self-love of men' had
been dealt two major blows. 'The first,' he argued, 'was when
they learnt that our earth was not the centre of the universe but
only a tiny fragment of a cosmic system of scarcely imaginable
vastness. This is associated in our minds with the name of Copernicus.' The second he credited to Darwin, who proved that
humans were not the pinnacle of creation, but instead descended
from the animal kingdom. Finally, he argued, 'human megalo-
mania will have suffered its third and most wounding blow from
the psychological research of the present time which seeks to
prove to the ego that it is not even master in its own house, but
must content itself with scanty information of what is going on
unconsciously in the mind.' As Copernicus displaced humans

from the centre of the cosmos, so Freud sought to displace consciousness from the centre of human activity. Because these moves destroyed illusions that so many held dear, they were attacked and resisted, and Freud claimed the Copernican mantle of the subversive iconoclast with pride.

Kant's Copernican revolution centred on the proper use of reason; Freud's on reason's inevitable failure. Both saw Copernicus and his work as worthy of emulation, yet this meant very different things to them. The same was true of the many figures you will encounter in this book who took up *On the Revolutions* in the two hundred years after it was written: to embrace it, modify it, or argue against it. As you read this book, you will see how our world slowly transformed from one in which we stood motionless at the very centre of creation to one in which we found ourselves moving around one small sun among a near infinitude of others. Revolutionary though these changes may seem, they were only the dawn of things, for the world would be unmoored yet again after Newton, as space and time themselves began to lose their traditional meanings and the deterministic physics that operated throughout the cosmos seemed to crumble at the microscopic level. But that is another story.

A Note on Translation

All translations are my own, aside from the following licensed translations, also noted after each relevant selection:

Three Copernican Treatises, trans. Edward Rosen (Dover, 1959).
On the Revolutions (*Nicholas Copernicus Complete Works, Volume 2*), trans. Edward Rosen (Johns Hopkins University Press, 1978).
Selections from Kepler's Astronomia Nova, trans. William H. Donahue (Green Lion Press, 2005).
The Galileo Affair: A Documentary History, trans. Maurice A. Finocchiaro (University of California Press, 1989).
Dialogue Concerning the Two Chief World Systems, Ptolemaic and Copernican, by Galileo Galilei, trans. Stillman Drake, 2nd edn (University of California Press, 1967).
The Principia: The Authoritative Translation: Mathematical Principles of Natural Philosophy, by Isaac Newton, trans. I. Bernard Cohen and Anne Whitman (University of California Press, 1999).

For my own translations, I have aimed at the general reader and have opted for language as straightforward and clear as possible. This is also why I have avoided footnotes, even when translations of particular words can be tricky, as other translators have discussed in some detail. *Orbis*, for example – one of the words in Copernicus's own title – can mean a three-dimensional, solid sphere on which the planets move, or a two-dimensional circle, or – by the time we get to Kepler – an orbit, the path a planet takes as it moves through space, based on distant but physical causes. In part this is a story of transformation, as the word itself changes its meaning, but at times it is quite difficult to decide which meaning individual authors intend. Further, some Latin words simply have no English equivalent; in those instances, I have tried to get as close as possible, like my use of 'immaterial emanation' to translate Kepler's Latin *species*. In most cases these translation difficulties or transformations are invisible to the reader, and I hope that the benefits of a clear and easy reading experience make up for that loss.

For works originally written in English, I have slightly modernized some of the spelling but have otherwise left them in their original form. Many of the excerpts are highly abridged, though I have chosen not to include the ellipses indicating the places of abridgement, hoping again for a smooth reading experience. Many, though not all, of the texts I have translated here are available in translation elsewhere, often in full and usually with a scholarly explanatory apparatus. In the Appendix at the end of the book I have provided a list of those translations so the interested reader can delve into each selection more thoroughly. Likewise, while I have avoided footnotes in the Introduction, many of the ideas I've discussed can be found in the many excellent books included in the 'General Overviews' section of the Appendix.

Timeline of Key Events

1450s Gutenberg prints the first copies of the Bible

1473 Copernicus born

1492 Columbus miscalculates the size of the Earth and ends up in the Americas

1496 *Epitome of Ptolemy's Almagest* by Peurbach and Regiomontanus appears

1514 Copernicus's ideas begin to circulate through his *Commentariolus* (*Little Commentary*) manuscript

1517 The Reformation begins with Luther's publication of his Ninety-five Theses

1522 Magellan's fleet circumnavigates the globe

1539 Georg Joachim Rheticus travels to study with Copernicus

1540 Rheticus publishes his *First Account* of the Copernican system

1543 Publication of *On the Revolutions*; death of Copernicus

1545 Council of Trent begins

1546 Tycho Brahe born

1551 Erasmus Reinhold publishes the *Prutenic Tables*

1551 The Jesuit Collegio Romano is founded

1563 Council of Trent ends

1564 Galileo Galilei born

1571 Johannes Kepler born

1572 Tycho's star, a bright supernova, appears in the sky

1576 Thomas Digges publishes a partial translation of Copernicus in English

1577 A bright comet appears and is observed by Tycho and others

1582 Gregorian calendar instituted by Pope Gregory XIII

1588 Tycho Brahe publishes his geo-heliocentric model

1596 René Descartes born

1600 Giordano Bruno burned at the stake in Rome

1600 Kepler begins to work for Tycho Brahe in Prague

1600 William Gilbert publishes *On the Magnet*

1601 Tycho Brahe dies

1608 Telescope invented in the Netherlands

1609 Kepler publishes *New Astronomy*

1609 Galileo constructs his first telescope and begins to observe the heavens

1610 Galileo publishes *The Starry Messenger* describing the moons of Jupiter

1616 *On the Revolutions* is placed on the Index of Forbidden Books until corrected

1619 Kepler publishes *The Harmony of the World*

1630 Kepler dies

1632 Galileo's publishes the *Dialogue on the Great World Systems*

1633 Galileo's trial and abjuration

1638 Galileo publishes the *Two New Sciences*

1642 Galileo dies; Isaac Newton born

1644 Descartes publishes the *Principles of Philosophy*

1651 Riccioli publishes the *New Almagest*

1665 First scientific journals, the French *Journal des sçavans* and the English *Philosophical Transactions of the Royal Society of London*, established

1686 Fontenelle publishes the *Conversations on the Plurality of Worlds*

1687 Newton publishes the *Mathematical Principles of Natural Philosophy*

1727 Newton dies

1838 Friedrich Bessel makes the first successful stellar parallax measurement, demonstrating the earth's orbital motion

1851 Foucault's pendulum experiment demonstrates the earth's rotational motion

THE DAWN OF
MODERN COSMOLOGY

Hartmann Schedel, Seventh Day from *Nuremberg Chronicle*, 1493

The Nuremberg Chronicle, *one of the most famous early printed books, was a history of the Christian world from its creation as described in the Bible to the time of its printing in Nuremberg in 1493. It was lavishly illustrated with woodcuts, including this image of the seventh day of creation, below. Here, the Aristotelian world view has been merged with the Christian. We see God, on his throne, resting from his creative work and surrounded by a choir of angels, with the four winds depicted at the corners. Beneath God's feet is the Aristotelian cosmos that he set in motion. At its centre is the stationary earth, surrounded by the other elements of water, air and fire. The moon, sun and five planets follow, as do the fixed stars and the Primum Mobile, the sphere that provides motion to the rest. All the heavenly bodies are embedded in solid crystalline spheres, which turn in eternal and unchanging circles around the earth. The average reader would have believed something much like this to be a true picture of the cosmos in 1493.*

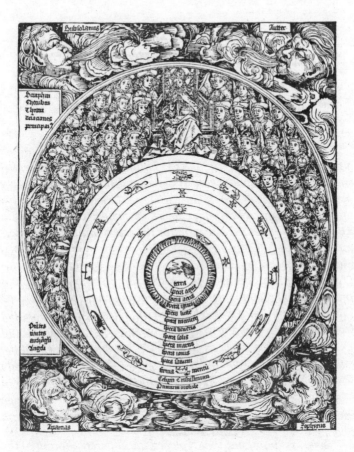

Nicolaus Copernicus,
Little Commentary, 1514[*]

Nicolaus Copernicus (1473–1543) studied astronomy, mathematics and medicine in both Poland and Italy, and ultimately obtained a degree in canon law. After assisting and caring for his uncle, the bishop of Warmia, Copernicus moved to Frombork in 1510, where he served as canon for the rest of his life. Throughout that period, he continued to pursue his astronomical interests, ultimately becoming convinced that a world system with a stationary central sun around which the earth revolved was preferable to Ptolemy's older geocentric model. He first described this new system in a short Latin work, later referred to as the Commentariolus, *or* Little Commentary, *likely written around 1514. Copernicus never published the work, which circulated among interested contemporaries in handwritten form. In the text, he highlighted what he believed to be a central fault of Ptolemaic theory: the fact that according to Ptolemy celestial objects do not move in uniform, perfect circles, 'as the rule of absolute motion requires'. His new theory corrected that error by positing seven 'assumptions' that replaced the apparent motion of the sun, the fixed stars and the retrograde motion of the planets with an annual revolution and diurnal rotation of the earth. These assumptions yielded a much larger universe than had previously been supposed, and an ordering of the planets that situated the earth as one of six bodies orbiting the sun – in other words, simply a planet, much like the others.*

[*] This translation is taken from *Three Copernican Treatises*, translated by Edward Rosen (Dover, 1959).

Our ancestors assumed, I observe, a large number of celestial spheres for this reason especially, to explain the apparent motion of the planets by the principle of regularity. For they thought it altogether absurd that a heavenly body, which is a perfect sphere, should not always move uniformly. They saw that by connecting and combining regular motions in various ways they could make any body appear to move to any position.

Callippus and Eudoxus, who endeavoured to solve the problem by the use of concentric spheres, were unable to account for all the planetary movements; they had to explain not merely the apparent revolutions of the planets but also the fact that these bodies appear to us sometimes to mount higher in the heavens, sometimes to ascend; and this fact is incompatible with the principle of concentricity. Therefore it seemed better to employ eccentrics and epicycles, a system which most scholars finally accepted.

Yet the planetary theories of Ptolemy and most other astronomers, although consistent with the numerical data, seemed likewise to present no small difficulty. For these theories were not adequate unless certain equants were also conceived; it then appeared that a planet moved with uniform velocity neither on its deferent nor about the centre of its epicycle. Hence a system of this sort seemed neither sufficiently absolute nor sufficiently pleasing to the mind.

Having become aware of these defects, I often considered whether there could perhaps be found a more reasonable arrangement of circles, from which every apparent inequality would be derived and in which everything would move uniformly about its proper centre, as the rule of absolute motion requires. After I had addressed myself to this very difficult and almost insoluble problem, the suggestion at length came to me how it could be solved with fewer and much simpler constructions than were formerly used, if some assumptions (which are called axioms) were granted me. They follow in this order.

Assumptions

1. There is no one centre of all the celestial circles or spheres.
2. The centre of the earth is not the centre of the universe, but only of gravity and of the lunar sphere.
3. All the spheres revolve about the sun as their mid-point, and therefore the sun is the centre of the universe.
4. The ratio of the earth's distance from the sun to the height of the firmament is so much smaller than the ratio of the earth's radius to its distance from the sun that the distance from the earth to the sun is imperceptible in comparison with the height of the firmament.
5. Whatever motion appears in the firmament arises not from any motion of the firmament, but from the earth's motion. The earth together with its circumjacent elements performs a complete rotation on its fixed poles with a daily motion, while the firmament and highest heaven abide unchanged.
6. What appear to us as motions of the sun arise not from its motion but from the motion of the earth and our sphere, with which we revolve about the sun like any other planet. The earth has, then, more than one motion.
7. The apparent retrograde and direct motion of the planets arises not from their motion, but from the earth's. The motion of the earth alone, therefore, suffices to explain so many apparent irregularities in the heaven.

Having set forth these assumptions, I shall endeavour briefly to show how uniformity of the motions can be saved in a systematic way. However, I have thought it well, for the sake of brevity, to omit from this sketch mathematical demonstrations, reserving these for my larger work. But in the explanation of the circles I shall set down here the lengths of the radii; and from these the reader who is not unacquainted with mathematics will readily perceive how closely this arrangement of circles agrees with the numerical data and observations.

Accordingly, let no one suppose that I have gratuitously asserted, with the Pythagoreans, the motion of the earth; strong proof will be found in my exposition of the circles. For the principal arguments by which the natural philosophers attempt to establish the immobility of the earth rest for the most part on the appearances; it is particularly such arguments that collapse here, since I treat the earth's immobility as due to an appearance.

The Order of the Spheres

The celestial spheres are arranged in the following order. The highest is the immovable sphere of the fixed stars, which contains and gives position to all things. Beneath it is Saturn, which Jupiter follows, then Mars. Below Mars is the sphere on which we revolve; then Venus; last is Mercury. The lunar sphere revolves about the centre of the earth and moves with the earth like an epicycle. In the same order also, one planet surpasses another in speed of revolution, according as they trace out greater or smaller circles. Thus Saturn completes its revolution in thirty years, Jupiter in twelve, Mars in two and one half, and the earth in one year; Venus in nine months; Mercury in three.

The Apparent Motions of the Sun

The earth has three motions. First, it revolves annually in a great circle about the sun in the order of the signs, always describing equal arcs in equal times; the distance from the centre of the circle to the centre of the sun is $1/25$ of the radius of the circle. The radius is assumed to have a length imperceptible in comparison with the height of the firmament; consequently the sun appears to revolve with this motion, as if the earth lay in the centre of the universe. However, this appearance is caused by the motion not of the sun but of the earth, so that, for example, when the earth is in the sign of Capricorn, the sun is seen diametrically opposite in Cancer, and so on. On account of the previously mentioned distance of the sun from the centre of the circle, this apparent motion of the sun is not uniform, the maximum inequality being $2\frac{1}{6}°$. The line drawn from the sun through the centre of the circle is invariably directed toward a point of the firmament about 10° west of the more brilliant of the two bright stars in the head of the Gemini; therefore, when the earth is opposite this point, and the centre of the circle lies between them, the sun is seen at its greatest distance from the earth. In this circle, then, the earth revolves together with whatever else is included within the lunar sphere.

The second motion, which is peculiar to the earth, is the daily rotation on the poles in the order of the signs, that is, from west to east. On account of this rotation the entire universe appears to revolve with enormous speed. Thus does the earth rotate together with its circumjacent waters and encircling atmosphere.

The third is the motion in declination. For, the axis of the daily rotation is not parallel to the axis of the great circle, but is inclined to it at an angle that intercepts a portion of a circumference, in our time about 23½°. Therefore, while the centre of the earth always remains in the plane of the ecliptic, that is, in the circumference of the great circle, the poles of the earth rotate, both of them describing small circles about centres equidistant from the axis of the great circle. The period of this motion is not quite a year and is nearly equal to the annual revolution on the great circle. But the axis of the great circle is invariably directed toward the points of the firmament which are called the poles of the ecliptic. In like manner the motion in declination, combined with the annual motion in their joint effect upon the poles of the daily rotation, would keep these poles constantly fixed at the same points of the heavens, if the periods of both motions were exactly equal. Now with the long passage of time it has become clear that this inclination of the earth to the firmament changes. Hence it is the common opinion that the firmament has several motions in conformity with a law not yet sufficiently understood. But the motion of the earth can explain all these changes in a less surprising way. I am not concerned to state what the path of the poles is. I am aware that, in lesser matters, a magnetized iron needle always points in the same direction. It has nevertheless seemed a better view to ascribe the changes to a sphere, whose motion governs the movements of the poles. This sphere must doubtless be sublunary.

Mercury runs on seven circles in all; Venus, on five; the earth, on three, and round it the moon on four; finally, Mars, Jupiter and Saturn on five each. Thus altogether, therefore, thirty-four circles suffice to explain the entire structure of the universe and the entire ballet of the planets.

Georg Joachim Rheticus,
First Account, 1540[*]

Georg Joachim Rheticus (1514–1574) was a twenty-five-year-old mathematics professor at the University of Wittenberg when he journeyed to Prussia in 1539 to study with Copernicus and learn more about his astronomical theories. Rheticus would ultimately stay with Copernicus for two and a half years, during which time he urged Copernicus to publish the major elaboration of the Little Commentary *that he had already written but hesitated to make public. To test how that work might be received and to spread Copernicus's ideas further, Rheticus wrote and published the* Narratio Prima, *or* First Account *of 1540 – the first published articulation of Copernicus's heliocentric theory. While noting the elimination of non-uniform motion in Copernicus's theory, as the* Little Commentary *had done, Rheticus particularly stressed the unity, simplicity and harmony of the Copernican system, in which everything was linked to the central sun, and all the appearances were explained by the motion of the earth alone. Referencing music, political rule and the human body, Rheticus argued that the heliocentric system should be preferred because only in it did 'the order and motions of the heavenly spheres agree in an absolute system'.*

My teacher has written a work of six books in which, in imitation of Ptolemy, he has embraced the whole of astronomy,

---

[*] This translation is taken from *Three Copernican Treatises*, translated by Edward Rosen (Dover, 1959).

stating and proving individual propositions mathematically and by the geometrical method.

Ptolemy's tireless diligence in calculating, his almost superhuman accuracy in observing, his truly divine procedure in examining and investigating all the motions and appearances, and finally his completely consistent method of statement and proof cannot be sufficiently admired and praised by anyone to whom Urania is gracious. In one respect, however, a burden greater than Ptolemy's confronts my teacher. For he must arrange in a certain and consistent scheme or harmony the series and order of all the motions and appearances, marshalled on the broad battlefield of astronomy by the observations of 2,000 years, as by famous generals. Ptolemy, on the other hand, had the observations of the ancients, to which he could safely entrust himself, for scarcely a quarter of this period. Time, the true god and teacher of the laws of the celestial state, discloses the errors of astronomy to us. For an imperceptible or unnoticed error at the foundation of astronomical hypotheses, principles and tables is revealed or greatly increased by the passage of time. Therefore my teacher must not so much restore astronomy as build it anew.

Ptolemy was able to harmonize satisfactorily most of the hypotheses of the ancients – Timocharis, Hipparchus and others – with every inequality in the motions known to him from so small an elapsed period of observation. Therefore he quite rightly and wisely – a praiseworthy action – selected those hypotheses which seemed to be in better agreement with reason and our senses, and which his greatest predecessors had employed. Nevertheless, the observations of all scholars and heaven itself and mathematical reasoning convince us that Ptolemy's hypotheses and those commonly accepted do not suffice to establish the perpetual and consistent connection and harmony of celestial phenomena and to formulate that harmony in tables and rules. It was therefore necessary for my teacher to devise new hypotheses, by the assumption of which he might geometrically and arithmetically deduce with sound logic systems of motion like those which the ancients and Ptolemy, raised on high, once perceived 'with the divine eye of the soul', and which careful observations reveal as

existing in the heavens to those today who study the remains of
the ancients.

The phenomena can be explained, as my teacher shows, by a
regular motion of the spherical earth; that is, by having the sun
occupy the centre of the universe, while the earth revolves
instead of the sun on the eccentric circle. Indeed, there is something divine in the circumstance that a sure understanding of
celestial phenomena must depend on the regular and uniform
motions of the terrestrial globe alone. My teacher saw that
only on this theory could all the circles in the universe be satisfactorily made to revolve uniformly and regularly about their
own centres, and not about other centres – an essential property of circular motion.

Since we see that this one motion of the earth satisfies an almost
infinite number of appearances, should we not attribute to God,
the creator of nature, that skill which we observe in the common
makers of clocks? For they carefully avoid inserting in the mechanism any superfluous wheel or any whose function could be
served better by another with a slight change of position. What
could dissuade my teacher, as a mathematician, from adopting a
serviceable theory of the motion of the terrestrial globe, when he
saw that on the assumption of this hypothesis there sufficed, for
the construction of a sound science of celestial phenomena, a single eighth sphere, and that motionless, the sun at rest in the centre
of the universe, and for the motions of the other planets, epicycles
on an eccentric or eccentrics on an eccentric or epicycles on an
epicycle? Moreover, the motion of the earth in its circle produces
the inequalities of all the planets except the moon; this one motion
alone seems to be the cause of every apparent inequality at a distance from the sun, in the case of the three superior planets, and
in the neighbourhood of the sun, in the case of Venus and Mercury. Finally, this motion makes it possible to satisfy each of the
planets by only one deviation in latitude of the deferent of the
planet. Hence it is particularly the planetary motions that require
such hypotheses.

My teacher was especially influenced by the realization that
the chief cause of all the uncertainty in astronomy was that the
masters of this science (no offence is intended to divine Ptolemy,

the father of astronomy) fashioned their theories and devices for correcting the motion of the heavenly bodies with too little regard for the rule which reminds us that the order and motions of the heavenly spheres agree in an absolute system. We fully grant these distinguished men their due honour, as we should. Nevertheless, we should have wished them, in establishing the harmony of the motions, to imitate the musicians who, when one string has either tightened or loosened, with great care and skill regulate and adjust the tones of all the other strings, until all together produce the desired harmony, and no dissonance is heard in any.

Under the commonly accepted principles of astronomy, it could be seen that all the celestial phenomena conform to the mean motion of the sun and that the entire harmony of the celestial motions is established and preserved under its control. Hence the sun was called by the ancients leader, governor of nature and king. But whether it carries on this administration as God rules the entire universe, a rule excellently described by Aristotle in the *On the Cosmos*, or whether, traversing the entire heaven so often and resting nowhere, it acts as God's administrator in nature, seems not yet altogether explained and settled. Which of these assumptions is preferable, I leave to be determined by geometers and philosophers (who are mathematically equipped). For in the trial and decision of such controversies, a verdict must be reached in accordance with not plausible opinions but mathematical laws (the court in which this case is heard). The former manner of rule has been set aside, the latter adopted. My teacher is convinced, however, that the rejected method of the sun's rule in the realm of nature must be revived, but in such a way that the received and accepted method retains its place. For he is aware that in human affairs the emperor need not himself hurry from city to city in order to perform the duty imposed on him by God; and that the heart does not move to the head or feet or other parts of the body to sustain a living creature, but fulfils its function through other organs designed by God for that purpose.

Now my teacher concluded that the mean motion of the sun must be the sort of motion that is not only established by the imagination, as in the case of the other planets, but is self-caused, since it appears to be truly 'both choral dancer and choral leader'. He then showed that his opinion was sound and not inconsistent with the truth, for he saw that by his hypotheses the efficient cause of the uniform motion of the sun could be geometrically deduced and proved. Hence the mean motion of the sun would necessarily be perceived in all the motions and appearances of the other planets in a definite manner, as appears in each of them. Thus the assumption of the motion of the earth on an eccentric provides a sure theory of celestial phenomena, in which no change should be made without at the same time re-establishing the entire system.

The Arrangement of the Universe

Aristotle says: 'That which causes derivative truths to be true is most true.' Accordingly, my teacher decided that he must assume such hypotheses as would contain causes capable of confirming the truth of the observations of previous centuries, and such as would themselves cause, we may hope, all future astronomical predictions of the phenomena to be found true.

First, surmounting no mean difficulties, he established by hypothesis that the sphere of the stars, which we commonly call the eighth sphere, was created by God to be the region which would enclose within its confines the entire realm of nature, and hence that it was created fixed and immovable as the place of the universe. Now motion is perceived only by comparison with something fixed; thus sailors on the sea, to whom 'land is no longer visible, only the sky on all sides and on all sides the water', are not aware of any motion of their ship when the sea is undisturbed by winds, even though they are borne along at such high speed that they pass over several long miles in an hour. Hence this sphere was studded by God for our sake with a large number of twinkling stars, in order that by comparison with them, surely fixed in place, we might observe the positions and motions of the other enclosed spheres and planets.

Then, in harmony with these arrangements, God stationed in the centre of the stage His governor of nature, king of the entire universe, conspicuous by its divine splendour, the sun 'to whose rhythm the gods move, and the world receives its laws and keeps the pacts ordained'. The other spheres are arranged in the following manner. The first place below the firmament or sphere of the stars falls to the sphere of Saturn, which encloses the spheres of first Jupiter, then Mars; the spheres of first Mercury, then Venus surround the sun; and the centres of the spheres of the five planets are located in the neighbourhood of the sun. Between the concave surface of Mars's sphere and the convex of Venus's, where there is ample space, the globe of the earth together with its adjacent elements, surrounded by the moon's sphere, revolves in a great circle which encloses within itself, in addition to the sun, the spheres of Mercury and Venus, so that the earth moves among the planets as one of them.

As I carefully consider this arrangement of the entire universe according to the opinion of my teacher, I realize that Pliny set down an excellent and accurate testament when he wrote 'to inquire what is beyond the universe or heaven, by which all things are overarched, is no concern of man, nor can the human mind form any conjecture concerning this question'. And he continues: 'The universe is sacred, without bounds, all in all; indeed, it is the totality, finite yet similar to the infinite, etc.' For if we follow my teacher, there will be nothing beyond the concave surface of the starry sphere for us to investigate, except insofar as Holy Writ has vouchsafed us knowledge, in which case again the road will be closed to placing anything beyond this concave surface. We will therefore gratefully admire and regard as sacrosanct all the rest of nature, enclosed by God within the starry heaven. In many ways and with innumerable instruments and gifts He has endowed us, and enabled us to study and know nature; we will advance to the point to which He desired us to advance, and we will not attempt to transgress the limits imposed by Him.

That the universe is boundless up to its concave surface, and truly similar to the infinite is known, moreover, from the fact that we see all the heavenly bodies twinkle, with the exception of the planets including Saturn, which, being the nearest of them to the firmament, revolves on the greatest circle. But this conclusion follows far more clearly by deduction from the hypotheses of my teacher. For the great circle which carries the earth has a perceptible ratio to the spheres of the five planets, and hence every inequality in the appearances of these planets is demonstrably derived from their relations to the sun. Every horizon on the earth, being a great circle of the universe, divides the sphere of the stars into equal parts. Equal periods in the revolutions of the sphere are shown to be measured by the fixed stars. Consequently it is quite clear that the sphere of the stars is, to the highest degree, similar to the infinite, since by comparison with it the great circle vanishes, and all the phenomena are observed exactly as if the earth were at rest in the centre of the universe.

Moreover, the remarkable symmetry and interconnection of the motions and spheres, as maintained by the assumption of the foregoing hypotheses, are not unworthy of God's workmanship and not unsuited to these divine bodies. These relations, I should say, can be conceived by the mind (on account of its affinity with the heavens) more quickly than they can be explained by any human utterance, just as in a demonstration they are usually impressed upon our minds, not so much by words as by the perfect and absolute ideas, if I may use the term, of these most delightful objects. Nevertheless, it is possible, in a general survey of the hypotheses, to see how the inexpressible harmony and agreement of all things manifest themselves.

For in the common hypotheses there appeared no end to the invention of spheres; moreover, spheres of an immensity that could be grasped by neither sense nor reason were revolved with extremely slow and extremely rapid motions. Some writers stated that the daily motion of all the lower spheres is caused by the highest movable sphere; but when a great storm of controversy raged over this question, they could not explain why a higher sphere should have power over a lower. Others, like Eudoxus and those who followed him, assigned to each planet a special sphere, the motion of which caused the planet to revolve about the earth once in a natural day. Moreover, ye immortal gods, what dispute, what strife there has been until now over the positions of the spheres of Venus and Mercury and their relation to the sun. But the case is still before the judge. Is there anyone

who does not see that it is very difficult and even impossible ever
to settle this question while the common hypotheses are accepted?
For what would prevent anyone from locating even Saturn
below the sun, provided that at the same time he preserved the
mutual proportions of the spheres and epicycle, since in these
same hypotheses there has not yet been established the common
measure of the spheres of the planets, whereby each sphere may
be geometrically confined to its place? I refrain from mentioning
here the vast commotion which those who defame this most
beautiful and most delightful part of philosophy have stirred up
on account of the great size of the epicycle of Venus, and on
account of the unequal motion, on the assumption of equants, of
the celestial spheres about their own centres.

However, in the hypotheses of my teacher, which accept, as
has been explained, the starry sphere as boundary, the sphere
of each planet advances uniformly with the motion assigned to
it by nature and completes its period without being forced into
any inequality by the power of a higher sphere. In addition, the
larger spheres revolve more slowly, and, as is proper, those that
are nearer to the sun, which may be said to be the source of
motion and light, revolve more swiftly. Hence Saturn, moving
freely in the ecliptic, revolves in thirty years, Jupiter in twelve,
and Mars in two. The centre of the earth measures the length
of the year by the fixed stars. Venus passes through the zodiac
in nine months, and Mercury, revolving about the sun on the
smallest sphere, traverses the universe in eighty days. Thus
there are only six moving spheres which revolve about the sun,
the centre of the universe. Their common measure is the great
circle which carries the earth, just as the radius of the spherical
earth is the common measure of the circles of the moon, the
distance of the sun from the moon, etc.

Who could have chosen a more suitable and more appropriate
number than six? By what number could anyone more easily
have persuaded mankind that the whole universe was divided
into spheres by God the Author and Creator of the world? For
the number six is honoured beyond all others in the sacred
prophecies of God and by the Pythagoreans and the other philosophers. What is more agreeable to God's handiwork than this

first and most perfect work should be summed up in this first and most perfect number? Moreover, the celestial harmony is achieved by the six aforementioned movable spheres. For they are all so arranged that no immense interval is left between one and another; and each, geometrically defined, so maintains its position that if you should try to move any one at all from its place you would thereby disrupt the entire system.

But now that we have touched on these general considerations, let us proceed to an exposition of the circular motions which are appropriate to the several spheres and to the bodies that cleave to and rest upon them. First we shall speak of the hypotheses for the motions of the terrestrial globe, on which we have our being.

The Motions Appropriate to the Great Circle and Its Related Bodies. The Three Motions of the Earth: Daily, Annual and the Motion in Declination

Following Plato and the Pythagoreans, the greatest mathematicians of that divine age, my teacher thought that in order to determine the causes of the phenomena circular motions must be ascribed to the spherical earth. He saw (as Aristotle also points out) that when one motion is assigned to the earth, it may properly have other motions, by analogy with the planets. He therefore decided to begin with the assumption that the earth has three motions, by far the most important of all.

For in the first place, having assumed the general arrangement of the universe described above, he showed that, enclosed by its poles within the lunar sphere, the earth, like a ball on a lathe, rotates from west to east, as God's will ordains; and that by this motion, the terrestrial globe produces day and night and the changing appearances of the heavens, according as it is turned toward the sun. In the second place, the centre of the earth, together with its adjacent elements and the lunar sphere, is carried uniformly in the plane of the ecliptic by the great circle which I have already mentioned more than once, in the order of the signs. In the third place, the equator and the axis of the earth have a variable inclination to the plane of the eclip-tic and move in the direction opposite to that of the motion of the centre, so that on account of this inclination of the earth's axis and the immensity of the starry sphere, no matter where the centre of the earth may be, the equator and the poles of the earth are almost invariably directed to the same points in the heavens. This result will ensue if the ends of the earth's axis, that is, the poles of the earth, are understood to move daily in precedence a distance almost exactly equal to the motion of the centre of the earth in consequence on the great circle, and to describe about the axis and poles of the greater circle or ecliptic small circles equidistant from them.

But to these motions we should add, in the opinion of my teacher, two librations of the poles of the earth, and the two motions, the one uniform and other unequal, with which the centre of the great circle advances in the ecliptic. We shall then have a true system of hypotheses for deducing in its entirety what the moderns call the doctrine of the first motion, which at present is derived from all sorts of motions of the starry sphere; and for determining the cause of the motions and phenomena of the sun and moon, as they have been carefully observed by scholars for the past 2,000 years. I may merely mention, since I shall have occasion to deal with the topic more fully below, that the motion of the great circle unquestionably affects the appearances of the other five planets. With so few motions and, as it were, with a single circle is so vast a subject comprehended.

The Second Part of the Hypotheses

The Motions of the Five Planets

When I reflect on this truly admirable structure of new hypoth-eses wrought by my teacher, I frequently recall that Platonic dialogue which indicates the qualities required in an astron-omer and then adds, 'No nature except an extraordinary one could ever easily formulate a theory.'

When I watched the work of learned men in the improvement

of the motions of Regiomontanus and his teacher Peurbach, I began to understand what sort of task and how great a difficulty it was to recall this queen of mathematics, astronomy, to her palace, as she deserved, and to restore the boundaries of her kingdom. But from the time that I became, by God's will, a spectator and witness of the labours which my teacher performs with energetic mind and has in large measure already accomplished, I realized that I had not dreamed of even the shadow of so great a burden of work. And it is so great a labour that it is not any hero who can endure and finally complete it. For this reason, I suppose, the ancients related that Hercules, sprung of Jupiter most high, no longer trusting his own shoulders, replaced the heavens upon Atlas, who, being long accustomed to the burden, resumed it with stout heart and undiminished vigour, as he had borne it in former days.

Moreover, divine Plato, master of wisdom as Pliny styles him, affirms not indistinctly in the *Epinomis* that astronomy was discovered under the guidance of God. Others perhaps interpret this opinion of Plato's otherwise. But when I see that my teacher always has before his eyes the observations of all ages together with his own, assembled in order as in catalogues; then when some conclusion must be drawn or contribution made to the science and its principles, he proceeds from the earliest observations to his own, seeking the mutual relationship which harmonizes them all; the results thus obtained by correct inference under the guidance of Urania he then compares with the hypotheses of Ptolemy and the ancients; and having made a most careful examination of these hypotheses, he finds that astronomical proof requires their rejection; he assumes new hypotheses, not indeed without divine inspiration and the favour of the gods; by applying mathematics, he geometrically establishes the conclusions which can be drawn from them by correct inference; he then harmonizes the ancient observations and his own with the hypotheses which he has adopted; and after performing all these operations he finally writes down the laws of astronomy – when, I say, I behold this procedure, I think that Plato must be understood as follows.

The mathematician who studies the motions of the stars is

surely like a blind man who, with only a staff to guide him, must make a great, endless, hazardous journey that winds through innumerable desolate places. What will be the result? Proceeding anxiously for a while and groping his way with his staff, he will at some time, leaning upon it, cry out in despair to heaven, earth and all the gods to aid him in his misery. God will permit him to try his strength for a period of years, that he may in the end learn that he cannot be rescued from threatening danger by his staff. Then God compassionately stretches forth His hand to the despairing man, and with His hand conducts him to the desired goal.

The staff of the astronomer is mathematics or geometry, by which he ventures at first to test the road and press on. For in the examination from afar of those divine objects so remote from us, of what avail is the strength of the human mind? Of what avail dim-sighted eyes? Accordingly, if God in His kindness had not endowed the astronomer with heroic ambitions and led him by the hand, as it were, along a road otherwise inaccessible to the human intellect, the astronomer would not be, I think, in any respect better circumstanced and more fortunate than the blind man, save that trusting in his reason and offering divine honours to his staff, he will one day rejoice in the recall of Urania from the underworld. When, however, he considers the matter aright, he will perceive that he is not more blessed than Orpheus, who was aware that Eurydice was following him as he danced his way up from Orcus; but when he reached the jaws of Avernus, she whom he dearly longed to possess disappeared from view and descended once more to the infernal regions. Let us then examine, as we set out to do, my teacher's hypotheses for the remaining planets, to see whether with unremitting devotion and under the guidance of God, he has led Urania back to the upper world and restored her to her place of honour.

With regard to the apparent motions of the sun and moon, it is perhaps possible to deny what is said about the motion of the earth, although I do not see how the explanation of precession is to be transferred to the sphere of the stars. But if anyone desires to look either to the principal end of astronomy and the order and harmony of the system of the spheres or to ease and elegance and a complete explanation of the causes of the phenomena, by the assumption of no other hypotheses will he demonstrate the apparent motions of the remaining planets more neatly and correctly. For all these phenomena appear to be linked most nobly together as by a golden chain; and each of the planets, by its position and order and every inequality of its motion, bears witness that the earth moves and that we who dwell upon the globe of the earth, instead of accepting its changes of position, believe that the planets wander in all sorts of motions of their own. And if it is possible anywhere else to see how God has left the universe for our discussion, it surely is eminently clear in this matter.

The Hypotheses for the Motions in Longitude of the Five Planets

What has been said thus far regarding the motion of the earth has been demonstrated by my teacher. Consequently, the entire inequality in the apparent motion of the planets which seems to occur in their positions with respect to the sun is caused by the annual motion of the earth on the great circle. Just as my teacher chose to employ an epicycle on an epicycle for the moon, so, for the purpose of demonstrating conveniently the order of the planets and the measurement of their motion, he has selected, for the three superior planets, epicycles on an eccentric, but for Venus and Mercury eccentrics on an eccentric.

If our eye were at the centre of the great circle, lines of sight drawn from it through the planets to the sphere of the stars would, as the lines of the true motions, be rotated in the ecliptic by the planets exactly as the schemes of the aforementioned circles and motions require, so that they would reveal the real inequalities of these motions in the zodiac. But we, as dwellers upon the earth, observe the apparent motions in the heavens from the earth. Hence we refer all the motions and phenomena to the centre of the earth as the foundation and inmost part of our abode, by drawing lines from it through the planets, as though our eye had moved from the centre of the great circle to the centre of the earth. Clearly, it is from this latter point that the inequalities of all the phenomena, as they are seen by us,

must be calculated. But if it is our purpose to deduce the true
and real inequalities in the motion of the planets, we must use
the lines drawn from the centre of the great circle, as has been
explained. To smooth our way through the topics in planetary
phenomena which remain to be discussed and to make the
whole treatise easier and more agreeable, let us imagine not
only the lines of true apparent motion drawn from the centre of
the earth through the planets to the ecliptic but also those
drawn from the centre of the great circle and therefore properly
called the lines of the inequality of motion.

When, as the earth advances with the motion of the great circle, it reaches a position where it is on a straight line between the
sun and one of the three superior planets, the planet will be seen
at its evening rising; and because the earth, when so situated, is at
its nearest to the planet, the ancients said that the planet was at
its nearest to the earth and in the perigee of its epicycle. But when
the sun approaches the line of the true and apparent place of the
planet – this occurs when the earth reaches the point opposite the
above-mentioned position – the planet begins to disappear by setting in the evening and to attain its greatest distance from the
earth, until the line of the true place of the planet passes also
through the centre of the sun. Then the sun lies between the
planet and the earth, and the planet is occulted. After occultation,
since the motion of the earth continues uninterrupted and since
the line of the true place of the sun withdraws from the line of the
true place of the planet, the planet reappears at its morning rising,
when it has attained the proper distance from the sun required by
the arc of vision.

Moreover, in the hypotheses of the three superior planets,
the great circle takes the place of the epicycle attributed to each
of the planets by the ancients. Hence the true apogee and perigee of the planet with respect to the great circle will be found
on the diameter of the great circle prolonged to meet the planet.
But the mean apogee and perigee will be found on the diameter
of the great circle that moves parallel to the line drawn from
the centre of the eccentric to the centre of the epicycle. Since in
the semicircle closer to the planet the earth approaches the
planet, and in the other, opposite semicircle recedes from it, in

the former semicircle the ends of the diameters of the great circle are in the perigees, but in the latter the apogees. For the former semicircle takes the place of the lower part of the epicycle, but the latter, the upper.

Imagine that conjunction of sun and planet is not far off. Let the centre of the earth be in the true place of the apogee of the planet with respect to the great circle; and let the line of the real inequality coincide with the line of the apparent place of the planet. However, as the earth in its motion moves away from this position, the line of the real inequality and the line of the true place of the planet begin to intersect in the planet. The former advances with the regular unequal motion of the planet in the order of the signs; and the latter, as it separates from the former, makes the planet seem to us to move more rapidly in the ecliptic than it really does with its own motion.

But when the earth reaches the part of the great circle that is nearer to the planet, the direction of its motion at once becomes westward, so that the apparent motion of the planet forthwith seems slower to us. Moreover, because the earth mounts toward the planet, the line of the true motion of the sun moves away from the planet, and the planet is thought to approach us, as though it were descending from its upper circumference. However, the motion of the planet seems to be direct, until the centre of the earth reaches the point on the great circle with respect to the planet where the angle through the line of the true place of the planet moves daily in precedence equals the diurnal angle of the real inequality in consequence. For there, since the two motions neutralize each other, the planet appears to remain at its first stationary point for a number of days, depending on the ratio of the great circle to the eccentric of the planet under consideration, the position of the planet on its circle, and the real rate of its motion. Then as the earth moves from this position nearer to the planet, we believe that the planet retrogrades and moves in precedence, since the regression of the line of the true place of the planet perceptibly exceeds the real motion of the planet. This apparent retrogradation continues until the earth reaches the true perigee of the planet with respect to the great circle, where the planet, at the midpoint of regression, is in opposition to the sun and nearest to the earth. When Mars is found in this position, it has, in addition to the common retrogradation or parallax caused by the great circle, another parallax caused by the sensible ratio of the radius of the earth to the distance of Mars, as careful observation will testify.

Finally, as the earth moves in consequence from this central conjunction with the planet, so to say, the westward regression diminishes exactly as it had previously increased, until, when the motions are again equal, the planet reaches its second stationary point. Then as the real motion of the planet exceeds the motion of the line of the true place of the planet, and as the earth advances, the situation is reached where the planet at length appears at the midpoint of its direction motion; and the earth again comes to the true apogee of the planet, whence we started its motion, and produces for us in order all the above-mentioned phenomena of each of the planets.

To the Reader, Copernicus, *On the Revolutions of the Heavenly Spheres*, 1543*

In 1541 Rheticus took the manuscript of On the Revolutions *to Nuremberg to be published, and he left it there in the hands of Andreas Osiander, a Lutheran theologian, who would manage the process of publication. The 'To the Reader' preface, though printed at the start of the book without attribution, was written by Osiander. In it, Osiander offered an instrumentalist reading of the ideas that followed in the body of the text, claiming that the goal of astronomy was simply to calculate celestial motions and make hypotheses about them, but that these hypotheses needed only to fit with the observations, and bore no direct relationship to the truth. He positioned Copernicus's hypotheses as simpler and more useful than those that came before, and urged readers to accept them without mistaking them for a true theory of the cosmos. Though this may have increased the initial acceptance of the book – which was indeed often approached in just this way by early readers – this was not a position that Copernicus himself adopted, as would have been clear to careful examiners of the text itself (as it was to Kepler, who was the first to publicly reveal the identity of the author of this 'To the Reader'). Copernicus, rather, stressed in his own preface to* On the Revolutions *that he considered his system to be a true and real description of the cosmos.*

* This translation is taken from *On the Revolutions* (*Nicholas Copernicus Complete Works, Volume 2*), translated by Edward Rosen (Johns Hopkins University Press, 1978).

To the Reader,
Concerning the Hypotheses of this Work:

There have already been widespread reports about the novel hypotheses of this work, which declares that the earth moves whereas the sun is at rest in the centre of the universe. Hence certain scholars, I have no doubt, are deeply offended and believe that the liberal arts, which were established long ago on a sound basis, should not be thrown into confusion. But if these men are willing to examine the matter closely, they will find that the author of this work has done nothing blameworthy. For it is the duty of an astronomer to compose the history of the celestial motions through careful and expert study. Then he must conceive and devise the causes of these motions or hypotheses about them. Since he cannot in any way attain to the true causes, he will adopt whatever suppositions enable the motions to be computed correctly from the principles of geometry for the future as well as for the past. The present author has performed both these duties excellently. For these hypotheses need not be true nor even probable. On the contrary, if they provide a calculus consistent with the observations, that alone is enough. Perhaps there is someone who is so ignorant of geometry and optics that he regards the epicycle of Venus as probable, or thinks that it is the reason why Venus sometimes precedes and sometimes follows the sun by forty degrees and even more. Is there anyone who is not aware that from this assumption it necessarily follows that the diameter of the planet at perigee should appear more than four times, and the body of the planet more than sixteen times, as great as at apogee? Yet this variation is refuted by the experience of every age. In this science there are some other no less important absurdities, which need not be set forth at the moment. For this art, it is quite clear, is completely and absolutely ignorant of the causes of the apparent non-uniform motions. And if any causes are devised by the imagination, as indeed very many are, they are not put forward to convince anyone that they are true, but merely to provide a reliable basis for computation. However, since different hypotheses are sometimes offered for one and the same motion (for example, eccentricity and an epicycle for the sun's motion), the astronomer will take as his first choice that hypothesis which is the easiest to grasp. The philosopher will perhaps rather seek the semblance of the truth. But neither of them will understand or state anything certain, unless it has been divinely revealed to him.

Therefore alongside the ancient hypotheses, which are no more probable, let us permit these new hypotheses also to become known, especially since they are admirable as well as simple and bring with them a huge treasure of very skilful observations. So far as hypotheses are concerned, let no one expect anything certain from astronomy, which cannot furnish it, lest he accept as the truth ideas conceived for another purpose, and depart from this study a greater fool than when he entered it. Farewell.

Copernicus, Prefatory Material from
On the Revolutions of the Heavenly Spheres, 1543*

Also appended to the start of On the Revolutions *were two letters: the first from the cardinal of Capua to Copernicus, urging him to publish his work, and the second Copernicus's dedication of the work to Pope Paul III. Both were likely intended to forestall theological objections to the work because of possible conflicts with scriptural claims of the earth's centrality. Copernicus pointed to many disagreements within traditional astronomy that his theory helped settle, gestured toward ancient precedents for his own work, and stressed its usefulness for the creation of tables and better calendars. He likewise asserted that astronomy was a discipline for experts, and that those without training in its technicalities did not have standing to criticize his work.*

Diligent reader, in this work, which has just been created and published, you have the motions of the fixed stars and planets, as these motions have been reconstituted on the basis of ancient as well as recent observations, and have moreover been embellished by new and marvellous hypotheses. You also have most convenient tables, from which you will be able to compute those motions with the utmost ease for any time whatever. Therefore buy, read and enjoy this work.

Let no one untrained in geometry enter here.

* This translation is taken from *On the Revolutions* (*Nicholas Copernicus Complete Works, Volume 2*), translated by Edward Rosen (Johns Hopkins University Press, 1978).

Letter of Nicholas Schönberg

Nicholas Schönberg, Cardinal of Capua, to Nicolaus
Copernicus, Greetings.

Some years ago word reached me concerning your proficiency,
of which everybody constantly spoke. At that time I began to
have a very high regard for you, and also to congratulate our
contemporaries among whom you enjoyed such great prestige.
For I had learned that you had not merely mastered the discoveries of the ancient astronomers uncommonly well but had also
formulated a new cosmology. In it you maintain that the earth
moves; that the sun occupies the lowest, and thus the central,
place in the universe; that the eighth heaven remain perpetually
motionless and fixed; and that, together with the elements
included in its sphere, the moon, situated between the heavens of
Mars and Venus, revolves around the sun in the period of a year.
I have also learned that you have written an exposition of this
whole system of astronomy, and have computed the planetary
motions and set them down in tables, to the greatest admiration
of all. Therefore with the utmost earnestness I entreat you, most
learned sir, unless I inconvenience you, to communicate this discovery of yours to scholars, and at the earliest possible moment
to send me your writings on the sphere of the universe together
with the tables and whatever else you have that is relevant to this
subject. Moreover, I have instructed Theodoric of Reden to have
everything copied in your quarters at my expense and dispatched
to me. If you gratify my desire in this matter, you will see that
you are dealing with a man who is zealous for your reputation
and eager to do justice to so fine a talent. Farewell.

Rome, 1 November 1536

To His Holiness, Pope Paul III: Nicolaus Copernicus's Preface to His Books *On The Revolutions*

I can readily imagine, Holy Father, that as soon as some people hear that in this volume, which I have written about the revolutions of the spheres of the universe, I ascribe certain motions to the terrestrial globe, they will shout that I must be immediately repudiated together with this belief. For I am not so enamoured of my own opinions that I disregard what others may think of them. I am aware that a philosopher's ideas are not subject to the judgement of ordinary persons, because it is his endeavour to seek the truth in all things, to the extent permitted to human reason by God. Yet I hold that completely erroneous views should be shunned. Those who know that the consensus of many centuries has sanctioned the conception that the earth remains at rest in the middle of the heaven as its centre would, I reflected, regard it as an insane pronouncement if I made the opposite assertion that the earth moves. Therefore I debated with myself for a long time whether to publish the volume which I wrote to prove the earth's motion or rather to follow the example of the Pythagoreans and certain others, who used to transmit philosophy's secrets only to kinsmen and friends, not in writing but by word of mouth, as is shown by Lysis' letter to Hipparchus. And they did so, it seems to me, not, as some suppose, because they were in some way jealous about their teachings, which would be spread around; on the contrary, they wanted the very beautiful thoughts attained by great men of deep devotion not to be ridiculed by those who are reluctant to exert themselves vigorously in any literary pursuit unless it is lucrative; or if they are stimulated to the non-acquisitive study of philosophy by the exhortation and example of others, yet because of their dullness of mind they play the same part among philosophers as drones among bees. When I weighed these considerations, the scorn which I had reason to fear on account of the novelty and unconventionality of my opinion almost induced me to abandon completely the work which I had undertaken.

But while I hesitated for a long time and even resisted, my

friends drew me back. Foremost among them was the cardinal of Capua, Nicholas Schönberg, renowned in every field of learning. Next to him was a man who loves me dearly, Tiedemann Giese, bishop of Chełmno, a close student of sacred letters as well as of all good literature. For he repeatedly encouraged me and, sometimes adding reproaches, urgently requested me to publish this volume and finally permit it to appear after being buried among my papers and lying concealed not merely until the ninth year but by now the fourth period of nine years. The same conduct was recommended to me by not a few other very eminent scholars. They exhorted me no longer to refuse, on account of the fear which I felt, to make my work available for the general use of students of astronomy. The crazier my doctrine of the earth's motion now appeared to most people, the argument ran, so much the more admiration and thanks would it gain after they saw the publication of my writings dispel the fog of absurdity by most luminous proofs. Influenced therefore by these persuasive men and by this hope, in the end I allowed my friends to bring out an edition of the volume, as they had long besought me to do.

However, Your Holiness will perhaps not be greatly surprised that I have dared to publish my studies after devoting so much effort to working them out that I did not hesitate to put down my thoughts about the earth's motion in written form too. But you are rather waiting to hear from me how it occurred to me to venture to conceive any motion of the earth, against the traditional opinion of astronomers and almost against common sense. I have accordingly no desire to conceal from Your Holiness that I was impelled to consider a different system of deducing the motions of the universe's spheres for no other reason than the realization that astronomers do not agree among themselves in their investigations of this subject. For, in the first place, they are so uncertain about the motion of the sun and moon that they cannot establish and observe a constant length even for the tropical year. Secondly, in determining the motions not only of these bodies but also of the other five planets, they do not use the same principles, assumptions and explanations of the apparent revolutions and motions. For while some employ only homocentrics, others utilize eccentrics and epicycles, and yet they do not quite reach their goal. For although those who put their faith in homocentrics showed that some non-uniform motions could be compounded in this way, nevertheless by this means they were unable to obtain any incontrovertible result in absolute agreement with the phenomena. On the other hand, those who devised the eccentrics seem thereby in large measure to have solved the problem of the apparent motions with appropriate calculations. But meanwhile they introduced a good many ideas which apparently contradict the first principles of uniform motion. Nor could they elicit or deduce from the eccentrics the principal consideration, that is, the structure of the universe and the true symmetry of its parts. On the contrary, their experience was just like someone taking from various places hands, feet, a head and other pieces, very well depicted, it may be, but not for the representation of a single person; since these fragments would not belong to one another at all, a monster rather than a man would be put together from them. Hence in the process of demonstration or 'method', as it is called, those who employed eccentrics are found either to have omitted something essential or to have admitted something extraneous and wholly irrelevant. This would not have happened to them, had they followed sound principles. For if the hypotheses assumed by them were not false, everything which follows from their hypotheses would be confirmed beyond any doubt. Even though what I am now saying may be obscure, it will nevertheless become clearer in the proper place.

For a long time, then, I reflected on this confusion in the astronomical traditions concerning the derivation of the motions of the universe's spheres. I began to be annoyed that the movements of the world machine, created for our sake by the best and most systematic Artisan of all, were not understood with greater certainty by the philosophers, who otherwise examined so precisely the most insignificant trifles of this world. For this reason I undertook the task of rereading the works of all the philosophers which I could obtain to learn whether anyone had ever proposed other motions of the universe's spheres than those expounded by the teachers of astronomy in the schools. And in fact first I found in Cicero that Hicetas supposed the earth to move. Later I also discovered in Plutarch that certain others were of this opinion. I have decided to set his words down here, so that they may be available to everybody:

> Some think that the earth remains at rest. But Philolaus the Pythagorean believes that, like the sun and moon, it revolves around the fire in an oblique circle. Heraclides of Pontus, and Ecphantus the Pythagorean make the earth move, not in a progressive motion, but like a wheel in a rotation from west to east about its own centre.

Therefore, having obtained the opportunity from these sources, I too began to consider the mobility of the earth. And even though the idea seemed absurd, nevertheless I knew that others before me had been granted the freedom to imagine any circles whatever for the purpose of explaining the heavenly phenomena. Hence I thought that I too would be readily permitted to ascertain whether explanations sounder than those of my predecessors could be found for the revolution of the celestial spheres on the assumption of some motion of the earth.

Having thus assumed the motions which I ascribe to the earth later on in the volume, by long and intense study I finally found that if the motions of the other planets are correlated with the orbiting of the earth, and are computed for the revolution of each planet, not only do their phenomena follow therefrom but also the order and size of all the planets and spheres, and heaven itself is so linked together that in no portion of it can anything be shifted without disrupting the remaining parts and the universe as a whole. Accordingly in the arrangement of the volume too I have adopted the following order. In the first book I set forth the entire distribution of the spheres together with the motions which I attribute to the earth, so that this book contains, as it were, the general structure of the universe. Then in the remaining books I correlate the motions of the other planets and of all the spheres with the movement of the earth so that I may thereby determine to what extent the motions and appearances of the other planets and spheres can be saved if they are correlated with the earth's motions. I have no doubt that acute and learned astronomers will agree with me if, as this discipline especially requires, they are willing to examine and consider, not superficially but thoroughly, what I adduce in this volume in proof of these matters. However, in order that the educated and uneducated alike may see that I do not run away from the judgement of anybody at all, I have preferred dedicating my studies to Your Holiness rather than to anyone else. For even in this very remote corner of the earth where I live you are considered the highest authority by virtue of the loftiness of your office and your love for all literature and astronomy too. Hence by your prestige and judgement you can easily suppress calumnious attacks although, as the proverb has it, there is no remedy for a backbite.

Perhaps there will be babblers who claim to be judges of astronomy although completely ignorant of the subject and, badly distorting some passage of Scripture to their purpose, will dare to find fault with my undertaking and censure it. I disregard them even to the extent of despising their criticism as unfounded. For it is not unknown that Lactantius, otherwise an illustrious writer but hardly an astronomer, speaks quite childishly about the earth's shape, when he mocks those who declared that the earth has the form of a globe. Hence scholars need not be surprised if any such persons will likewise ridicule me. Astronomy is written for astronomers. To them my work too will seem, unless I am mistaken, to make some contribution also to the Church, at the head of which Your Holiness now stands. For not so long ago under Leo X the Lateran Council considered the problem of reforming the ecclesiastical calendar. The issue remained undecided then only because the lengths of the year and month and the motions of the sun and moon were regarded as not yet adequately measured. From that time on, at the suggestion of that most distinguished man, Paul, bishop of Fossombrone, who was then in charge of this matter, I have directed my attention to a more precise study of these topics. But what I have accomplished in this regard, I leave to the judgement of Your Holiness in particular and of all other learned astronomers. And lest I appear to Your Holiness to promise more about the usefulness of this volume than I can fulfil, I now turn to the work itself.

Copernicus, Book One, *On the Revolutions of the Heavenly Spheres*, 1543 *

While much of On the Revolutions *was technical and mathematical, Book One offered an overview of Copernicus's main ideas in non-mathematical terms. Copernicus insisted that circular motion was the most perfect and the most suitable for celestial objects, and thus eliminated the Ptolemaic equant; he likewise argued that as the earth was a sphere, rotational motion was natural to it. Along with a daily rotation, which replaced the daily rotational motion of the celestial sphere in the Ptolemaic system, Copernicus gave the earth two other motions: an annual revolution around the sun, replacing the sun's annual motion around the ecliptic; and a motion of its axis that ensured its fixed orientation relative to the stars. The earth's annual motion explained many of the appearances of the planets that had seemed arbitrary in the older system; retrograde motion, for example, was now understood to be simply an appearance caused when the earth overtook another planet as both circled the sun. Further, Copernicus emphasized another advantage he believed his system had over Ptolemy's: in his formulation, the order of the planets was no longer arbitrary, and Copernicus could assign them not only definite positions but also specific distances from the sun. Finally, Copernicus insisted on the vastness of the cosmos and the immense distance of the fixed stars, as compared to the Ptolemaic system; this could theoretically account for the lack of an observed*

* This translation is taken from *On the Revolutions* (*Nicholas Copernicus Complete Works, Volume 2*), translated by Edward Rosen (Johns Hopkins University Press, 1978).

stellar parallax, which should appear if the earth were in motion,
but did not.

Introduction

Among the many various literary and artistic pursuits which
invigorate men's minds, the strongest affection and utmost zeal
should, I think, promote the studies concerned with the most
beautiful objects, most deserving to be known. This is the nature
of the discipline which deals with the universe's divine revolutions, the asters' motions, sizes, distances, risings and settings,
as well as the causes of the other phenomena in the sky, and
which, in short, explains its whole appearance. What indeed is
more beautiful than heaven, which of course contains all things
of beauty? This is proclaimed by its very names, *caelum* and
mundus, the latter denoting purity and ornament, the former a
carving. On account of heaven's transcendent perfection most
philosophers have called it a visible god. If then the value of the
arts is judged by the subject matter which they treat, that art
will be by far the foremost which is labelled astronomy by some,
astrology by others, but by many of the ancients, the consummation of mathematics. Unquestionably the summit of the
liberal arts and most worthy of a free man, it is supported by
almost all the branches of mathematics. Arithmetic, geometry,
optics, surveying, mechanics and whatever others there are all
contribute to it.

Although all the good arts serve to draw man's mind away
from vices and lead it toward better things, this function can be
more fully performed by this art, which also provides extraordinary intellectual pleasure. For when a man is occupied with
things which he sees established in the finest order and directed
by divine management, will not the unremitting contemplation
of them and a certain familiarity with them stimulate him to
the best and to admiration for the Maker of everything, in
whom are all happiness and every good? For would not the
godly Psalmist in vain declare that he was made glad through
the work of the Lord and rejoiced in the works of His hands,

were we not drawn to the contemplation of the highest good by this means, as though by a chariot?

The great benefit and adornment which this art confers on the commonwealth (not to mention the countless advantages to individuals) are most excellently observed by Plato. In the *Laws*, Book VII, he thinks that it should be cultivated chiefly because by dividing time into groups of days as months and years, it would keep the state alert and attentive to the festivals and sacrifices. Whoever denies its necessity for the teacher of any branch of higher learning is thinking foolishly, according to Plato. In his opinion it is highly unlikely that anyone lacking the requisite knowledge of the sun, moon and other heavenly bodies can become and be called godlike.

However, this divine rather than human science, which investigates the loftiest subjects, is not free from perplexities. The main reason is that its principles and assumptions, called 'hypotheses' by the Greeks, have been a source of disagreement, as we see, among most of those who undertook to deal with this subject, and so they did not rely on the same ideas. An additional reason is that the motion of the planets and the revolution of the stars could not be measured with numerical precision and completely understood except with the passage of time and the aid of many earlier observations, through which this knowledge was transmitted to posterity from hand to hand, so to say. To be sure, Claudius Ptolemy of Alexandria, who far excels the rest by his wonderful skill and industry, brought this entire art almost to perfection with the help of observations extending over a period of more than four hundred years, so that there no longer seemed to be any gap which he had not closed. Nevertheless very many things, as we perceive, do not agree with the conclusions which ought to follow from his system, and besides certain other motions have been discovered which were not yet known to him. Hence Plutarch too, in discussing the sun's tropical year, says that so far the motion of the heavenly bodies has eluded the skill of the astronomers. For, to use the year itself as an example, it is well known, I think, how different the opinions concerning it have always been, so that many have abandoned all hope that an exact determination of it could be found. The situation is the same with regard to other heavenly bodies.

Nevertheless, to avoid giving the impression that this difficulty is an excuse for indolence, by the grace of God, without whom we can accomplish nothing, I shall attempt a broader inquiry into these matters. For, the number of aids we have to assist our enterprise grows with the interval of time extending from the originators of this art to us. Their discoveries may be compared with what I have newly found. I acknowledge, moreover, that I shall treat many topics differently from my predecessors, and yet I shall do so thanks to them, for it was they who first opened the road to the investigation of these very questions.

The Universe Is Spherical – Chapter 1

First of all, we must note that the universe is spherical. The reason is either that, of all forms, the sphere is the most perfect, needing no joint and being a complete whole, which can be neither increased nor diminished; or that it is the most capacious of figures, best suited to enclose and retain all things; or even that all the separate parts of the universe, I mean the sun, moon, planets and stars, are seen to be of this shape; or that wholes strive to be circumscribed by this boundary, as is apparent in drops of water and other fluid bodies when they seek to be self-contained. Hence no one will question the attribution of this form to the divine bodies.

The Earth Too Is Spherical – Chapter 2

The earth also is spherical, since it presses upon its centre from every direction. Yet it is not immediately recognized as a perfect sphere on account of the great height of the mountains and depth of the valleys. They scarcely alter the general sphericity of the earth, however, as is clear from the following considerations. For a traveller going from any place toward the north, that pole of the daily rotation gradually climbs higher, while the opposite pole drops down an equal amount. More stars in the north are seen not to set, while in the south certain stars are no longer seen to rise. Thus Italy does not see Canopus, which is visible in Egypt; and Italy does see the River's last star, which is unfamiliar to our area in the colder region. Such stars, conversely, move higher in the heavens for a traveller heading southward, while those which are high in our sky sink down. Meanwhile, moreover, the elevations of the poles have the same ratio everywhere to the portions of the earth that have been traversed. This happens on no other figure than the sphere. Hence the earth too is evidently enclosed between poles and is therefore spherical. Furthermore, evening eclipses of the sun and moon are not seen by easterners, nor morning eclipses by westerners, while those occurring in between are seen later by easterners but earlier by westerners.

The waters press down into the same figure also, as sailors are aware, since land which is not seen from a ship is visible from the top of its mast. On the other hand, if a light is attached to the top of the mast, as the ship draws away from land, those who remain ashore see the light drop down gradually until it finally disappears, as though setting. Water, furthermore, being fluid by nature, manifestly always seeks the same lower levels as earth and pushes up from the shore no higher than its rise permits. Hence whatever land emerges out of the ocean is admittedly that much higher.

How Earth Forms a Single Sphere with Water – Chapter 3

Pouring forth its seas everywhere, then, the ocean envelops the earth and fills its deeper chasms. Both tend toward the same centre because of their heaviness. Accordingly, there had to be less water than land, to avoid having the water engulf the entire earth and to have the water recede from some portions of the land and from the many islands lying here and there, for the preservation of living creatures. For what are the inhabited countries and the mainland itself but an island larger than the others?

We should not heed certain Peripatetics [Aristotelians] who declared that the entire body of water is ten times greater than all the land. For, according to the conjecture which they accepted, in the transmutation of the elements as one unit of earth dissolves, it becomes ten units of water. They also assert that the earth bulges out to some extent as it does because it is not of equal weight everywhere on account of its cavities, its centre of gravity being different from its centre of magnitude. But they err through ignorance of the art of geometry. For they do not realize that the water cannot be even seven times greater and still leave any part of the land dry, unless earth as a whole vacated the centre of gravity and yielded that position to water, as if the latter were heavier than itself. For spheres are to each other as the cubes of their diameters. Therefore, if earth were the eighth part to seven parts of water, earth's diameter could not be greater than the distance from centre to the circumference of the waters. So far are they from being as much as ten times greater.

Moreover, there is no difference between the earth's centres of gravity and magnitude. This can be established by the fact that from the ocean inward the curvature of the land does not mount steadily in a continuous rise. If it did, it would keep the sea water out completely and in no way permit the inland seas and such vast gulfs to intrude. Furthermore, the depth of the abyss would never stop increasing from the shore of the ocean outward, so that no island or reef or any form of land would be encountered by sailors on the longer voyages. But it is well known that almost in the middle of the inhabited lands barely fifteen furlongs remain between the eastern Mediterranean and the Red Sea. On the other hand, in his *Geography* Ptolemy extended the habitable area halfway around the world. Beyond that meridian, where he left unknown land, the moderns have added Cathay and territory as vast as sixty degrees of longitude, so that now the earth is inhabited over a greater stretch of longitude than is left for the ocean. To these regions, moreover, should be added the islands discovered in our time under the rulers of Spain and Portugal, and especially America, named after the ship's captain who found it. On account of its still undisclosed size it is thought to be a second group of inhabited countries. There are also many other islands, heretofore unknown. So little reason have we to marvel at the existence of antipodes or antichthones. Indeed, geometrical reasoning about the location of America compels us to believe that it is diametrically opposite the Ganges district of India.

From all these facts, finally, I think it is clear that land and water together press upon a single centre of gravity; that the earth has no other centre of magnitude; in that, since earth is heavier, its gaps are filled with water; and that consequently there is little water in comparison with land, even though more water perhaps appears on the surface.

The earth together with its surrounding waters must in fact have such a shape as its shadow reveals, for it eclipses the moon with the arc of a perfect circle. Therefore the earth is not flat, as Empedocles and Anaximenes thought; nor drum-shaped, as Leucippus; nor bowl-shaped, as Heraclitus; nor hollow in another way, as Democritus; nor again cylindrical, as Anaximander; nor does its lower side extend infinitely downward, the thickness diminishing toward the bottom, as Xenophanes taught; but it is perfectly round, as the philosophers hold.

The Motion of the Heavenly Bodies Is Uniform, Eternal and Circular or Compounded of Circular Motions – Chapter 4

I shall now recall to mind that the motion of the heavenly bodies is circular, since the motion appropriate to a sphere is rotation in a circle. By this very act the sphere expresses its form as the simplest body, wherein neither beginning nor end can be found, nor can the one be distinguished from the other, while the sphere itself traverses the same points to return upon itself.

In connection with the numerous spheres, however, there are many motions. The most conspicuous of all is the daily rotation, which the Greeks call *nuchthēmeron*, that is, the interval of a day and a night. The entire universe, with the exception of the earth, is conceived as whirling from east to west in this rotation. It is recognized as the common measure of all motions, since we even compute time itself chiefly by the number of days.

Secondly, we see other revolutions as advancing in the opposite direction, that is, from west to east; I refer to those of the sun, moon and five planets. The sun thus regulates the year for us, and the moon the month, which are also very familiar periods of time. In like manner each of the other five planets completes its own orbit.

Yet these motions differ in many ways. In the first place, they do not swing around the same poles as the first motion, but run obliquely through the zodiac. Secondly, these bodies are not seen moving uniformly in their orbits, since the sun and moon are observed to be sometimes slow, at other times faster in their course. Moreover, we see the other five planets also retrograde at times, and stationary at either end. And whereas the sun always advances along its own direct path, they wander in various ways, straying sometimes to the south and sometimes to the north; that is why they are called 'planets'. Furthermore, they are at times nearer to the earth, when they are said to be in perigee; at other times they are further away, when they are said to be in apogee.

We must acknowledge, nevertheless, that their motions are circular or compounded of several circles, because these non-uniformities recur regularly according to a constant law. This could not happen unless the motions were circular, since only the circle can bring back the past. Thus, for example, by a composite motion of circles the sun restores to us the inequality of days and nights as well as the four seasons of the year. Several motions are discerned herein, because a simple heavenly body cannot be moved by a single sphere non-uniformly. For this non-uniformity would have to be caused either by an inconstancy, whether imposed from without or generated from within, in the moving force or by an alteration in the revolving body. From either alternative, however, the intellect shrinks. It is improper to conceive any such defect in objects constituted in the best order.

It stands to reason, therefore, that their uniform motions appear non-uniform to us. The cause may be either that their circles have poles different from the earth's or that the earth is not at the centre of the circles on which they revolve. To us who watch the course of these planets from the earth, it happens that our eye does not keep the same distance from every part of their orbits, but on account of their varying distances these bodies seem larger when nearer than when further away (as has been proved in optics). Likewise, in equal arcs of their orbits their motions will appear unequal in equal times on account of the observer's varying distance. Hence I deem it above all necessary that we should carefully scrutinize the relation of the earth to the heavens lest, in our desire to examine the loftiest objects, we remain ignorant of things nearest to us, and by the same error attribute to the celestial bodies what belongs to the earth.

Does Circular Motion Suit the Earth?
What Is Its Position? – Chapter 5

Now that the earth too has been shown to have the form of a sphere, we must in my opinion see whether also in this case the form entails the motion, and what place in the universe is occupied by the earth. Without the answers to these questions it is impossible to find the correct explanation of what is seen in the heavens. To be sure, there is general agreement among the authorities that the earth is at rest in the middle of the universe. They hold the contrary view to be inconceivable or downright silly. Nevertheless, if we examine the matter more carefully, we shall see that this problem has not yet been solved and is therefore by no means to be disregarded.

Every observed change of place is caused by a motion of either the observed object or the observer or, of course, by an unequal displacement of each. For when things move with equal speed in the same direction, the motion is not perceived, as between the observed object and the observer, I mean. It is the earth, however, from which the celestial ballet is beheld in its repeated performances before our eyes. Therefore, if any motion is ascribed to the earth, in all things outside it the same motion will appear, but in the opposite direction, as though they were moving past it. Such in particular is the daily rotation, since it seems to involve the entire universe except the earth and what is around it. However, if you grant that the heavens have no part in this motion but that the earth rotates from west to east, upon earnest consideration you will find that this is the actual situation concerning the apparent rising and setting of the sun, moon, stars and planets. Moreover since the heavens, which enclose and provide the setting for everything, constitute the space common to all things, it is not at first blush clear why motion should not be attributed rather to the enclosed than to the enclosing, to the thing located in space rather than to the framework of space. This opinion was indeed maintained by Heraclides and Ecphantus, the Pythagoreans, and by Eficetas of Syracuse, according to Cicero. They rotated the earth in the middle of the universe, for they ascribed the setting of the stars to the earth's interposition, and their rising to its withdrawal.

If we assume its daily rotation, another and no less important question follows concerning the earth's position. To be sure, heretofore there has been virtually unanimous acceptance of the belief that the middle of the universe is the earth. Anyone who denies that the earth occupies the middle or centre of the universe may nevertheless assert that its distance is insignificant in comparison with that of the sphere of the fixed stars, but perceptible and noteworthy in relation to the spheres of the sun and the other planets. He may deem this to be the reason why their motions appear non-uniform, as conforming to a centre other than the centre of the earth. Perhaps he can produce a not inept explanation of the apparent non-uniform motion. For the fact that the same planets are observed nearer to the earth and further away necessarily proves that the centre of the earth is not the centre of their circles. It is less clear whether the approach and withdrawal are executed by the earth or the planets.

It will occasion no surprise if, in addition to the daily rotation, some other motion is assigned to the earth. That the earth rotates, that it also travels with several motions, and that it is one of the heavenly bodies are said to have been the opinions of Philolaus the Pythagorean. He was no ordinary astronomer, inasmuch as Plato did not delay going to Italy for the sake of visiting him, as Plato's biographers report.

But many have thought it possible to prove by geometrical reasoning that the earth is in the middle of the universe; that being like a point in relation to the immense heavens, it serves as their centre; and that it is motionless because, when the universe moves, the centre remains unmoved, and the things nearest to the centre are carried most slowly.

The Immensity of the Heavens Compared to the Size of the Earth – Chapter 6

The massive bulk of the earth does indeed shrink to insignificance in comparison with the size of the heavens. This can be ascertained from the fact that the boundary circles (for that is the translation of the Greek term 'horizons') bisect the entire sphere of the heavens. This could not happen if the earth's size or distance from the universe's centre were noteworthy in comparison with the heavens. For, a circle that bisects a sphere passes through its centre, and is the greatest circle that can be described on it.

Thus, let circle ABCD be a horizon, and let the earth, from which we do our observing, be E, the centre of the horizon, which separates what is seen from what is not seen. Now, through a dioptra or horoscopic instrument or water level placed at E, let the first point of the Crab be sighted rising at point C, and at that instant the first point of the Goat is perceived to be setting at A. Then A, E and C are on a straight line through the dioptra. This line is evidently a diameter of the ecliptic, since six visible signs form a semicircle, and E, the centre, is identical with the horizon's centre. Again, let the signs shift their position until the first point of the Goat rises at B. At that time the Crab will also be observed setting at D. BED will be a straight line and a diameter of the ecliptic. But, as we have already seen, AEC also is a diameter of the same circle. Its centre, obviously, is the intersection. A horizon, then, in this way always bisects the ecliptic, which is a great circle of the sphere. But on a sphere, if a circle bisects any great circle, the bisecting circle is itself a great circle. Consequently, a horizon is one of the great circles, and its centre is clearly identical with the centre of the ecliptic.

Yet a line drawn from the earth's surface must be distinct from the line drawn from the earth's centre. Nevertheless, because these lines are immense in relation to the earth, they become like parallel lines. Because their terminus is enormously remote they appear to be a single line. For in comparison with their length the space enclosed by them becomes imperceptible, as is demonstrated in optics. This reasoning certainly makes it quite clear that the heavens are immense by comparison with the earth and present the aspect of an infinite magnitude, while on the testimony of the senses the earth is related to the heavens as a point to a body, and a finite to an infinite magnitude.

But no other conclusion seems to have been established. For it does not follow that the earth must be at rest in the middle of the universe. Indeed, a rotation in twenty-four hours of the enormously vast universe should astonish us even more than a rotation of its least part, which is the earth. For, the argument that the centre is motionless, and what is nearest the centre moves the least, does not prove that the earth is at rest in the middle of the universe.

To take a similar case, suppose you say that the heavens rotate but the poles are stationary, and what is closest to the poles moves the least. The Little Bear, for example, being very close to the pole, is observed to move much more slowly than the Eagle or the Little Dog because it describes a smaller circle. Yet all these constellations belong to a single sphere. A sphere's movement, vanishing at its axis, does not permit an equal motion of all its parts. Nevertheless these are brought round in equal times, though not over equal spaces, by the rotation of the whole sphere. The upshot of the argument, then, is the claim that the earth as a part of the celestial sphere shares in the same nature and movement so that, being close to the centre, it has a slight motion. Therefore, being a body and not the centre, it too will describe arcs like those of a celestial circle, though smaller, in the same time. The falsity of this contention is clearer than daylight. For it would always have to be noon in one place, and always midnight in another, so that the daily risings and settings could not take place, since the motion of the whole and the part would be one and inseparable.

But things separated by the diversity of their situations are subject to a very different relation: those enclosed in a smaller orbit revolve faster than those traversing a bigger circle. Thus Saturn, the highest of the planets, revolves in thirty years; the moon, undoubtedly the nearest to the earth, completes its course in a month; and to close the series, it will be thought, the earth rotates in the period of a day and a night. Accordingly, the same question about the daily rotation emerges again. On the other hand, likewise still undetermined is the earth's position, which has been made even less certain by what was said above. For that proof establishes no conclusion other than the heavens' unlimited size in relation to the earth. Yet how far this immensity extends is not at all clear. At the opposite extreme are the very tiny indivisible bodies called 'atoms'. Being imperceptible, they do not immediately constitute a visible body when they are taken two or a few at a time. But they can be multiplied to such an extent that in the end there are enough of them to combine in a perceptible magnitude. The same may be said also about the position of the earth. Although it is not in the centre of the universe, nevertheless its distance therefrom is still insignificant, especially in relation to the sphere of the fixed stars.

Why the Ancients Thought That the Earth Remained at Rest in the Middle of the Universe as Its Centre – Chapter 7

Accordingly, the ancient philosophers sought to establish that the earth remains at rest in the middle of the universe by certain other arguments. As their main reason, however, they adduce heaviness and lightness. Earth is in fact the heaviest element, and everything that has weight is borne toward it in an effort to reach its inmost centre. The earth being spherical, by their own nature heavy objects are carried to it from all directions at right angles to its surface. Hence, if they were not checked at its surface, they would collide at its centre, since a straight line perpendicular to a horizontal plane at its point of tangency with a sphere leads to the centre. But things brought to the middle, it seems to follow, come to rest at the middle. All the more, then, will the entire earth be at rest in the middle, and as the recipient of every falling body it will remain motionless thanks to its weight.

In like manner, the ancient philosophers analyse motion and its nature in a further attempt to confirm their conclusion. Thus, according to Aristotle, the motion of a single simple body is simple; of the simple motions, one is straight and the other is circular; of the straight motions, one is upward and the other is downward. Hence every simple motion is either toward the middle, that is, downward; or away from the middle, that is, upward; or around the middle, that is, circular. To be carried downward, that is, to seek the middle, is a property only of earth and water, which are considered heavy; on the other hand, air and fire, which are endowed with lightness, move upward and away from the middle. To these four elements it seems reasonable to assign rectilinear motion, but to the heavenly bodies, circular motion around the middle. This is what Aristotle says.

Therefore, remarks Ptolemy of Alexandria, if the earth were to move, merely in a daily rotation, the opposite of what was said above would have to occur, since a motion would have to be exceedingly violent and its speed unsurpassable to carry the entire circumference of the earth around in twenty-four hours. But things which undergo an abrupt rotation seem utterly unsuited to gather [bodies to themselves], and seem more likely, if they have been produced by combination, to fly apart unless they are held together by some bond. The earth would long ago have burst asunder, he says, and dropped out of the skies (a quite preposterous notion); and, what is more, living creatures and any other loose weights would by no means remain unshaken. Nor would objects falling in a straight line descend perpendicularly to their appointed place, which would meantime have been withdrawn by so rapid a movement. Moreover, clouds and anything else floating in the air would be seen drifting always westward.

The Inadequacy of the Previous Arguments
and a Refutation of Them – Chapter 8

For these and similar reasons forsooth the ancients insist that the earth remains at rest in the middle of the universe, and that this is its status beyond any doubt. Yet if anyone believes that the earth rotates, surely he will hold that its motion is natural, not violent. But what is in accordance with nature produces effects contrary to those resulting from violence, since things to which force or violence is applied must disintegrate and cannot long endure. On the other hand, that which is brought into existence by nature is well ordered and preserved in its best state. Ptolemy has no cause, then, to fear that the earth and everything earthly will be disrupted by a rotation created through nature's handiwork, which is quite different from what art or human intelligence can accomplish.

But why does he not feel this apprehension even more for the universe, whose motion must be the swifter, the bigger the heavens are than the earth? Or have the heavens become immense because the indescribable violence of their motion drives them away from the centre? Would they also fall apart if they came to a halt? Were this reasoning sound, surely the size of the heavens would likewise grow to infinity. For the higher they are driven by the power of their motion, the faster that motion will be, since the circumference of which it must make the circuit in the period of twenty-four hours is constantly expanding; and, in turn, as the velocity of the motion mounts, the vastness of the heavens is enlarged. In this way the speed will increase the size, and the size the speed, to infinity. Yet according to the familiar axiom of physics that the infinite cannot be traversed or moved in any way, the heavens will therefore necessarily remain stationary.

But beyond the heavens there is said to be no body, no space, no void, absolutely nothing, so that there is nowhere the heavens can go. In that case it is really astonishing if something can be held in check by nothing. If the heavens are infinite, however, and finite at their inner concavity only, there will perhaps be more reason to believe that beyond the heavens there is nothing. For, every single thing, no matter what size it attains, will be inside them, but the heavens will abide motionless. For, the chief contention by which it is sought to prove that the universe is finite is its motion. Let us therefore leave the question whether the universe is finite or infinite to be discussed by the natural philosophers.

We regard it as a certainty that the earth, enclosed between poles, is bounded by a spherical surface. Why then do we still hesitate to grant it the motion appropriate by nature to its form rather than attribute a movement to the entire universe, whose limit is unknown and unknowable? Why should we not admit, with regard to the daily rotation, that the appearance is in the heavens and the reality in the earth? This situation closely resembles what Virgil's Aeneas says: 'Forth from the harbour we sail, and the land and the cities slip backward.'

For when a ship is floating calmly along, the sailors see its motion mirrored in everything outside, while on the other hand they suppose that they are stationary, together with everything on board. In the same way, the motion of the earth can unquestionably produce the impression that the entire universe is rotating.

Then what about the clouds and the other things that hang in the air in any manner whatsoever, or the bodies that fall down, and conversely those that rise aloft? We would only say that not merely the earth and the watery element joined with it have this motion, but also no small part of the air and whatever is linked in the same way to the earth. The reason may be either that the nearby air, mingling with earthy or watery matter, conforms to the same nature as the earth, or that the air's motion, acquired from the earth by proximity, shares without resistance in its unceasing rotation. No less astonishingly, on the other hand, is the celestial movement declared to be accompanied by the uppermost belt of air. This is indicated by those bodies that appear suddenly, I mean, those that the Greeks called 'comets' and 'bearded stars'. Like the other heavenly bodies, they rise and set. They are thought to be generated in that region. That part of the air, we can maintain, is unaffected by the earth's motion on account of its great distance from the earth. The air closest to the earth will accordingly seem to be still. And so will the things suspended in it, unless they are tossed to and fro, as indeed they are, by the wind or some other disturbance. For what else is the wind in the air but the wave in the sea?

We must in fact avow that the motion of falling and rising bodies in the framework of the universe is twofold, being in every case a compound of straight and circular. For, things that sink of their own weight, being predominantly earthy, undoubtedly retain the same nature as the whole of which they are parts. Nor is the explanation different in the case of those things, which, being fiery, are driven forcibly upward. For also fire here on the earth feeds mainly on earthy matter, and flame is defined as nothing but blazing smoke. Now it is a property of fire to expand what it enters. It does this with such great force that it cannot be prevented in any way by any device from bursting through restraints and completing its work. But the motion of expansion is directed from the centre to the circumference. Therefore, if any part of the earth is set afire, it is carried from the middle upwards. Hence the statement that the motion of a simple body is simple holds true in particular for circular motion, as long as the simple body abides in its natural place and with its whole. For when it is in place, it has none but circular motion, which remains wholly within itself like a body at rest. Rectilinear motion, however, affects things which leave their natural place or are thrust out of it or quit it in any manner whatsoever. Yet nothing is so incompatible with the orderly arrangement of the universe and the design of the totality as something out of place. Therefore rectilinear motion occurs only to things that are not in proper condition and are not in complete accord with their nature, when they are separated from their whole and forsake its unity.

Furthermore, bodies that are carried upward and downward, even when deprived of circular motion, do not execute a simple, constant and uniform motion. For they cannot be governed by their lightness or by the impetus of their weight. Whatever falls moves slowly at first but increases its speed as it drops. On the other hand, we see this earthly fire (for we behold no other), after it has been lifted up high, slacken all at once, thereby revealing the reason to be the violence applied to the earthy matter. Circular motion, however, always rolls along uniformly, since it has an unfailing cause. But rectilinear motion has a cause that quickly stops functioning. For when rectilinear motion brings bodies to their own place, they cease to be heavy or light, and their motion ends. Hence, since circular motion belongs to wholes, but parts have rectilinear motion in addition, we can say that 'circular' subsists with 'rectilinear' as 'being alive' with 'being sick'. Surely Aristotle's division of simple motion into three types, away from the middle, toward the middle, and around the middle, will be construed merely as a logical exercise. In like manner we distinguish line, point and surface, even though one cannot exist without another, and none of them without body.

As a quality, moreover, immobility is deemed nobler and more divine than change and instability, which are therefore better suited to the earth than to the universe. Besides, it would seem quite absurd to attribute motion to the framework of space or that which encloses the whole of space, and not, more appropriately, to that which is enclosed and occupies some space, namely, the earth. Last of all, the planets obviously approach closer to the earth and recede further from it. Then the motion of a single body around the middle, which is thought to be the centre of the earth, will be both away from the middle and also toward it. Motion around the middle, consequently, must be interpreted in a more general way, the sufficient condition being that each such motion encircle its own centre. You see, then, that all these arguments make it more likely that the earth moves than that it is at rest. This is especially true of the daily rotation, as particularly appropriate to the earth. This is enough, in my opinion, about the first part of the question.

Can Several Motions Be Attributed to the Earth? The Centre of the Universe – Chapter 9

Accordingly, since nothing prevents the earth from moving, I suggest that we should now consider also whether several motions suit it, so that it can be regarded as one of the planets. For, it is not the centre of all the revolutions. This is indicated by the planets' apparent non-uniform motion and their varying distances from the earth. These phenomena cannot be explained by circles concentric with the earth. Therefore, since there are many centres, it will not be by accident that the further question arises whether the centre of the universe is identical with the centre of terrestrial gravity or with some other point. For my part I believe that gravity is nothing but a certain natural desire, which the divine providence of the Creator of all things has implanted in parts, to gather as a unity and a whole by combining in the form of a globe. This impulse is present, we may suppose, also in the sun, the moon and the other brilliant planets, so that through its operation they remain in that spherical shape which they display. Nevertheless, they swing round their circuits in diverse ways. If, then, the earth too moves in other ways, for example, about a centre, its additional motions must likewise be reflected in many bodies outside it. Among these motions we find the yearly revolution. For if this is transformed from a solar to a terrestrial movement, with the sun acknowledged to be at rest, the risings and settings which bring the zodiacal signs and fixed stars into view morning and evening will appear in the same way. The stations of the planets, moreover, as well as their retrogradations and [resumptions of] forward motion will be recognized as being, not movements of the planets, but a motion of the earth, which the planets borrow for their own appearances. Lastly, it will be realized that the sun occupies the middle of the universe. All these facts are disclosed to us by the principle governing the order in which the planets follow one another, and by the harmony of the entire universe, if only we look at the matter, as the saying goes, with both eyes.

The Order of the Heavenly Spheres – Chapter 10

Of all things visible, the highest is the heaven of the fixed stars. This, I see, is doubted by nobody. But the ancient philosophers wanted to arrange the planets in accordance with the duration of the revolutions. Their principle assumes that of objects moving equally fast, those further away seem to travel more slowly, as is proved in Euclid's *Optics*. The moon revolves in the shortest period of time because, in their opinion, it runs on the smallest circle as the nearest to the earth. The highest planet, on the other hand, is Saturn, which completes the biggest circuit in the longest time. Below it is Jupiter, followed by Mars.

With regard to Venus and Mercury, however, differences of opinion are found. For, these planets do not pass through every elongation from the sun, as the other planets do. Hence Venus and Mercury are located above the sun by some authorities, like Plato's Timaeus, but below the sun by others, like Ptolemy and many of the moderns. Al-Bitruji places Venus above the sun, and Mercury below it.

According to Plato's followers, all the planets, being dark bodies otherwise, shine because they receive sunlight. If they were below the sun, therefore, they would undergo no great elongation from it, and hence they would be seen halved or at any rate less than fully round. For, the light which they receive would be reflected mostly upward, that is, toward the sun, as we see in the new or dying moon. In addition, they argue, the sun must sometimes be eclipsed by the interposition of these planets, and its light cut off in proportion to their size. Since this is never observed, these planets do not pass beneath the sun at all, according to those who follow Plato.

On the other hand, those who locate Venus and Mercury below the sun base their reasoning on the wide space which they notice between the sun and the moon. For the moon's greatest distance from the earth is $64\frac{1}{6}$ earth-radii. This is contained, according to them, about eighteen times in the sun's least distance from the earth, which is 1,160 earth-radii. Therefore between the sun and the moon there are 1,096 earth-radii. Consequently, to avoid having so vast a space remain empty, they announce that the same numbers almost exactly fill up the apsidal distances, by which they compute the thickness of those spheres. Thus the moon's apogee is followed by Mercury's perigee. Mercury's apogee is succeeded by the perigee of Venus, whose apogee, finally, almost reaches the sun's perigee. For between the apsides of Mercury they calculate about $177\frac{1}{2}$ earth-radii. Then the remaining space is very nearly filled by Venus's interval of 910 earth-radii.

Therefore they do not admit that these heavenly bodies have any opacity like the moon's. On the contrary, these shine either with their own light or with the sunlight absorbed throughout their bodies. Moreover, they do not eclipse the sun, because it rarely happens that they interfere with our view of the sun, since they generally deviate in latitude. Besides, they are tiny bodies in comparison with the sun. Venus, although bigger than Mercury, can occult barely a hundredth of the sun. So says Al-Battani of Raqqa, who thinks that the sun's diameter is ten times larger [than Venus's], and therefore so minute a speck is not easily descried in the most brilliant light. Yet in his *Paraphrase* of Ptolemy, Ibn Rushd reports having seen something blackish when he found a conjunction of the sun and Mercury indicated in the tables. And thus these two planets are judged to be moving below the sun's sphere.

But this reasoning also is weak and unreliable. This is obvious from the fact that there are 38 earth-radii to the moon's perigee, according to Ptolemy, but more than 49 according to a more accurate determination, as will be made clear below. Yet so great a space contains, as we know, nothing but air and, if you please, also what is called 'the element of fire'. Moreover, the diameter of Venus's epicycle which carries it 45° more or less to either side of the sun, must be six times longer than the line drawn from the earth's centre to Venus's perigee, as will be demonstrated in the proper place. In this entire space which would be taken up by that huge epicycle of Venus and which, moreover, is so much bigger than what would accommodate the earth, air, ether, moon and Mercury, what will they say is contained if Venus revolved around a motionless earth?

Ptolemy argues also that the sun must move in the middle between the planets which show every elongation from it and those which do not. This argument carries no conviction because its error is revealed by the fact that the moon too shows every elongation from the sun.

Now there are those who locate Venus and then Mercury below the sun, or separate these planets [from the sun] in some other sequence. What reason will they adduce to explain why Venus and Mercury do not likewise traverse separate orbits divergent from the sun, like the other planets, without violating the arrangement [of the planets] in accordance with their [relative] swiftness and slowness? Then one of two alternatives will have to be true. Either the earth is not the centre to which the order of the planets and spheres is referred, or there really is no principle of arrangement nor any apparent reason why the highest place belongs to Saturn rather than to Jupiter or any other planet.

In my judgement, therefore, we should not in the least disregard what was familiar to Martianus Capella, the author of an encyclopedia, and to certain other Latin writers. For according to them, Venus and Mercury revolve around the sun as their centre. This is the reason, in their opinion, why these planets diverge no further from the sun than is permitted by the curvature of their revolutions. For they do not encircle the earth, like the other planets, but 'have opposite circles'. Then what else do these authors mean but that the centre of their spheres is near the sun? Thus Mercury's sphere will surely be enclosed within Venus's, which by common consent is more than twice as big, and inside that wide region it will occupy a space adequate for itself. If anyone seizes this opportunity to link Saturn, Jupiter and Mars also to that centre, provided he understands their spheres to be so large that together with Venus and Mercury the earth too is enclosed inside and encircled, he will not be mistaken, as is shown by the regular pattern of their motions.

For [these outer planets] are always closest to the earth, as is well known, about the time of their evening rising, that is, when they are in opposition to the sun, with the earth between them and the sun. On the other hand, they are at their furthest from the earth at the time of their evening setting, when they become invisible in the vicinity of the sun, namely, when we have the sun between them and the earth. These facts are enough to show that their centre belongs more to the sun, and is identical with the centre around which Venus and Mercury likewise execute their revolutions.

But since all these planets are related to a single centre, the space between Venus's convex sphere and Mars's concave sphere must be set apart as also a sphere or spherical shell, both of whose surfaces are concentric with those spheres. This [intercalated sphere] receives the earth together with its attendant, the moon, and whatever is contained within the moon's sphere. Mainly for the reason that in this space we find quite an appropriate and adequate place for the moon, we can by no means detach it from the earth, since it is incontrovertibly nearest to the earth.

Hence I feel no shame in asserting that this whole region engirdled by the moon, and the centre of the earth, traverse this grand circle amid the rest of the planets in an annual revolution around the sun. Near the sun is the centre of the universe. Moreover, since the sun remains stationary, whatever appears as a motion of the sun is really due rather to the motion of the earth. In comparison with any other spheres of the planets, the distance from the earth to the sun has a magnitude which is quite appreciable in proportion to those dimensions. But the size of the universe is so great that the distance earth–sun is imperceptible in relation to the sphere of the fixed stars. This should be admitted, I believe, in preference to perplexing the mind with an almost infinite multitude of spheres, as must be done by those who kept the earth in the middle of the universe. On the contrary, we should rather heed the wisdom of nature. Just as it especially avoids producing anything superfluous or useless, so it frequently prefers to endow a single thing with many effects.

All these statements are difficult and almost inconceivable, being of course opposed to the beliefs of many people. Yet, as we proceed, with God's help I shall make them clearer than sunlight, at any rate to those who are not unacquainted with the science of astronomy. Consequently, with the first principle remaining intact, for nobody will propound a more suitable principle than that the size of the spheres is measured by the length of the time, the order of the spheres is the following, beginning with the highest.

The first and the highest of all is the sphere of the fixed stars, which contains itself and everything, and is therefore immovable. It is unquestionably the place of the universe, to which the motion and position of all the other heavenly bodies are compared. Some people think that it also shifts in some way.

A different explanation of why this appears to be so will be adduced in my discussion of the earth's motion.

This is followed by the first of the planets, Saturn, which completes its circuit in thirty years. After Saturn, Jupiter accomplishes its revolution in twelve years. Then Mars revolves in two years. The annual revolution takes the series' fourth place, which contains the earth, as I said, together with the lunar sphere as an epicycle. In the fifth place Venus returns in nine months. Lastly, the sixth place is held by Mercury, which revolves in a period of eighty days.

At rest, however, in the middle of everything is the sun. For in this most beautiful temple, who would place this lamp in another or better position than that from which it can light up the whole thing at the same time? For, the sun is not inappropriately called by some people the lantern of the universe, its mind by others, and its ruler by still others. [Hermes] the Thrice Greatest labels it a visible god, and Sophocles' Electra, the all-seeing. Thus indeed, as though seated on a royal throne, the sun governs the family of planets revolving around it. Moreover, the earth is not deprived of the moon's attendance. On the contrary, as Aristotle says in a work on animals, the moon has the closest kinship with the earth. Meanwhile the earth has intercourse with the sun, and is impregnated for its yearly parturition.

In this arrangement, therefore, we discover a marvellous symmetry of the universe, and an established harmonious linkage between the motion of the spheres and their size, such as can be found in no other way. For this permits a not inattentive student to perceive why the forward and backward arcs appear greater in Jupiter than in Saturn and smaller than in Mars, and on the other hand greater in Venus than in Mercury. This reversal in direction appears more frequently in Saturn than in Jupiter, and also more rarely in Mars and Venus than in Mercury. Moreover, when Saturn, Jupiter and Mars rise at sunset, they are nearer to the earth than when they set in the evening or appear at a later hour. But Mars in particular, when it shines all night, seems to equal Jupiter in size, being distinguished only by its reddish colour. Yet in the other configurations it is found barely among the stars of the second magnitude, being recognized by those who track it with assiduous observations. All these phenomena proceed from the same cause, which is in the earth's motion.

Yet none of these phenomena appears in the fixed stars. This proves their immense height, which makes even the sphere of the annual motion, or its reflection, vanish from before our eyes. For, every visible object has some measure of distance beyond which it is no longer seen, as is demonstrated in optics. From Saturn, the highest of the planets, to the sphere of the fixed stars there is an additional gap of the largest size. This is shown by the twinkling lights of the stars. By this token in particular they are distinguished from the planets, for there had to be a very great difference between what moves and what does not move. So vast, without any question, is the divine handiwork of the most excellent Almighty.

Proof of the Earth's Triple Motion – Chapter 11

In so many and such important ways, then, do the planets bear witness to the earth's mobility. I shall now give a summary of this motion, insofar as the phenomena are explained by it as a principle. As a whole, it must be admitted to be a threefold motion.

The first motion, named *nuchthēmeron* by the Greeks, as I said, is the rotation which is the characteristic of a day plus a night. This turns around the earth's axis from west to east, just as the universe is deemed to be carried in the opposite direction. It describes the equator, which some people call the 'circle of equal days', in imitation of the designation used by the Greeks, whose term for it is *isēmerinos*.

The second is the yearly motion of the centre, which traces the ecliptic around the sun. Its direction is likewise from west to east, that is, in the order of the zodiacal signs. It travels between Venus and Mars, as I mentioned, together with its associates. Because of it, the sun seems to move through the zodiac in a similar motion. Thus, for example, when the earth's centre is passing through the Goat, the sun appears to be traversing the

Crab; with the earth in the Water Bearer, the sun seems to be in the Lion, and so on, as I remarked.

To this circle, which goes through the middle of the signs, and to its plane, the equator and the earth's axis must be understood to have a variable inclination. For if they stayed at a constant angle, and were affected exclusively by the motion of the centre, no inequality of days and nights would be observed. On the contrary, it would always be either the longest or shortest day or the day of equal daylight and darkness, or summer or winter, or whatever the character of the season, it would remain identical and unchanged.

The third motion in inclination is consequently required. This also is a yearly revolution, but it occurs in the reverse order of the signs, that is, in the direction opposite to that of the motion of the centre. These two motions are opposite in direction and nearly equal in period. The result is that the earth's axis and equator, the largest of the parallels of latitude on it, face almost the same portion of the heavens, just as if they remained motionless. Meanwhile the sun seems to move through the obliquity of the ecliptic with the motion of the earth's centre, as though this were the centre of the universe. Only remember that, in relation to the sphere of the fixed stars, the distance between the sun and the earth vanishes from our sight forthwith.

Copernicus, Heliocentric System from
On the Revolutions, 1543

Neither Copernicus's Little Commentary *nor Rheticus's* First
Account *included diagrams (though Kepler's teacher Michael
Maestlin would add some to the latter in later printings). The first
image depicting the new heliocentric system was this one from*
On the Revolutions. *This image was not drawn to scale, nor
did it depict Copernicus's conception of the actual complexity of
the planetary motions; rather, it offered a highly simplified and
schematic look at the ordering of celestial bodies according to
Copernicus, from the central sun to the outer realm of the fixed
stars, which bounded the cosmos. The circles here may represent
cross-sections of the planetary spheres, the three-dimensional
bodies in which the planets and stars moved – which Copernicus
accepted as real, though he did not take a clear stand on the question of their solidity.*

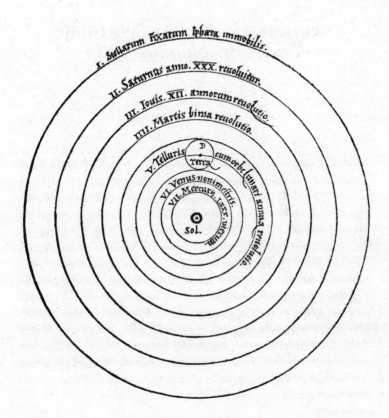

Copernicus, Tusi Couple from
On the Revolutions, 1543

The thirteenth-century Persian astronomer Nasir al-Din al-Tusi developed a mathematical device that became known as the Tusi couple. The device used two circles, one with a diameter twice the size of the other; their combined motions allowed a point on the circumference of the smaller circle to move in a straight line. Al-Tusi and others of his school were critical of some of the devices employed by Ptolemaic astronomy, like the equant point, and preferred to rely on perfect circles instead; the Tusi couple was one solution to the problem of how to account for latitudinal motion of the planets using only perfect circles. In the diagram here, Copernicus clearly depicts a Tusi couple, which was central to his own elimination of Ptolemaic equants, though he does not mention its origin. Transmission from Islamic astronomy seems likely (though its precise route is uncertain) since Copernicus uses very similar letters to label his diagram as those in al-Tusi's earlier manuscript.

cludit tandem quã diximus intortã lineam FKILGMINF. Itacǫ
manifestum est, quòd in una reuersione obliquitatis bis præce
dentium biscǫ sequentium limitem terræ polus attingit.

Quomodo motus reciprocus siue librationis ex
circularibus constet. Cap. IIII.

Vod igitur iste motus apparentijs consentiat am-
modo declarabimus. Interim uero quæret aliquis,
quo nam modo possit illarum librationum æquali-
tas intelligi, cum à principio dictum sit, motum cçle
stem æqualẽ esse, uel ex æqualibus ac circularibus cõpositum.
Hic aũt utrobicǫ duo motus
in uno apparẽt sub utriscǫ ter
minis, qbus necesse est cessa-
tionẽ interuenire. Fatebimur
quidem geminatos esse, at ex
ǫqualibus hoc modo demon
strant. Sit recta linea A B, quç
quadrifariã secetur in C D E si
gnis, & in D describãtur circu
li homocentri, ac in eodẽ pla
no A D B, & C D E, & in circũfe-
rentia interioris circuli assu-
mat utcũcǫ F signũ, & in ipso
F cẽtro, interuallo uero F D cir
culus describatur G H D, qui

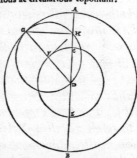

secet A B rectã lineã in H signo, & agat dimetiẽs D F G. Ostẽdendũ
est, cǫ geminis motibus circuloru G H D & C F E cõcurrẽtibus in-
uicẽ H mobile p eandẽ rectam lineã A B hinc inde reciprocãdo re
pat. Quod erit, si intelligat H moueri in diuersam partẽ, & duplo
magis ipso F. Quoniã idẽ angulus, q sub C D F in cẽtro circuli C F E
& circũferẽtia ipsius G H D cõsistẽs cõpræhẽdit utrãcǫ circuferen
tiã circuloru ǫqliũ G H dupla ipsi F C, posito cǫ aliquãdo in cõiun
ctiõe rectaru linearu A C D & D F G mobile H fuerit in G cõgruente
cũ A, & F in C. Nũc aũt in dextras ptes p F C motũ est centrũ F &
ipsum H p G H circumferentiã in sinistras duplo maiores ipsi C F.

r iij   uel

Erasmus Reinhold, *Prutenic Tables*, 1551

Copernicus had asserted that the theories and models of his book would be useful for practical purposes. Erasmus Reinhold (1511–1553), astronomy professor at the University of Wittenberg, was one of the first to realize that claim by constructing astronomical tables based on Copernicus's work. These tables were named after Reinhold's patron, Duke Albert of Prussia – they were thus called the Prussian, or, as more commonly, the Prutenic Tables. *Reinhold's tables were one of the first and fastest ways that Copernicus's theories spread. Through Reinhold, they were disseminated as mathematical models that offered improved and simpler calculations for planetary positions, eclipses and calendars, as well as for astrology. Yet they did not spread as a true physical system, for Reinhold still adhered to the older geocentric model. Still, Reinhold did celebrate Copernicus's work, writing in the* Prutenic Tables *that 'all posterity will, with a grateful heart, celebrate the name of Copernicus, by whose labour and zeal the doctrine of celestial motions itself, nearly collapsed, has again been restored. His great light, kindled with the help of God, led him to find and disclose many things that until this age were either unknown or hidden'.*

1 Sexagena.

Gradus	Subtrahe Eccentri. par ' ''	Dif. A ' ''	Scru. Prop. ' ''	Dif. A	Adde Parallaxis orbis p2r ' ''	Dif. A ' ''	Excessus parall. par ' ''	Dif. A ' ''	Gradus
30	11 3 3	1 10	20 21	29	30 59 4	16 1	5 1 58	4 48	30
31	11 4 2	1 0	20 50	29	31 14 51	15 47	5 6 51	4 53	29
32	11 4 49	0 47	21 19	29	31 30 25	15 34	5 11 48	4 57	28
33	11 5 24	0 35	21 48	29	31 45 45	15 20	5 16 49	5 1	27
34	11 5 48	0 24	22 18	30	32 0 51	15 6	5 21 53	5 4	26
35	11 5 59	0 11	22 48	30	32 15 42	14 51	5 27 2	5 9	25
36	11 5 59	0 0	23 19	31	32 30 18	14 36	5 32 15	5 13	24
37	11 5 46	0 13	23 49	30	32 44 38	14 20	5 37 33	5 18	23
38	11 5 22	0 24	24 20	31	32 58 42	14 4	5 43 55	5 22	22
39	11 4 45	0 37	24 51	32	33 12 29	13 47	5 48 21	5 31	21
40	11 3 57	0 48	25 33	32	33 25 59	13 30	5 53 52	5 37	20
41	11 2 56	1 1	25 55	32	33 39 11	13 12	5 59 29	5 42	19
42	11 1 42	1 14	26 27	32	33 52 4	12 53	6 5 11	5 47	18
43	11 0 17	1 25	26 59	33	34 4 38	12 34	6 10 58	5 51	17
44	10 58 39	1 38	27 32	33	34 16 53	12 15	6 16 49	5 57	16
45	10 56 49	1 50	28 5	33	34 28 47	11 54	6 22 46	5 57	15
46	10 54 46	2 3	28 38	33	34 40 20	11 33	6 28 49	6 3	14
47	10 52 31	2 15	29 11	33	34 51 31	11 11	6 34 57	6 8	13
48	10 50 3	2 28	29 44	34	35 2 20	10 49	6 41 11	6 14	12
49	10 47 23	2 40	30 18	34	35 12 45	10 25	6 47 31	6 20	11
50	10 44 20	2 53	30 52	34	35 22 45	10 0	6 53 57	6 26	10
51	10 41 25	3 5	31 26	34	35 32 20	9 35	7 0 29	6 32	9
52	10 38 7	3 18	32 0	34	35 41 30	9 10	7 7 7	6 38	8
53	10 34 37	3 30	32 34	34	35 50 13	8 43	7 13 51	6 44	7
54	10 30 54	3 43	33 8	35	35 58 28	8 15	7 20 41	6 50	6
55	10 26 59	3 55	33 42	35	36 6 13	7 45	7 27 39	6 58	5
56	10 22 51	4 8	34 18	34	36 13 28	7 15	7 34 44	7 5	4
57	10 18 31	4 20	34 52	35	36 20 12	6 44	7 41 56	7 12	3
58	10 13 58	4 33	35 27	35	36 26 25	6 13	7 49 13	7 17	2
59	10 9 12	4 46	36 2	35	36 32 3	5 38	7 56 38	7 25	1
60	10 4 14	4 58	36 37		36 37 7	5 4	8 4 9	7 31	0
	Adde	S A		S	Subtrahe	S		S	Gradus

4 Sexagenæ

S. 3.

Robert Recorde, *The Castle of Knowledge*, 1556

Robert Recorde (c.1510–1558) was a physician and a fellow of mathematics at Oxford. His 1556 Castle of Knowledge *was an introduction to the study of astronomy written as a set of dialogues between a master and his student. Published only thirteen years after Copernicus's* On the Revolutions, *it is the first book that discusses Copernicus's heliocentric theory in English. Copernicus is mentioned only very briefly, in the last of the volume's four books. There, Recorde treated Copernicus's theory as a claim about the physical world, rather than an instrumental fiction as Osiander's anonymous preface had suggested. While he exhibited more openness to the earth's diurnal rotation ('circular motion') than annual revolution ('direct motion out of the centre of the world'), Recorde ultimately did not commit himself to one position or the other, but rather deferred the issue, which he judged too advanced for an introductory discussion.*

SCHOLAR: I hear all learned men say Ptolemy is the father of that Art [astronomy], and proves all his words by strong and invincible reasons.

MASTER: No man can worthily praise Ptolemy, his travail being so great, his diligence so exact in observations and conference with all nations, and all ages, and his reasonable examination of all opinions with demonstrable confirmation of his own assertion. Yet must you and all men take heed, that both in him and in all men's works, you be not abused by their authority, but evermore attend to their reasons, and examine them well, ever regarding more what is said, and how it is proved, then who says it: for authority oftentimes deceives many men.

But as for the quietness of the earth I need not spend any time in proving it, since that opinion is so fixed in most men's heads, that they account it mere madness to bring the question in doubt. And therefore it is as much folly to travail to prove that which no man denies, as it were with great study to dissuade that thing, which no man does covet nor any man allows: or to blame that which no man praises, nor any man likes.

SCHOLAR: Yet sometime it chances, that the opinion most generally received, is not most true.

MASTER: And so do some men judge of this matter, for not only Eraclides Ponticus, a great philosopher, and two great clerks of Pythagoras' school, Philolaus and Ecphantus, were of the contrary opinion, but also Nicias Syracusius, and Aristarchus Samius, seem with strong arguments to approve it: but the reasons are difficult for this Introduction, & therefore I will omit them till another time. And so will I do the reasons that Ptolemy, Theon, & others do allege, to prove the earth to be without motion: and the rather, because those reasons do not proceed so demonstrably, but they may be answered fully, of him that holds the contrary. I mean, concerning circular motion: marry direct motion out of the centre of the world, seems more easily confuted, and by the same reason, which being before alleged for proving the earth to be in the middle and centre of the world.

SCHOLAR: I perceive it well: for as if the earth were always out of the centre of the world, those former absurdities would at all times appear: so if at any time the earth should move out of his place, those inconveniences would then appear.

MASTER: This is truly to be gathered: howbeit, Copernicus a man of great learning, of much experience, and of wonderful diligence in observation, has renewed the opinion of Aristarchus Samius, and affirms that the earth not only moves circularly about his own centre, but also may be, yea and is, continually out of the precise centre of the world 38 hundred thousand miles: but because the understanding of that controversy depends upon profounder knowledge than in this Introduction may be uttered conveniently, I will let it pass till some other time.

SCHOLAR: Nay sir in good faith, I desire not to hear such vain fantasies, so far against common reason, and repugnant to the consent of all the learned multitude of Writers, and therefore let it pass for ever, and a day longer.

MASTER: You are too young to be a good judge in so great a matter: it passes far your learning, and theirs also that are much better learned than you, to improve his suppositions by good arguments, and therefore you were best to condemn no thing that you do not well understand: but another time, as I said, I will so declare his supposition, that you shall not only wonder to hear it, but also peradventure be as earnest then to credit it as you are now to condemn it.

Tycho Brahe, *On the New Star*, 1573

In 1572 a supernova appeared in the constellation Cassiopeia that was so bright it was visible by day for two weeks, and by night for fourteen months. It sparked attention and debate among astronomers across Europe. Among them was the Danish nobleman and astronomer Tycho Brahe (1546–1601), whose measurements of its position with respect to other stars revealed no observable parallax, suggesting that the new star was extremely distant, beyond not only the realm of the moon but also of the planets. The old Aristotelian wisdom, broadly accepted by everyone including Tycho himself until the observations of the new star, had insisted on a dichotomy of heavens and earth: in this view, change only occurred below the sphere of the moon, while the celestial realm was immutable. Tycho's recorded observations and arguments to the contrary in his On the New Star, *like the earlier models of Copernicus, forced some readers to question their assumptions about the nature of the heavens and the ways in which they differed from the earth. The image shows Tycho's depiction of the new star (at I) and its position with respect to Cassiopeia.*

Last year, on 11 November, in the evening after sunset, I was contemplating the stars in the clear sky, as is my custom, when I observed a star nearly overhead that was new and unusual, shining more conspicuously than the others. And since I have known all the stars perfectly almost since childhood (for there is no great difficulty in this science), it was quite clear that no star had existed previously in that place in the sky, not even a small one, still less one so conspicuously bright. But all philosophers agree, and the facts themselves clearly declare, that there is no alteration of either generation or corruption in the ethereal region of the celestial world, but that the heavens and the ethereal bodies contained in them do not increase or diminish, or vary in number, or magnitude, or brightness, or in any other way. Rather, they are always the same, similar to themselves in all ways, enduring with no wearing out over the years. Moreover, observations of all experts made over thousands of years testify that all the stars have remained the same in number, position, order, motion and quantity. But this star is located in the eighth sphere, together with the other fixed stars, and has noticeably changed both its magnitude and its colour, and is not in the region of the elements, below the sphere of the moon. We have proven, through infallible demonstrations, that this new body was generated in the ether itself, contrary to the opinions and decrees of all the philosophers.

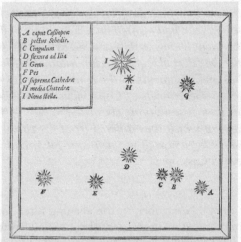

Distantiam verò huius stellæ à fixis aliquibus in hac Cassiopeiæ constellatione, exquisito instrumento, & omnium minutorum capaci, aliquoties observaui. Inueni autem eam distare ab ea, quæ est in pectore, Schedir appellata B, 7. partibus & 55. minutis: à superiori
verò

Thomas Digges, *A Perfect Description of the Celestial Orbs*, 1576

Thomas Digges (1546–1595) was the son of the English mathematician Leonard Digges. He was tutored by the polymath John Dee, who was acquainted with the Copernican system and had written a preface to John Feild's Ephemeris, *the first English version of the* Prutenic Tables *based on Copernicus's heliocentric model. Neither Dee nor Feild endorsed the Copernican system as true, and Digges was the first to issue his full support for heliocentrism in English. He did so in 1576 in a newly issued edition of his father's* Prognostication Everlasting. *That book had endorsed the Ptolemaic system and included a diagram of the heavens based upon it. For his new edition, which was widely read, Thomas Digges added an appendix entitled* A Perfect Description of the Celestial Orbs *in which he argued for the reality and superiority of the Copernican system and offered some refutations of the older arguments for a stationary earth. He also included a translation into English of parts of Book I of* On the Revolutions, *with some comments and additions.*

To the Reader:

Having of late (gentle reader) corrected and reformed sundry faults that by negligence in printing have crept into my father's General Prognostication, among other things I found a description or Model of the world, and situation of Spheres Celestial and Elementary, according to the doctrine of Ptolemy, whereunto all Universities (led thereto chiefly by the authority of

Aristotle) sithens have consented. But in this our age, one rare wit (seeing the continual errors that from time to time more & more have been discovered, besides the infinite absurdities in their Theorickes, which they have been forced to admit that would not confess any mobility in the ball of the earth) has by long study, painful practice, and rare invention delivered a new Theorick or model of the world, showing that the Earth rests not in the Centre of the whole world, but only in the Centre of this our mortal world or Globe of Elements, which environed and enclosed in the Moon's Orb, and together with the whole Globe of mortality, is carried yearly round about the Sun, which like a king in the midst of all reigns and gives laws of motion to the rest, spherically dispersing his glorious beams of light through all this sacred Celestial Temple. And the Earth itself to be one of the Planets, having his peculiar & strays courses turning every twenty-four hours round upon his own Centre, whereby the Sun and great Globe of fixed stars seem to sway about and turn, albeit indeed they remain fixed. So many ways is the sense of mortal men abused, but reason and deep discourse of wit having opened these things to Copernicus, & the same being with demonstrations Mathematical, most apparently by him to the world delivered.

I thought it convenient, together with the old Theorick, also to publish this, to the end such noble English minds (as delight to reach above the baser sort of men) might not be altogether defrauded of so noble a part of Philosophy. And to the end it might manifestly appear that Copernicus meant not as some have fondly accused him, to deliver these grounds of the Earth's mobility only as Mathematical principles, feigned & not as Philosophical truly averred. I have also from him delivered both the Philosophical reasons by Aristotle and others produced to maintain the Earth's stability, and also their solutions and insufficiency, wherein I cannot a little commend the modesty of that grave Philosopher Aristotle, who seeing (no doubt) the insufficiency of his own reasons in seeking to confute the Earth's motion, uses these words: 'these things are explained as much as our ability allows'. Howbeit his disciples have not with like sobriety maintained the same.

Thus much for my own part in this case I will only say: there is no doubt but of a true ground truer effects may be produced than of principles that are false, and of true principles falsehood or absurdity cannot be inferred. If, therefore, the Earth be situate immoveable in the Centre of the world, why find we not Theorickes upon that ground to produce effects as true and certain as these of Copernicus? Why cast we not away those equant circles and motions irregular, seeing our own Philosopher Aristotle himself, the light of our Universities, hath taught us: 'A simple motion is appropriate for a simple body.' But if contrary, it be found impossible (the Earth's stability being granted) but that we must necessarily fall into these absurdities, and cannot by any means avoid them. Why shall we so much dote in the appearance of our senses, which many ways may be abused, and not suffer our selves to be directed by the rule of Reason, which the great GOD hath given us as a Lamp to lighten the darkness of our understanding, and the perfect guide to lead us to the golden branch of Verity amid the Forest of errors.

A Perfect Description of the Celestial Orbs according to the most ancient doctrine of the Pythagoreans, lately revived by Copernicus and by Geometrical Demonstrations approved.

We need not to be ashamed to confess this whole globe of Elements enclosed with the Moon's sphere, together with the earth as the Centre of the same, to be by this great Orb together with the other Planets about the Sun turned, making by his revolution our year. And whatsoever seem to us to proceed by the moving of the Sun, the same to proceed indeed by the revolution of the earth, the Sun still remaining fixed & immoveable in the midst. And the distance of the earth from the Sun to be such, as being compared with the other Planets, makes evident alterations and diversity of Aspects, but if it be referred to the Orb of stars fixed, then has it no proportion sensible, but as a point or a Centre to a circumference, which should far more reasonable to be granted, than to fall into such an infinite multitude of absurd imaginations, as they

were fain to admit that will needs willfully maintain the earth's
stability in the Centre of the world. But rather herein to direct ourselves by that wisdom we see in all God's natural works, where we
may behold one thing rather endued with many virtues and effects,
than any superfluous or unnecessary part admitted. And all these
things, although they seem hard strange & incredible, yet to any
reasonable man that has his understanding ripened with Mathematical demonstration, Copernicus, in his *Revolutions*, according
to his promise, has made them more evident and clear than the
Sun's beams. These grounds therefore admitted, which no man
reasonably can impugn – that the greater orb requires the longer
time to run his Period – the orderly and most beautiful frame of
the Heavens does ensue.

In the midst of all is the Sun. For in so stately a temple as this,
who would desire to set his lamp in any other better or more
convenient place than this, from whence uniformly it might distribute light to all, for not unfitly it is of some called the lamp or
light of the world, of others the mind, of others the Ruler of the
world. Trismegistus calls him the visible God. Thus does the
Sun, like a king sitting in his throne, govern his courts of inferior
powers. Neither is the Earth defrauded of the service of the
Moon, but as Aristotle says, of all other the Moon with the earth
has highest alliance, so here are they matched accordingly.

In this form of frame may we behold such a wonderful Symmetry of motions and situations, as in no other can be proposed.
The times whereby we the Inhabitants of the earth are directed,
are constituted by the revolutions of the earth. The circulation of
her Centre causes the year, the conversion of her circumference
makes the natural day, and the revolution of the Moon produces
the month. By the only view of this Theoricke the cause & reason
is apparent why in Jupiter the progressions and Retrogradations
are greater than in Saturn, and less than in Mars, why also in
Venus they are more than in Mercury. And why such changes
from Direct to Retrograde, Stationary, &c. happen notwithstanding more rifely in Saturn than in Jupiter & yet more rarely
in Mars, why in Venus not so commonly as in Mercury.

Herein can we never sufficiently admire this wonderful &
incomprehensible huge frame of God's work proposed to our

senses, seeing first this ball of the earth wherein we move, to the common sort seems great, and that, in respect of the Moon's Orb, is very small, but compared with the Great Orb wherein it is carried, it scarcely retains any sensible proportion, so marvellously is that Orb of Annual motion greater than this little dark star wherein we live. But that Great Orb, being as is before declared, but as a point in respect of the immensity of that immoveable heaven, we may easily consider what little portion of God's frame our Elementary corruptible world is, but never sufficiently of that fixed Orb garnished with lights innumerable and reaching up in Spherical altitude without end.

But because the world hath so long a time been carried with an opinion of the Earth's stability, as the contrary cannot but be now very impersuasible, I have thought good out of Copernicus also to give a taste of the reasons philosophical alleged for the Earth's stability, and their solutions, that such as are not able with Geometrical eyes to behold the secret perfection of Copernicus' Theorick, may yet by these familiar natural reasons be induced to search further, and not rashly to condemn for fantastical, so ancient doctrine revived, and by Copernicus so demonstratively approved.

What reasons moved Aristotle and others that followed him to think the earth to rest immoveable as a Centre to the whole world.

The most effectual reasons that they produce to prove the Earth's stability in the middle or lowest part of the world, is that of Gravity and Levity. For of all other the Element of the earth (say they) is most heavy, and all ponderous things are carried unto it, striving, as it were, to sway even down to the inmost part thereof. For the earth being round, into the which all weighty things on every side fall, making right angles on the surface must needs if they were not stayed on the surface pass to the Centre, seeing every right line that falls perpendicularly upon the Horizon, in that place where it touches the earth must needs pass by the Centre. And those things that are carried toward that Medium, it is likely that there also they would rest. So much therefore, the rather shall the Earth rest in the middle, and (receiving all things into itself that fall) by his own weight shall be most immoveable. Again, they seek to prove it by reason of motion and his nature, for of one and the same simple body, the motion must also be simple, says Aristotle. Of simple motions there are two kinds: Right and Circular. Right are either up or down: so that every simple motion is either downward toward the Centre, or upward from the Centre – or Circular – about the Centre. Now unto the earth and water, in respect of their weight, the motion downward is convenient to seek the Centre. To air and fire, in regard of their lightness, upward and from the Centre. So it is mete to these elements to attribute the right or straight motion, and to the heavens only it is proper circularly about this mean or Centre to be turned round. Thus much Aristotle. If therefore, says Ptolemy of Alexandria, the earth should turn but only by the daily motion, things quite contrary to these should happen. For his motion should be most swift and violent that, in twenty-four hours, should let pass the whole circuit of the earth, and those things which by sudden turning are stirred, are altogether unmeet to collect, but rather to disperse things united, unless they should by some firm fastening be kept together. And long ere this, the Earth, being dissolved in pieces, should have been scattered through the heavens, which were a mockery to think of, and much more, beasts and all other weights that are loose could not remain unshaken. And also, things falling should not light on the places perpendicular under them, neither should they fall directly thereto, the same being violently in the mean carried away. Clouds also and other things hanging in the air, should always seem to us to be carried toward the West.

The Solutions of these
Reasons with their insufficiency.

These are the causes and such other, wherewith they approve the Earth to rest in the middle of the world, and that out of all question. But he that will maintain the Earth's mobility may say that this motion is not violent, but natural. And these things, which are naturally moved, have effects contrary to such as are violently carried. For such motions wherein force and violence is used, must needs be dissolved and cannot be of long continuance, but those which by nature are caused, remain still in their perfect estate and are concerned and kept in their most excellent constitution. Without cause therefore did Ptolemy fear lest the Earth and all earthly things should be torn in pieces by this revolution of the Earth, caused by the working of nature, whose operations are far different from those of Art or such as human intelligence may reach unto.

But why should he not much more think and misdoubt the same of the world, whose motion must of necessity be so much more swift and vehement than this of the Earth, as the Heaven is greater than the Earth. Is therefore the Heaven made so huge in quantity that it might with unspeakable vehemency of motion be severed from the Centre, lest happily resting it should fall, as some Philosophers have affirmed? Surely, if this reason should take place, the magnitude of the Heaven should infinitely extend. For the more the motion should violently be carried higher, the greater should the swiftness be, by reason of the increasing of the circumference, which must of necessity in twenty-four hours be passed over, and in like manner by increase of the motion the Magnitude must also necessarily be augmented. Thus should the swiftness increase the Magnitude, and the Magnitude the swiftness, infinitely. But, according to that ground of nature, whatsoever is infinite can never be passed over. The Heaven therefore, of necessity must stand and rest fixed.

But, say they, without the Heaven there is no body, no place, no emptiness; no, not any thing at all, whether heaven should or could further extend. But this surely is very strange that nothing should have such efficient power to restrain something, the same having a very essence and being. Yet if we would thus confess that the Heaven were indeed infinite upward, and only finite downward in respect of his spherical concavity, much more perhaps might that saying be verified, that without the Heaven is nothing, seeing everything in respect of the infiniteness thereof had place sufficient within the same. But then must it of necessity remain immoveable. For the chiefest reason that has moved some to think the Heaven limited was Motion, which they thought without controversy to be indeed in it. But whether the world have his bounds, or be indeed infinite and without bounds, let us leave that to be discussed of Philosophers. Sure we are that the Earth is not infinite, but hath a circumference limited. Seeing, therefore, all Philosophers consent that limited bodies may have Motion, and infinite cannot have any, why do we yet stagger to confess motion in the Earth, being most agreeable to his form and nature, whose bounds also and circumference we know, rather than to imagine that the whole world should sway and turn, whose end we know not, nor possibly can of any mortal man be known.

And therefore, the true Motion indeed to be in the Earth, and the apparent only in Heaven: And that these appearances are no otherwise than if the Virgilian Aeneas should say: 'We set out from port, and the land and cities recede.' For a ship carried in a smooth Sea with such tranquillity does pass away, that all things on the shores and the Seas to the sailors seem to move, and themselves only quietly to rest with all such things as are aboard with them, so surely may it be in the Earth, whose Motion being natural and not forcible, of all other is most uniform and unperceivable, whereby to us that sail therein, the whole world may seem to roll about.

But what shall we then say of Clouds and other things hanging or resting in the air or tending upward, but that not only the Earth and Sea making one globe, but also no small part of the air is likewise circularly carried, and in like sort all such things as are derived from them or have any manner of alliance with them. Either for that the lower Region of the air, being mixed with Earthly and watery vapours, follow the same nature of the Earth. Either that it be gained and gotten from the Earth by reason of Vicinity or Contiguity. Which if any man marvel at, let him consider how the old Philosophers did yield the same reason for the revolution of the highest Region of the air, wherein we may sometime behold Comets carried circularly no otherwise than the bodies Celestial seem to be, and yet hath that Region of the air less convenience with the Orbs Celestial, than this low part with the Earth. But we affirm that part of the air in respect of this great distance to be destitute of this motion terrestrial, and that this part of the air that is next to the Earth does appear most still and quiet by reason of his uniform natural accompanying of the Earth, and likewise things that hang therein unless by winds or other violent accident they be tossed to and fro. For the wind in the air, is nothing else but as the waves in the Sea.

And of things ascending and descending in respect of the world, we must confess them to have a mixed motion of right & circular, albeit it seem to us right & straight, not otherwise than if in a ship under sail a man should softly let a plummet down from the top along by the mast even to the deck. This plummet, passing always by the straight mast, seems also to fall in a right line, but being by discourse of reason weighed, his motion is sound mixed of right and circular. For such things as naturally fall downward being of earthly nature there is no doubt, but as parts they retain the nature of the whole.

Hereto we may adjoin, that the condition of immobility is more noble and divine than that of change, alteration, or instability, & therefore more agreeable to Heaven than to this Earth, where all things are subject to continual mutability. And seeing by evident proof of Geometrical mensuration we find that the Planets are sometimes nigher to us and sometimes more remote, and that therefore even the maintainers of the Earth's stability are enforced to confess that the Earth is not their Orb's Centre, this motion around the middle must in more general sort be taken, and that it may be understood that every Orb has his peculiar Medium and Centre, in regards whereof this simple and uniform motion is to be considered. Seeing therefore that these Orbs have several Centres, it may be doubted whether the Centre of this earthly gravity be also of the world. For Gravity is nothing else but a certain proclivity or natural coveting of parts to be coupled with the whole, which by divine providence of the Creator of all is given & impressed into the parts, that they should restore themselves into their unity and integrity, concurring in spherical form. Which kind of propriety or affection it is likely also that the Moon and other glorious bodies want not to knit & combine their parts together, and to maintain them in their round shape, which bodies not withstanding are by sundry motions sundry ways conveighed. Thus, as it is apparent by these natural reasons that the mobility of the Earth is more probable and likely than the stability. So, if it be Mathematically considered, and with Geometrical mensurations every part of every Theorick examined, the discreet student shall find that Copernicus, not without great reason did propone this ground of the Earth's Mobility.

Thomas Digges, Frontispiece, *A Perfect Description of the Celestial Orbs*, 1576

In contrast to his father's depiction of a geocentric cosmos, Thomas Digges included in his own appendix a diagram of the heliocentric universe, the first to appear in a book written in English. It included one crucial change from the diagram of the heliocentric model in Copernicus's On the Revolutions, *however. Though Copernicus had switched the positions of the sun and the earth, he had retained the eighth sphere of fixed stars that had formed the outermost boundary of the traditional cosmos, though he had argued that it was significantly further away than had previously been supposed, to account for the absence of any detectable parallax. Digges did not believe that the cosmos had any fixed outer boundary at all, arguing instead for a universe that was infinite. In this image the outer boundary of the eighth sphere is gone, and instead the 'innumerable' fixed stars extend 'infinitely up'.*

❧ A perfit description of the Cœlestiall Orbes,
according to the most auncient doctrine of the
Pythagoreans, &c.

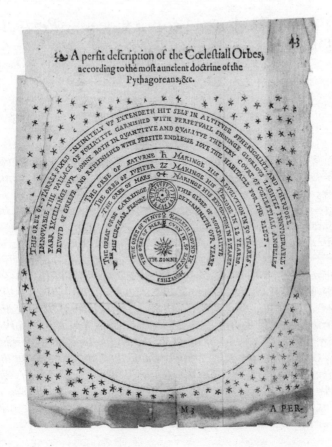

THIS ORBE OF STARRES FIXED INFINITELY VP EXTENDETH HIT SELF IN ALTITVDE SPHERICALLYE, AND THERFORE IMMOVABLE THE PALLACE OF FOELICITYE GARNISHED WITH PERPETVALL SHININGE GLORIOVS LIGHTES INNVMERABLE, FARR EXCELLINGE OVR SONNE BOTH IN QVANTITYE AND QVALITYE THE VERY COVRT OF CŒLESTIALL ANGELLES DEVOYD OF GRIEFE AND REPLENISHED WITH PERFITE ENDLESSE IOYE THE HABITACLE FOR THE ELECT.

THE ORBE OF SATVRNE ♄ MAKINGE HIS REVOLVTION IN 30 YEARES.

THE ORBE OF IVPITER ♃ MAKINGE HIS PERIODE IN 12 YEARES.

THE ORBE OF MARS ♂ MAKINGE HIS REVOLVTION IN 2 YEARES.

THE GREAT ORBE CARRIINGE THIS GLOBE OF MORTALITYE
IN HIS CIRCVLAR PERIODE DETERMINETH OVR YEARE.

THE ORBE OF VENVS ♀ ROVLLETH ROVND IN 9 MONITHES.

THE ORBE OF MERCVRY IN 88 DAYES.

THE SONNE

Diego de Zúñiga of Salamanca, *Commentary on Job*, 1584

Diego de Zúñiga (1536–1597) was a Spanish Augustinian scholar. He was educated at the University of Salamanca, which in 1561 had introduced the optional teaching of Copernicus, and in 1595 had made that teaching required. In his 1584 Commentary on Job, *Zúñiga used the exegesis of Job 9:6 to argue both that Scripture itself was better explained using the Copernican system, and also that the Copernican system was superior astronomically to the Ptolemaic. Later, however, in his* Philosophia Prima Pars *of 1597, Zúñiga reversed his position and argued that the heliocentric theory was philosophically and astronomically absurd. Zúñiga's defence of Copernican theory is more famous than his rejection, however, because his* Commentary on Job *had the distinction of being the only text singled out alongside Copernicus's* On the Revolutions *for 'prohibition until corrected' by the Congregation of the Index's 1616 decree in Rome (with Foscarini's letter on Copernican theory, the only other text mentioned by the decree, prohibited entirely).*

Job 9.6: He who moves the earth from its place, and its columns are shaken.

Marginal note: The motion of the earth is not contrary to Scripture.

This passage depicts another effect of God in order to demonstrate his especially great power along with his infinite wisdom.

While it indeed seems difficult, it might be greatly illuminated by the opinion of the Pythagoreans who believe that the earth is moved by its own nature, and that the motion of the planets, which differ so greatly in slowness and speed, cannot be explained otherwise. Philolaus held this opinion, as did Heraclides of Pontus, as Plutarch reports in his book *On the Doctrines of the Philosophers*. Numa Pompilius followed them, and so did divine Plato later in life, which I marvel at even more. He said then that thinking differently would be highly absurd, as Plutarch recounts in his *Numa*. And Hippocrates, in his book on airs, says that the earth is a vehicle. In our own time Copernicus indicates the paths of the planets using this opinion.

Nor is there a doubt that the positions of the planets are far better and more certainly obtained from this doctrine than from the *Almagest* of Ptolemy and the opinions of others. For it is certain that Ptolemy could neither explain the motion of the equinoxes nor determine a fixed and stable beginning of the year, which he himself admits in the third book and second chapter of the *Almagest* and leaves for future astronomers to discover, who would be able to compare more observations at greater intervals than he could. And although the Alphonsines and Thabit ben Core attempted to explain it, they nevertheless both accomplished nothing. For the positions of the Alphonsines disagree with each other, as Ricius proves. On the other hand, while Thabit's method is more precise and he relates a more stable beginning of the year, which Ptolemy wanted, nevertheless it is already evident that the equinoxes have progressed longer than he believed they could have. Moreover, the sun is known to be much closer to us than was once thought, by more than forty thousand stadia, and neither Ptolemy nor other astrologers recognized the reason for its motion. Yet the reasons for these things are explained and demonstrated very thoroughly by Copernicus from the motion of the earth, and everything else fits more aptly.

His opinion is not at all refuted by what Solomon says in Ecclesiastes: 'but the earth remains [stands still] for ever.' This signifies only that although there are varied and changing future ages and generations of men, nevertheless the earth is one and persists in the same manner. For the full passage is this:

'Generations pass and generations come, but the earth remains for ever.' Hence the context does not include the question of whether the earth is immobile (as the philosophers relate).

On the other hand, the fact that the Sacred Scriptures (in this chapter of Ecclesiastes and in many other places) mention the motion of the sun does not negate Copernicus's desire for it to stand immobile at the centre of the universe. For the motion of the earth is assigned to the sun in the words of both Copernicus and those who follow him, such that they often call the path of the earth the path of the sun. Finally, there are no places in the Sacred Scriptures which say as openly that the earth does not move as the one that says here that it does. For the passage here is readily connected to this opinion we are describing when it declares, in words that show the wonderful power and wisdom of God, who stimulates and manages the motion of the earth (which by nature is extremely heavy), that 'its pillars are shaken'. What is signified in this teaching is that it is moved from its foundations. To those who will still not approve this opinion of ancient and recent philosophers, one may – though not as favourably – note that the occasional violent shaking of the earth may agree with the motion of the earth. Or rather, that what is signified is the very great reverence that the earth exhibits for God, and the fear with which it venerates him and obeys his orders, such that at his command the whole earth trembles and departs.

Giordano Bruno,
Ash Wednesday Supper, 1584

Giordano Bruno (1548–1600) was born in Nola, Naples (and is therefore often called 'the Nolan', as below). He studied philosophy and theology and became a Dominican friar, but fled Italy under suspicion of heresy, travelling through Europe and ultimately writing the Ash Wednesday Supper *in Italian while in London. It is a dialogue in five parts between three principal characters: Theophilus, a supporter of Copernican astronomy (and a stand-in for Bruno himself); Smith (an Englishman sympathetic to Theophilus's arguments); and Prudencio (a supporter of Aristotle and Ptolemaic astronomy). A fourth character, Frulla (who doesn't appear in the selection below), occasionally interjects some comic relief. Bruno's support for Copernicus was primarily rooted in philosophical speculation rather than mathematics or astronomical observation, and though he understood Copernicus to be offering a real system of the cosmos, he argued that Copernicus did not go nearly far enough. Bruno used the dialogue to highlight some of the implications he believed Copernicus had missed, among them an infinite universe in which each star was itself a solar system with possible inhabitants, and an earth that had a soul responsible for its movement. Bruno emphasized that if the universe were truly infinite, then neither the earth nor the sun was central, for there was no centre at all of which to speak. After Bruno returned to Italy he was arrested and tried for heresy, and ultimately burned at the stake in Rome – for his controversial theological beliefs and not for his heliocentrism, a doctrine that had not yet been formally banned by the Church.*

THEOPHILUS: Recently, two visitors came to the Nolan on behalf of a royal squire, to convey to him how much the squire wanted to converse with him so that he could understand that Copernicus of his, and other paradoxes of his new philosophy. To which the Nolan replied that with respect to judgement and determination he saw not with the eyes of Copernicus or Ptolemy, but rather with his own. Although with respect to observations, he knew that he owed much to these and other diligent mathematicians, who over time have successively added light to light and have given us principles sufficient to make the kind of judgements only possible after the industry of many ages. He added that those men are like interpreters who translate words from one language to another; it is not they themselves but others who deepen the understanding.

SMITH: Please, let me understand: what is your opinion of Copernicus?

THEOPHILUS: He had an earnest, diligent, careful and mature intellect. He was a man who was not inferior to any astronomer who came before him, except for his place later in time; a man who, with respect to natural learning, was far superior to Ptolemy, Hipparchus, Eudoxus and all the others who walked in their footsteps; someone who had to free himself from some of the false presuppositions – I do not want to say blindness – of the common and popular philosophy. However, he did not go all that far from them. Since he was a scholar of mathematics, rather than of nature, he was unable to dig deep and penetrate far enough so that he could remove and uproot all the unfit and empty principles. Thus he could not completely untangle all the difficult contradictions and free himself and others from many vain inquiries in order to settle attention on constant and certain things.

Despite that, who can fully praise the virtue of this German? With little regard for the foolish multitude, he stood firm against the torrent of opposing belief, and although almost unarmed with true arguments, he reclaimed what wretched and rusty fragments he could from the hands of antiquity, repolished them, and joined and soldered them so greatly with his mathematical (rather than physical) discourse, that a cause once ridiculed, wretched and vilified became honoured, esteemed and more probable than the contrary, and certainly more convenient and accessible for theory and the calculation of causes.

Thus this German, although he only had sufficient means to resist falsehood, rather than to defeat and suppress it, still stood firm; he determined inwardly and acknowledged outwardly that this globe necessarily moves with respect to the universe, rather than that the universe itself, with so many innumerable bodies more magnificent and much larger than it, should – against nature, reason and sense – acknowledge this globe as the centre and base of their revolutions and influences. Therefore, who will be so rude and discourteous towards the work of this man as to place in oblivion so much that he has done? He was ordained by the gods as a dawn that must precede the emerging sun of the ancient and true philosophy, buried for so many centuries in the dark caverns of blind, malicious, wayward and envious ignorance. Though we recognize what he could not do, when we compare him to the masses – who flock together, run about, and move ahead rashly by heeding unreasonable and ignoble beliefs – this man ought to be counted among those who, with their blessed intellect, were able to stand upright and advance through the faithful guidance of the eye of divine intelligence.

The Nolan accomplished something quite different. He freed the human mind, which was locked in an artificial prison of turbulent air, peering through holes from which one could scarcely behold the distant stars. Its wings were maimed, so it could not fly and open the veil of clouds to see what was really up there and free itself from the chimeras that issued from the mud and caves of the earth, as though they were Mercuries and Apollos descending from heaven. With many forms of imposture, they filled the whole world with infinite folly, beastliness and vice, while pretending they were great virtue, piety and discipline. They stifled that light which made the souls of our ancient fathers divine and heroic, and approved and confirmed the shadowy darkness of sophists and asses. Thus human reason, oppressed for so long, and in its lucid intervals sometimes weeping over its base condition, turned back to the wise divine mind, which always whispers in its inner ear with words like these: 'Mistress, who shall for me to heaven ascend, / To bring back my forsaken wit?'

Behold now the man who traversed the air, penetrated the sky, navigated the stars, passed beyond the borders of the world and made vanish the fantastical walls of the first, eighth, ninth and tenth spheres, and whatever others could have been added to these based on the reports of foolish mathematicians and the blind vision of popular philosophers. Using his full sense and reason and with the key of very diligent inquiry, he opened whatever cloisters of truth are able to be opened by us. He stripped bare the parts of nature that had been veiled, gave eyes to moles and illuminated the blind, who could not fix their eyes and admire its image in the many mirrors that they set against it on all sides. He loosened the tongues of the mute, who did not know and did not dare to express complicated opinions, and healed the lame, who could not make that spiritual progress that is denied to ignoble and destructible things. He makes their view no less than if they were the very inhabitants of the sun, the moon and other named stars. He shows how similar or dissimilar, greater or worse are those bodies which we see far away from this one that is close to us and to which we are united. And he opens our eyes to see this deity, our mother, who feeds and nourishes us on her back, produces us from her womb, and always gathers us there again – and who is not to be thought of as nothing other than a body without soul or life, or as the dregs of bodily substances.

In this way we know that if we were on the moon, or on other stars, we would be in a place quite similar to this, and perhaps even worse than it. Likewise, there can be other bodies just as good or even better for themselves and for the happiness of their own living creatures. Thus we perceive how many stars, how many planets, how many gods – of which there are hundreds of thousands – assist in the service and contemplation of the first, universal, infinite and eternal one. Our reason is no longer imprisoned in the shackles of eight, nine and ten fantastical movable and moving spheres. We know that there is only one heaven, an immense ethereal region, where these magnificent lights keep their own distances in order to participate in perpetual life. These flaming bodies are ambassadors, who announce the excellence of the glory and majesty of God. In this way we are stimulated to discover the infinite effect of the infinite cause, the true and living traces of the infinite force. We have a doctrine that teaches not to seek the divinity at a remove from us; rather, it is near us, and indeed, within us, more than we are within ourselves. In just this way the inhabitants of other worlds do not seek it near us, since it is near and within themselves. Know that the moon is no more a heaven to us than we are to the moon.

PRUDENCIO: Be that as it may, I do not want to deviate from the opinion of the ancients; for as the wise man says, in antiquity is wisdom.

THEOPHILUS: And he adds, in many years is prudence. If you understood well what you are saying, you would see that the opposite of what you think should be inferred from your foundation. That is to say, that we are older and have been around longer than our predecessors with respect to those things that pertain to certain judgements, like the subject under discussion. The judgement of Eudoxus, who lived shortly after the rebirth of astronomy – if it was not reborn in him – could not be as fully developed as that of Callippus, who lived thirty years after the death of Alexander the Great, for the latter was able to add additional observations alongside the added years. For the same reason, Hipparchus had to know more than Callippus, because he saw the changes that happened up to 196 years after the death of Alexander. Menelaus, the Roman geometer, who saw the difference in the motions 462 years after Alexander died, rightly understood more than Hipparchus. Still more could be seen by Machometto Aracense [al-Battani] 1,202 years after that. Copernicus, almost in our times, saw even more than that, 1,849 years after Alexander. But some of these men, though nearer to us, have not been wiser than those who came before, and the majority of those in our times are no cleverer. This is because both those in the past and those today do not learn from the years of others, and, still worse, both live as though dead in their own years.

PRUDENCIO: Say what you please and cast matters in whatever form and direction you like: I am a friend of antiquity. And with respect to your opinions and paradoxes, I do not believe that so many and such wise men have been as ignorant as you and other friends of novelty think.

THEOPHILUS: Well, Master Prudencio, if this popular opinion of yours is true insofar as it is ancient, then it was certainly false when it was new. And before the philosophy that corresponds to your way of thinking, there was another, that of the Chaldeans, the Egyptians, the Magi, the Orphics, the Pythagoreans and others of earlier memory, which corresponds to ours. Against that philosophy, the first to rebel were the senseless and vain logicians and mathematicians, not so much enemies of antiquity as strangers to truth. Let us therefore put aside the discussion of the old and the new, because there is nothing old which was not once new, as your Aristotle well noted. Your Aristotle noted that just as there are vicissitudes in other things, there are also changes in opinions and their diverse ramifications; thus to esteem philosophies based on their antiquity is like wanting to decide if night or day came first. The thing to consider, rather, is whether we are in the day, and whether the light of truth is above our horizon, or above that of our adversaries opposite us; whether we are in darkness, or they are. And finally, whether we, who begin to renew the ancient philosophy, are the morning that ends the night, or the evening that ends the day.

SMITH: But what do you say about the fact that most people of our time think exactly the opposite, especially with respect to that doctrine?

THEOPHILUS: I am not surprised, because it is usually the case that the less people know, the more they believe they know, and those who are complete fools think that they know everything.

SMITH: Tell me, how can these people be corrected?

THEOPHILUS: By using some form of argument to remove their confidence that they know, and stripping from them, as much as possible, with crafty persuasion, their foolish opinion, so that in the end they are open to listening.

SMITH: When the Nolan spoke to Dr Nundinio and Dr Torquato about Copernicus, what did they say?

THEOPHILUS: Dr Nundinio said in Latin: 'I want to explain what we said. We believe that Copernicus was not of the opinion that the earth moves, which is an unfit and impossible thing, but that he attributed motion to it, rather than to the eighth sphere, for the convenience of the computations.' The Nolan said that if Copernicus had said that the earth moved for this reason alone, and not for any other, then he had understood little enough of things. But it is certain that Copernicus did understand what he said, and he proved it with all his efforts.

SMITH: Could it be that those people ignorantly tossed out this judgement on the opinion of Copernicus, if they could not marshal it from some of his propositions?

THEOPHILUS: You should know that this statement started out with Dr Torquato, who – though I can believe that he flipped through all of Copernicus – retained from the book only the name of its author, its printer, the place where it was printed, the year and the number of pages. And as someone not ignorant in grammar, he would have understood a short preliminary letter attached to it by some ignorant and presumptuous ass. That man (as if he wanted to help the author by excusing him, or perhaps ultimately to help the other asses find in this book their lettuce and fruit, so that they would not have to leave it undigested) warned them to read the book in this way at the very start, and to consider his words: [Theophilus quotes the 'To the Reader' by Osiander; see page 27].

See what a good door-keeper he is! Consider how well he opens the door for you to enter into participation of that most esteemed knowledge, without which knowing how to calculate, measure, and use geometry and perspective are nothing other than pastimes for clever fools! Consider how faithfully he serves the master of the house. For to Copernicus it was not enough to say only that the earth moves: he also openly asserts and confirms, when writing to the pope, that the opinions of philosophers are very far from popular opinion, which is unworthy of being followed and very worthy of being deserted, since it is contrary to the progression of truth.

And many other clear arguments emerge from his writing, notwithstanding that in the end it seems that he somehow wants agreement as much from those who understand his philosophy as from those others who are pure mathematicians. Thus he writes that if they do not like his supposition because it seems unsuitable, it would still behove them to concede to him the freedom of putting the earth in motion in order to construct more solid demonstrations than those made by the ancients, who felt free to feign many sorts and models of circles in order to determine the appearances of the stars. From these words we surely do not conclude that he doubted that which he constantly acknowledged and proved in his book by sufficiently responding to the arguments of those who supposed the opposite. There he acts not only as a mathematician, who makes suppositions, but also as a physicist, who demonstrates the motion of the earth. But certainly it amounts little to the Nolan what Copernicus, Nicetas of Syracuse the Pythagorean, Philolaus, Heraclites of Pontus, Ecphantus the Pythagorean, Plato in the *Timaeus* (although timidly and inconstantly, because he came to it more by faith than by science), and the divine Cusanus in his second book *On Learned Ignorance*, and other very singular people, have said, taught and established earlier: because he has other, more proper and solid principles, which are held not because of authority but because of human sense and reason. Thus he holds them to be as certain as any other thing about which one can be certain.

SMITH: This makes sense. But pray tell more about the argument of that man who stands at the doorway of Copernicus's book, for he seems to think that it seems at least probable, if not actually true, that the star of Venus ought to vary in size as much as it does in distance.

THEOPHILUS: That fool, who keenly fears that some people will become deluded by the doctrine of Copernicus – I don't know whether he could have brought a more unsuitable point than that one, which he puts forth with such solemnity, deeming it sufficient to demonstrate that those who think about this are very ignorant when it comes to optics and geometry. I would like to know where this fiend gained his skill in optics and geometry, as he shows how ignorant he is in those subjects, and are all those from whom he learned.

I would like to know how we can infer the proximity or remoteness of luminous bodies from their size? And conversely, how, from the distance or proximity of those bodies, we can infer a proportional change in their size? I would like to know with what principle of perspective or optics we can definitively determine the correct distance, or the major and minor variations, from the variation in its diameter? I would like to understand if we are mistaken when we draw this conclusion: that we cannot infer the true size or distance of a luminous body from its apparent magnitude. For just as the same rules do not apply for opaque and luminous bodies, so the same rules do not apply for a body that is less luminous and one that is more luminous, so as to enable us to judge their size or their distance.

The size of a man's head cannot be seen two miles away, but a lamp much smaller than it, or some other similar light, would be seen without much difference (but with some difference), sixty miles away, as the candles of Valona can be seen from Otranto in Puglia, between which there is a large expanse of the Ionian Sea. Everyone who has sense and reason knows that if the lamps were lit with twice as powerful a light, they would be seen at a distance of 140 miles, as the first one would be seen at a distance of 70 miles, without variation in size – if the power of the light were tripled, at 210 miles; if quadrupled, at 280; and similarly for other increases of proportions and degrees. For it is more the quality and intensity of the light, rather than the magnitude of the shining body, that maintains the diameter and size of the body for us.

Next Nundinio said that it cannot be probable that the earth moves, being that it is the middle and centre of the universe, which has to be the fixed and constant foundation of all motion. The Nolan replied that the same thing can be said by someone who holds the sun to be the middle of the universe, and therefore fixed and immobile, as Copernicus and many others understood, all of whom believe that the universe had a limited circumference. It happens that this type of reasoning (if it is even reasoning) offers nothing against them, and presupposes its own principles. And it offers nothing against the Nolan, who wants the world to be infinite, and thus for there to be no body in it which would simply and necessarily be the middle of it, or at the far end, or in between the two. Instead, there would be certain relations to other bodies, and boundaries selected intentionally.

SMITH: What do you think of this?

THEOPHILUS: That it is very well said! For just as no natural body is known to be perfectly round and consequently to have a perfect centre, so too, of the sensible motions that we see physically in natural bodies, none of them is perfectly circular and regular around some centre, despite how much some people want them to be, inventing wads and fillings of unequal orbs of diverse diameters, and other plasters and medicines, in order to heal nature until it comes to serve Master Aristotle or someone else with the conclusion that all motion is continuous and regular around the centre. But we, who look not at fantastic shadows but at things themselves, we, who see an aerial, ethereal, spiritual, liquid body, capable of motion and of rest, though immense and infinite (which we must affirm at least, because we see no end to it, either sensibly or rationally), we know with certainty that, being the effect and creation of an infinite cause and principle, it must be infinitely infinite, in its bodily capacity and in its mode of being. And I am certain that it will be impossible, both for Nundinio and also for all professors of understanding, to find a semi-probable reason for there to be a boundary to this body of the universe, and consequently, one for the stars, which are contained in its space, to be finite in number, and also for there to be a naturally determined centre and middle to it.

SMITH: But did Nundinio add something to that? Did he bring some argument, or probability, to infer that the universe was first, finite; second, that the earth was at its middle; and third, that that centre is fully and completely immobile and without local motion?

THEOPHILUS: Nundinio, like someone who says what he says out of faith and custom, and denies things because of their unusualness or novelty, as is common for those who contemplate little and are not in charge of their own actions, whether rational or natural, remained stupid and astonished, like someone to whom a new ghost suddenly appears.

The Nolan briefly mentioned that there were innumerable earths similar to this one. Dr Nundinio asked about the quality of those other globes. He wanted to know what material made up those bodies that are thought to be of the fifth essence, an unalterable and incorruptible material, which at its densest makes up the stars.

The Nolan replied to him that the other globes that are earths are not different at all from this one in species, except for being larger or smaller, just as in other species of animals there are inequalities due to individual differences. But he thinks that those spheres which are made of fire (like the sun is for now) are different in species, just as hot is from cold and something luminous on its own from something luminous by way of light from another. It was here that Nundinio began to show his teeth, widen his jaws, squeeze his eyes, wrinkle his eyebrows, flare his nostrils and croak like a capon from his windpipe. Through this laughter, he wanted others to think that he well understood that he was right, and that the Nolan said ridiculous things.

Asked why he was laughing, he replied that this talk and imagining that there are other earths which have the same properties and accidents seems like it is taken from the *True History* of Lucian. The Nolan replied that when Lucian said that the moon was another earth, inhabited and cultivated like this one, he said it to mock the philosophers who assert that there are many earths (and particularly the moon, whose similarity to this globe of ours is all the more evident the closer it is to us). But he was wrong, and showed himself to be commonly ignorant and blind. Because if we consider carefully, we will find that the earth and so many other bodies, which are called stars, are the principal members of the universe, as they give life and nourishment to things which take material from them and renew themselves with it. They thus have a great deal of life within themselves. This is how they move toward other things, through the spaces suitable to them, with a structured and natural will that comes from a principle within them. There are no other extrinsic motors which transport these bodies as though they are affixed to moving fantastical spheres. For if there were, the motion would be violent and outside the nature of the moving object. The motor would be imperfect, and the motion and the moved object would be hasty and laborious, and there would also be many other inconveniences. Consider too that just as the male moves to the female, and the female to the male, and every plant and animal, whether more or less expressly, moves to its vital principle, so too do the sun and other stars. A magnet moves to iron, straw to amber, and finally everything finds that which is similar to it, and escapes that which is opposed to it. Everything happens according to the sufficient internal principle through which things become naturally enlivened, rather than via an external principle of the kind that moves things either against or beyond their own nature. Thus the earth and other stars move according to their individual local differences via an internal principle which is their own souls. 'Do you believe,' said Nundinio, 'that this soul is sensitive?' 'Not only sensitive,' replied the Nolan, 'but also intelligent; not only intelligent like we are, but perhaps even more so.' Here Nundinio was silent and did not laugh.

Finally, he said this, something which fills many little books: that if it were true that the earth moved towards the east, it would be necessary that clouds of the air always appear to move towards the west, because of the very fast and rapid motion of this globe, which has to complete a great rotation in the period of twenty-four hours. To this the Nolan replied that the area through which the clouds and wind move is part of the earth, because by 'earth' he means (and this must be so) the whole machine and the whole animal, which consists of dissimilar parts. Thus the rivers, the stones, the seas, the vaporous and turbulent air, which is enclosed in the highest mountains, all belong to the earth as members of it, just as the air which is in the lungs and other cavities of animals – which they breathe, and which expands their arteries and fulfils other functions necessary for life – belongs to them. The clouds, then, move as accidents within the body of the earth, all the way to its bowels, much like the waters.

SMITH: You have well satisfied me and wonderfully disclosed many secrets of nature which are hidden under this key. You responded to the argument taken from the winds and clouds. From that response, we can undertake another response to Aristotle in the second book of *On the Heavens*, where he says that it would be impossible for a stone thrown high up to come down along the same straight perpendicular line, but that it would necessarily be left far behind toward the west due to the very fast motion of the earth. And with this projection being directed back toward the earth, from its motion there necessarily follows a change in every relationship of straightness and obliquity, because there is a difference between the motion of a ship and the motion of those things that are on the ship. For if this were not true, it would follow that when the ship races through the sea, no one could ever throw anything straight from one side of it to the other, nor would it be possible for anyone to jump up and return standing to the place from where he took off.

THEOPHILUS: Consequently, all the things that are on the earth move with the earth. If, therefore, something was thrown on to the earth from a place outside the earth, it would lose the straightness of its motion because of the earth's motion.

This is apparent on a ship passing through a river. If anyone on the bank of the river threw a stone straight ahead at it, his throw would miss because of the speed of the ship on its course. But if someone were on the mast of the ship, then however fast it moves, he would not miss his target. Suppose there are two men, one inside the moving ship, and the other outside it. Both of them have their hand raised to the same height in the air, and from that same height both drop a stone without giving it any forward force. The first one would hit the prefixed target without missing it by any amount or deviating from the straight line, while the second one would be left behind. This is only because the stone that leaves the hand of the first man on the ship, and which consequently moves with the ship's motion, has an impressed virtue which the other stone, coming from the hand of the man outside the ship does not have, even if both stones have the same weight, the same air to pass through, and start (if possible) from the same point, with the same force. The only conclusion we can draw from this difference is that things that are connected to the ship or similarly belong to it move along with it. One of the stones brings with it the virtue of the mover that moves with the ship, while the other does not participate in that motion. From this it is clearly evident that the virtue of going straight depends entirely on the force of the original impressed virtue, rather than on the point where motion ends, or where it begins, or the medium through which it moves.

SMITH: But the Holy Scripture (whose sense must be much recommended as something that proceeds from superior intelligences which do not err) in many places hints at and supposes the opposite [of the earth's motion]!

THEOPHILUS: Now as to that, believe me, if the gods intended to teach us the theory of natural things, as they have done us the favour of teaching us the practice of moral things, I would be more inclined to approach things based on faith in their revelations rather than on the certainty of my own reasons and feelings. But as everyone can see very clearly, the divine books do not treat demonstrations and speculations about natural things as if they were philosophy, told over in the service of our intellect. Rather, the practice of moral actions is set by laws for the benefit of our mind and affection. The divine lawmaker, having this as his purpose, does not care to speak about truths that would give no profit to the common people by allowing them to withdraw from evil and cling to good. Instead, he leaves these thoughts to contemplative men, and speaks to the common people in such a way, according to their own forms of understanding and speaking, that allows them to understand what is important.

Do these venerable men want it to be that when Moses said that God made two great luminaries among all the others, the sun and the moon, these must be understood as absolutely so, with all the others being smaller than the moon? Or do they actually think this follows a common sense and an ordinary mode of speaking and understanding? Are there not many stars that are bigger than the moon? Can they not be bigger than the sun? Does the earth lack something – is it not a more beautiful and larger luminary than the moon, as it receives the great splendour of the sun in its oceans and seas, and may well appear as a very luminous body to other worlds, called stars, just as they appear to us as so many twinkling torches? Of course, he does not call the earth a great or small luminary, though he says such about the sun and the moon. And this was said well and truly for the kind of speech it was, as he had to make himself understood according to the common way of speaking and understanding, and not as someone who, whether madman or fool, uses knowledge or wisdom. To speak in terms of truth when it is not necessary, and to wish that the common and foolish people, from whom practise is required, under-stood it, would be like wishing that the hand had an eye, when hands are made by nature not to see, but to do.

Torquato then inquired about the ratio between the motion of the planets and the motion of the earth. The Nolan replied that he had not come to lecture, nor to teach, but to respond; and that the symmetry, order and measure of the motion of the heavens he presupposed to be that known by the ancients and moderns; that he does not dispute this, and his purpose is not to quarrel with mathematicians in order to challenge their measurements and theories, which he accepts and believes. Rather, his purpose pertains to the nature and verification of the foundations of those motions. The Nolan also said: 'If I take the time to answer this question, we would be here all night without debating or laying out the grounds for our chal-lenges to the common philosophy. The fact is that both sides tolerate all the same hypotheses in order to draw conclusions about the true reason and the quantity of all the motions: and in this we agree. So why agonize over the matter? See if from the given observations and granted verifications you can infer something against our position, and then you are free to press ahead with your condemnations.'

Then, to complicate the business, they asked the Nolan to explain what exactly he wanted to defend, against which the Dr Torquato mentioned earlier would argue. The Nolan replied that he had explained too much already, and that if the arguments of his adversaries were scarce, this was not due to a lack of material, as should be clear even to a blind man. But again he asserted that the universe is infinite, and that it consists of an immense ethereal region; that it is actually one heaven, which is called the embrace of space, in which there are many stars fixed in it just like the earth; and that the moon, the sun and the innumerable other bodies are in that ethereal region just as the earth is. These are great living creatures, of which many are visible on all sides because of the clear light which emanates from their bodies. Some are effectively hot, like the sun and other innumerable fires, and others are cold, like the earth, the moon, Venus and other innumerable earths. In order to communicate with one another and to partake in each other's vital principle, some of these complete their revolutions at certain spaces and with certain distances around others, as is evident in these seven which turn around the sun. The earth is one of these; it also moves in twenty-four hours from the west to the east, which causes the appearance of the motion of the universe around it. This is what we call mundane or diurnal motion. And this imagined motion [of the universe around it] is very false, and is unnatural and impossible. For it is possible, suitable, true and necessary that the earth moves around its own centre, in order to participate in light and darkness, day and night, heat and cold; and around the sun, in order to participate in spring, summer, autumn and winter; and around the poles and the opposite hemispheric points, for the renewal of ages and the change of its countenance. This is in order that in areas where there was sea there would one day be dry land; where the torrid zone once was would become a cold area; and where the tropical zone was would become equinoctial. Ultimately, these things happen so that there will be continual change in all things. The same things that happen here happen on other stars too, which are not without reason called worlds by the true ancient philosophers.

The other stars are not fixed in the sky, just as this star, the earth, is not fixed in the same firmament, which is the air. And the place where the tail of the Bear is found is no worthier of being called the eighth sphere than the earth, where we are, is. Because these separate bodies are in the same ethereal region, like the same great space and field, separated from one another by suitable intervals. Consider the reason why there are judged to be seven heavens for the planets, and only one for all the other stars. The varied motion seen in those seven, and the regular one in all the other stars, which perpetually maintain the same distance and order, make it seem like the latter all share one motion and one fixed sphere, and that there are not more than eight sensible spheres for all the luminaries, which are nailed to them. But what if we approach things more clearly and with better regulated reason, in the same way that we know that the apparent motion of the universe proceeds from the rotation of the earth? If we judge that the nature of this earthly body in the middle of the air is similar to the nature of all the other bodies, then we will first be able to believe and then demonstratively prove the opposite of that dream, that fantasy, which first established that unfit account and has generated and still generates so many innumerable others. The way that error occurred was that, as we turn our eyes in every direction from the centre of the horizon, we can judge greater and lesser distance from, between and in those things which are closest to us; but from a certain limit beyond that, they all appear equally distant. Thus looking at the stars in the firmament, we recognize the differences in the motions and distances of some of the nearest stars, but the ones that are more distant and very far off appear motionless and equally distant with respect to longitude. Therefore, the fact that we do not see much motion in those stars, and that they do not seem to move away from or approach one another, is not because they do not actually do so as the nearer ones do.

Because there is no reason why the former would not have the same accidents as the latter, or why both types of bodies would not receive the same virtues, which would cause them to move around each other. Further, they should not be called fixed because they actually keep the same distance between us and them, but rather because their motion is undetectable to us. This can be seen in the example of a very distant ship: if it turns thirty or forty paces, it will appear to be standing still just as if it did not move at all. Similarly, we should consider the even more distant large and luminous bodies, many of which – possibly an immeasurable number – may be as large or as bright as the sun, but whose great circles and motions cannot be seen by us. So if some of these stars vary in their proximity, we cannot discern it except through very long observations, which have not yet been initiated or pursued. Because no one has believed this motion possible, nor sought or presumed it, and we know that the beginning of an investigation is the knowledge or recognition that the thing in question exists, that it is possible and appropriate, and that profit can be derived from it.

SMITH: I would know very well what to answer to people who say that it would be difficult for the earth to move because it is such a large body and so dense and heavy. Yet I would like to hear your way of answering, for I see that you are so resolved in your reasons.

THEOPHILUS: The way to reply is this: that the same could be said of the moon, the sun and of other very large and innumerable bodies, which those adversaries assume to be quickly circling the earth with such immense obits. And yet they think it a big deal for the earth to turn around its own centre in twenty-four hours, and around the sun in one year. Know that neither the earth nor any other body is absolutely heavy or light. Nothing is heavy or light in its own place, nor are these differences and qualities a principal part of bodies, or particular to isolated individuals in the universe. Rather, they pertain to parts that are divided from the whole, and that are found wandering outside their own region. These are as naturally forced towards their place of keeping as iron is towards the magnet – not determinately from below, or above, or to the right, but from any local difference, whatever it is. The parts of the earth in the air come towards us, because this is their sphere; but if that sphere were in an opposite place, they would move from us straight there. The same is true of water and fire. Water is not heavy in its place, and does not burden those things which are deep in the sea. Our arms, head and other limbs are not heavy to our torso, and nothing naturally constituted causes an act of violence in its natural place. Gravity and levity are not actually seen in things that are in their own places and natural dispositions, but only in things that have a certain impetus with which they force themselves to the place suitable for them. It is, however, absurd to call any body naturally heavy or light, being that these qualities are not conditions of things that are in their natural constitution, but only of things outside it – which never happens to a whole sphere, but sometimes to parts of one.

Tycho Brahe, *On Recent Celestial Phenomena*, 1588

Aristotle had taught that comets were meteorological phenomena located in the upper atmosphere. Scholars like Girolamo Cardano (1501–1576) had already begun to argue in the 1530s, after a series of comets were observed, that they were much further away than that, and thus that they seemed to contradict the Aristotelian position on celestial immutability. When another bright comet appeared in 1577 and remained visible for two months, Tycho Brahe used his observations to make this case yet again, much as he had for the supernova of 1572. He also worked out the comet's orbital path, which moved across the planetary spheres. Though traditionally these spheres were understood to be solid, Tycho argued that the spheres themselves were not real and the heavens were fluid, so that the crossing of the paths of celestial objects was not a problem. He also used this 1588 work to argue for a new, third world system that accepted much of Copernicus's innovations while retaining a central, immobile earth as in the Ptolemaic system. In the Tychonic system, the moon and sun orbit the earth at the centre, while the five planets orbit the sun. Tycho's system was observationally and mathematically identical to Copernicus's, the primary difference being a physical one – what actually moved and what remained stationary.

It is clear and indisputable that our comet had nothing in common with the elementary world. Rather, it has shown in its departure that it was far higher than the moon, and fully ethereal, with its tail clearly looking back at a specific star, perpetually maintaining an Olympic proportion.

Now, it remains – and seems particularly appropriate – for us to assign to it a certain place in the very large space of the ether itself, so that it will be clear between which orbs of the planetary region it has directed its path. For the ethereal world encompasses incredible vastness, to such a degree that if we suppose this elementary one, from the centre of the earth to the nearest bounds of the moon, to be almost 52 semi-diameters of the earth (each of which has 860 of our German, or common, miles), this will be contained in the remaining space of the planetary region – that is, from the earth to the outermost distance of the star Saturn – 235 times. In this very vast interstice, seven planets ceaselessly rehearse their wonderful and nearly divine periodic motions. I can say nothing about that immense distance of the eighth sphere, which without doubt is a far more enormous space than that to the furthest distance of Saturn. Indeed, if it were necessary to consult the Copernican hypothesis, that space between Saturn and the fixed stars would exceed several times the distance of the sun from the earth (which is nevertheless such that it surpasses the semi-diameter of the elementary world about twenty times). For otherwise (according to his speculation) the annual revolution of the earth in the great sphere would not be imperceptible with respect to the eighth sphere, as it ought to be.

Since the region of the celestial world is of such great and incredible magnitude, and since in the preceding it was at least generally demonstrated that this comet kept itself within the limits of the space of the ether, it will seem that a deeper explanation of the whole matter is not satisfactory unless we also distinguish in particular in which part of that very wide ether, and next to which spheres of the planets, it has established its course, and by what route it completes it. In order that this might be more rightly and clearly enough understood, let me disclose here a little more deeply, as the matter itself demands, my thoughts from four years ago concerning the very disposition of the celestial revolutions or the framework of the whole system of the world, although I had decided hitherto to spare them in this astronomical work.

When I turned my attention to the old Ptolemaic distribution of the celestial spheres, I considered that they were not pleasing enough, and that so many epicycles – by which are justified the relationships of the planets with the sun and their stations and retrogradations, with some part of the apparent inequality – were redundant. Still more, I considered that these hypotheses sin against the very first principles of the art, as long as they unsuitably admit as possible uniform circular motion, not around something's own centre, as is proper, but around the unconnected centre of another, namely an eccentric (which, on account of this, they commonly call an equant). At the same time, I considered the newly introduced innovation of that remarkable Copernicus, whose ideas are in the model of Aristarchus of Samos (as is well known from the book of Archimedes to the Sicilian king Gelon concerning the number of grains of sand), in which he very ingeniously guards against those things which are redundant and contrary in the Ptolemaic arrangement, and commits no offence against mathematical principles. However, he decided that the great, lethargic body of the earth was moved about, though it is less well suited for motion on a disordered path (especially three of them) than the ethereal lights. Thus he opposes not only physical principles, but also the authority of the Sacred Scripture, which several times confirms the stability of the earth (as we will assert more extensively elsewhere). For this reason, I will not speak now about the very vast space between the sphere of Saturn and the eighth sphere, which by this reasoning is rendered entirely empty of stars, and of the other more problematic things that accompany this speculation. But since, I say, I had examined both these hypotheses and found them admitting non-trivial absurdities of this sort, I began to contemplate inwardly and more deeply whether some method of hypothesizing could be discovered which would agree completely and properly both with mathematics and with physics while evading theological censure, and at the same time satisfying entirely the celestial appearances. And finally, almost as if unexpectedly aided, an order came to me by which the form of the celestial revolutions would be most suitably arranged, such that all these irregularities are avoided. I will now communicate it briefly to devotees of celestial philosophy.

I think it beyond all conceivable doubt that the earth, which we inhabit, occupies the centre of the universe and is not carried around in an annual motion, as Copernicus preferred. This accords with ancient astronomers and the accepted opinions of natural philosophers, and is attested to as well in the Sacred Scripture. But I do not concur with Ptolemy and the ancients in the belief that the centre of all the spheres of the planetary region is near the earth. Rather, I judge that the celestial circuits are administered thus: so that both the lamps of the world, which aid in the judgement of times, and also the remotest eighth sphere, containing all the others, regard the earth as the centre of their revolutions. However, I profess that additional circles lead the five planets around the sun itself, as their leader and king, and that they always observe it as the midpoint of their advancing revolutions, such that the centres of the spheres which they mark out around it are also carried around annually on its circuit. I have found this to hold not only for Venus and Mercury, on account of their small digressions from the sun, but also for the three other superior planets. And in this way all that apparent inequality of motions in the three more distant planets (which enclose in the vastness of their revolutions around the sun both the earth and the whole elementary world ending with the moon) – which were connected to epicycles by the ancients, and to the annual motion of the earth by Copernicus – is explained very suitably through such an association of the centre of their spheres with the annual revolution of the sun. Ample opportunity is thereby extended to explain their stations and retrogradations, and their approach to the earth and retreat from it, and their apparent variations in magnitude, and other things of the sort tolerated either with the pretext of epicycles or with the assumption of the motion of the earth. All these things are furnished as well from the smaller circuits of Venus and Mercury around the sun itself but not around the earth, when the former method of epicycles is understood as fictitious and the ancient judgement concerning the disposition of these planets above and below the sun is dissolved. And hence a clear cause is rendered for why the simple motion of the sun is intermingled by necessity with the motions of all five of the planets in a fixed and particular course. Thus all the celestial appearances align themselves to the pattern of the sun, and it is like Apollo (which is the name it was assigned by the ancients) in the middle of the muses, regulating the whole harmony of the planetary choir.

Indeed, as far as the rest of the more particular differences of the apparent inequality – which the ancients imagined as happening through eccentrics and equants, and Copernicus through an epicycle on the circumference of the eccentric and equal to its changing revolution – those too can be easily saved in our hypothesis, whether through a circle of appropriate size in an eccentric sphere around the sun, or through a double circle in some concentric sphere. Thus it happens that no less than in the Copernican system, all circular motions regard their proper centre, with the Ptolemaic disarray repudiated. We will explain in what way all these things happen fully and more particularly in a work we have decided to develop (God willing) on the revival of astronomy, where we will deal with the hypothesis of celestial motions with due competence and demonstrate that all the appearances of the planets fit extremely well with each other, and that they correspond more accurately than with other ones used until now.

In order that our new discovery of the arrangement of the celestial spheres be better understood, I present an illustration of it:

Indeed, I have established a fuller explanation of this new arrangement of the celestial spheres, which includes powerful corollaries of the whole present effort, to be added around the end of the work. There it will be shown first through the motion of comets and clearly proved that the very machine of heaven is not a hard and impenetrable body, densely filled with various real spheres, as was believed hitherto by the majority of people, but is rather extremely limpid and simple. Likewise, the circuits of the planets are free and without the aid or conveyance of any real spheres, but rather are directed divinely by an inner knowledge, and clearly endure no obstacles whatsoever. From here too it will be apparent that no absurdity occurs in this ordering

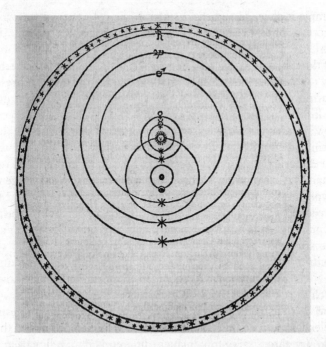

New Hypothesis of a system of the world recently devised by the
author, in which both the ancient Ptolemaic redundancies and
disorder and also the recent Copernican physical absurdity in the
motion of the earth, are excluded, and all things correspond very
neatly with the celestial appearances

of the celestial spheres from the consequence that Mars as an
evening star is nearer to the Earth than the sun itself. For this
does not admit any real and inappropriate penetration of the
spheres (since they are not in reality in the heavens, but are proposed practically for the sake of teaching and understanding the
matter). Nor can the bodies of any planets themselves ever come
to meet one another, or disturb, for any reason, the harmony of
motions which each of them observes, even as the imaginary
orbs of Mercury, Venus and Mars are intermingled with the
sun's, and cross it. This will be shown more extensively and

clearly in that place, near (as I said) the colophon of the work, and especially in our astronomical volume, where we will deal expertly with these things.

Now, however, we will borrow from this new plan of ethereal revolutions itself at least that part which satisfies the present business of assigning a place to this comet, and a hypothesis that helps to order its appearances.

With the foundations of these celestial revolutions having been cast, I say that everything corresponds most fittingly with the apparent motion of this comet, if we understand that it itself too – as if some strange and special planet – has found the centre of its circuit, no less than the other planets, at the sun, and has marked out around it a certain portion of its own orbit, in which it goes beyond not only the sphere of Mercury but also the sphere of Venus. Thus it can depart from the sun up to a sixth of the heavens, while Venus is removed from it not much beyond an eighth part of them. In truth, the comet advanced such that if it were assumed to be at the inmost part of its orbit and closest to earth when it was connected to the mean motion of the sun, then it follows that it has proceeded from there through the progression of signs (differently than happens in the case of Venus and Mercury) in the direction of the apogee of its sphere, with the centre of its revolution agreeing continually with the simple motion of the sun. So that all these things may be perceived more clearly, at this point we will now reveal to the eyes a suitable arrangement for fashioning the spheres.

A signifies the globe of the earth, which is in the centre of the universe, around which the moon revolves nearby in the sphere BEFD, in which the whole region of the elements is enclosed. Moreover, the fact that the comet was by no means found within these limits of the lunar sphere was demonstrated by us sufficiently in Chapter 6. Above, let CHIG be the annual sphere of the sun, revolved around the earth, in which the sun is represented near C, at which too are found the centres of all the spheres of the remaining five planets, according to our innovation of the celestial hypothesis (about which I have spoken). And because the star of Mercury is revolved around that same sun most closely in the sphere LKMN, and a little above this is

DE COMETA ANNI 1577.

motui coniunctus fuisset, in infima Orbis sui parte & Terris proxima constituto assumatur, atq; hinc per consequentiā Signorum, aliter quàm in Venere & Mercurio viuunit, versus eiusdē Orbis Apogæum perrexisse, centro hui9 reuolutionis Solis simplici motui perpetuò concurrente, admittatur. Quæ omnia vt rectiùs percipiantur, nunc orbiū huc aliquid facientiū oportunā dispositionem oculis subijciemus.

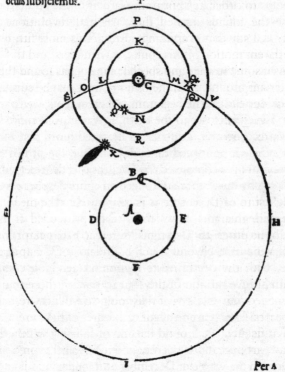

the star of Venus in the sphere OPQR, it is fitting that the comet is carried around in an equal manner on a sphere just a bit greater around the sun, which includes only the spheres of Mercury and Venus, but not the moon along with the earth (as the star of Mars does in its revolution) since it does not wander from the sun in a digression greater than 60 degrees. This same sphere, which we assign to the comet, is signified by the circle

STVX, such that the comet is at that moment near X, about as it was at the time when it was first noticed by us. Its motion in this sphere follows the order of the signs, other than the way that Venus and Mercury are revolved, such that it goes around from X through S to T. Moreover, the centre of its sphere observes a simple motion agreeing constantly with the sun. And with this arrangement of the circuit of the comet between the celestial spheres granted, I declare that it is possible to adequately account for its apparent motion as seen by us, dwelling on A, the earth.

However, it must be noticed that the comet, led in this same circular path around the sun, has not always exhibited a uniform advance. Instead, at the beginning, when in the lower part of its sphere, which is closer to the earth, it was moved more slowly. Afterwards, however, it increased its speed more and more. By that principle it happened that while around the 9th or 10th of November it only completed ten-twelfths of a degree on its circle in one day, by the 20th it discharged an entire degree in a day. At the beginning of December it increased that motion bit by bit from one degree and five-twelfths all the way until in the days immediately after 20 December, when it had been through a degree and a half, beyond which boundary it did not stretch its motion, but slowly returned to a more relaxed state – still, thus far, with a slow variation, such that up to the 26th day of January, when it was last seen by us, only five-twenty-fourths had disappeared from one and a half of a degree of the motion proper to it. For its advance around the end of January within the natural day was once more a degree and five-twelfths, precisely as it was through the whole of December and January. It did not alter its daily progress in its sphere except at the most by five-twenty-fourths – so little, in so much time, was its revolution around the sun lacking in perfect uniformity. Indeed, in November it generally showed more of the same, by a little more rapid a variation. All this is seen more fully in the fourth row of the table which we will add to the end of the following chapter.

Indeed, I admit (a fact that would be more fitting) that if the comet had completed equal arcs in an equal time interval through the whole of its duration on its own sphere, then a simple uniformity of revolution would be more properly maintained, with the same regularity through which the planets themselves steadily observe a perpetual uniformity in their circuits. Granted that this very inequality which happens to the comet in its own revolution could be bounded and corrected, whether by a twist through the centre of its sphere around the sun circularly with the requisite method in the opposite direction, or through an addition to the circular circumference of the same, by which the motion may here be curbed and there let loose. However, because by tying the motion in such knots this business would acquire more obscurity and veils than light and keenness, I did not wish to construct a more muddled arrangement of the various motions to keep them uniform, especially since it would be very ill-fitting for such quickly vanishing bodies as comets to be guilty of such artificially composed and multi-formed layers of motions. Thus I have preferred to retain those same daily steps of the comet in its sphere around the sun that experience itself supplied to us, notwithstanding the fact that at the start they wandered a bit more slowly, while afterwards they returned more quickly in successive advances, especially since it followed just about a constant uniformity through the greatest and most productive time of its appearance. For in December and January, two whole months, the uniform motion was not varied by more than five-twenty-fourths (as I indicated previously), which is truly very small and of almost no importance. In November alone, and at least through half of the month, it admitted a sensible alteration. Thus only approximately one-fifth a part of its whole duration was guilty of inequality, while the remaining four were almost exempt from it.

Nor is it true that on account of this not very great or significant inequality, anyone should suppose that the certainty of our hypotheses is shaken. For it is probable that comets, just as they do not have bodies that are truly perfect and consummate for perpetual duration (as do the rest of the stars, coeval with the beginning of the world), also do not observe so absolute and constant a uniform path in their circuits. Rather, it is as if they are a kind of mimes, imitating in a certain way the uniform regularity of the planets, but not fully following it. Comets of some years later, which will be no less in the ethereal region of the world, have shown this too fairly clearly. Therefore, whether this comet of ours completed its circuit in a manner not completely and exquisitely circular, but a little more oblong (the sort of figure which is commonly called an ovoid), or whether it advances along a perfectly circular path, but with a motion through it that is slower at the start, and afterwards slowly increased: all the same, it is revolved around the sun, though it admits some inequality, and is not confused and disorderly.

Johannes Kepler, *Cosmographic Mystery*, 1596

Johannes Kepler (1571–1630) was a young man of twenty-three when he left his studies at the University of Tübingen to teach mathematics at the Lutheran high school in Graz. Already a convinced Copernican inspired by his Tübingen teacher Michael Maestlin, Kepler believed that there were certain questions Copernicus had not adequately answered, including the reason for the number of planets (now six) and for the distances between them. Kepler believed that true theories would not only adequately match observed phenomena but also illuminate their causes, and he sought a rational explanation, built into nature, that would bolster Copernicus's model. One year later, in 1595, he believed he had found that explanation in the five Platonic solids, or regular polyhedra. Kepler argued that nesting the solids one within the other would create a series of six concentric spheres (thus explaining the number of planets), with the ratios between them explaining the planetary distances. This showed, according to Kepler, both that nature itself was fundamentally geometrical and that geometry was the key to explaining the Copernican system rather than simply describing it, as Copernicus had done. Also noteworthy in this book, often referred to by its Latin title, Mysterium Cosmographicum, *is Kepler's discussion of the slow progress of astronomy and the reasons why its theories might not perfectly match the observed data (as his own Platonic solid theory, while coming close, did not).*

Forerunner of cosmological discussions, containing the
cosmographic mystery, on the wonderful proportion of the
celestial spheres, and on the genuine and particular causes of
the number, magnitude and periodic motions of the heavens,
demonstrated through the five regular geometric solids

Greetings, friendly reader
What kind of world this is, what reason and plan God had for
    creating it,
Whence God selected the numbers, what rule there is for so great a
    mass,
What might bring about six circuits, which intervals fall between
    which sphere,
Why there is so great a gap between Jupiter and Mars, by no means
    in the first spheres,
Pythagoras teaches all these things here to you with the five figures.
Certainly he has shown by this example that we can be born again
Having gone astray for two thousand years, until Copernicus,
Distinguished as a better observer of the world.
But you, do not neglect the fruits discovered in the shells.

Preface to the Reader

Reader, my purpose is to demonstrate in this little book that
the best and greatest Creator, in the creation of this mobile
world and the arrangement of the heavens, looked back at
those five regular solids, highly celebrated from Pythagoras and
Plato continuously until us, and that he adjusted the number of
heavenly bodies, the proportions and the plan of motions to
the nature of those solids. But before I may allow you to come
to that subject itself, I will discuss with you something both
about how this little book came to be and also about my reason
for preparing it, which I believe relate both to your understanding and my reputation.

    Six years ago, when I was at Tübingen learning from the very
illustrious Michael Maestlin, and troubled by the innumerable
motions of the customary opinion about the world, I was so

delighted by Copernicus, whom Maestlin made frequent mention of in his lectures, that I not only frequently defended his
claims in the disputations of natural philosophy candidates, but
also composed a meticulous disputation on the first motion,
arguing that it happens due to the revolution of the earth. I
already ascribed the motion of the sun to that same earth, as
Copernicus had done mathematically, but I preferred natural
philosophical, or rather metaphysical, methods. To that effect I
collected, little by little, partly from the mouth of Maestlin, and
partly through my own battle, the mathematical advantages that
Copernicus has over Ptolemy. (Joachim Rheticus could have
freed me from this labour, as he pursued all this briefly and
clearly in his *First Account*.) Meanwhile, while engaged with
that perpetual labour but as a sideline alongside theology, it happened that at the right moment I came to Graz, and there I
succeeded Georg Stadius after he died; my duties there involved
me more directly in those studies. As I explained the principles of
astronomy there, all the things that I had either heard before
from Maestlin or attempted myself were of great use to me.

Then, in the year 1595, when I really wanted to take a break
from lecturing and to finish my official duties, I ruminated over
this material with all my mental vigour. And there were three
principal things whose causes (that is, why they were this way
and not some other way) I sought determinedly: the number,
size and motion of the spheres. The fact that I dared this was
brought about by that beautiful harmony of resting things: of
the sun, the fixed stars and the intermediate area, with God the
Father, and the Son, and the Holy Spirit. (I will pursue this likeness further in my *Cosmographia*.) Since, therefore, things that
were at rest were like this, I had no doubt that mobile things
would reveal themselves thus. At the start I approached the
matter with numbers, and considered whether one sphere was
double another, or triple, or quadruple, or what have you, and
how much any one was separated from any other according
to Copernicus. I wasted a great amount of time in that labour,
as though I played a game, because no regularity was found –
neither in the proportions themselves, nor in their increases;
and I gained nothing useful from it, except that I engraved very

deeply on my memory what the distances themselves were that Copernicus had recorded.

Because this did not succeed, I attempted an approach by another path, and yet another. Nearly a whole summer was lost in this torment. Finally, I happened nearer to the actual matter by a certain unpredictable chance. I believe that what happened to me was divinely inspired, as I obtained by chance that which I was unable to achieve by any labour. And I believe it even more because I had always begged God that if Copernicus had spoken the truth, then things should progress thusly. Therefore, on the 9th/19th of July, in the year 1595, as I was about to point out to my students the jumps of the great conjunctions through the eight signs, and how they crossed gradually from one trigon into another, I inscribed many triangles, or almost triangles, in the same circle, as if the end of one was the beginning of another. Consequently, the sides of the triangles cut into each other mutually at points which sketched out a small circle. The radius of the circle inscribed in the triangle is half the radius of the circumscribed one.

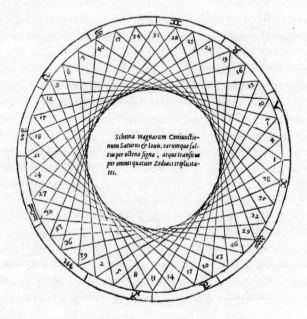

Schema magnarum Coniunctio-
num Saturni & Iouis, earumque sal-
tus per octena signa ; atque transitus
per omnes quatuor Zodiaci triplicita-
tes.

The ratio between the two circles seemed to the eye almost the same as that between Saturn and Jupiter, and the triangle was formed first, just as Saturn and Jupiter are the first planets. I immediately attempted the second distance, between Mars and Jupiter, in a quadrangular shape, the third in a five-sided one, and the fourth in a six-sided one. And since the eyes still cried out in protest, in the second distance, between Jupiter and Mars, I added a square to the triangle and the five-sided figure. Following up each one is unending.

Still, the end of this useless attempt was the same as the beginning of the final and favourable one. I reflected truly that if indeed I wanted to preserve an order among the figures in this way, I would never arrive all the way at the sun, nor would I have a reason why there are six moving spheres, rather than twenty or a hundred. And yet the figures were pleasing, inasmuch as they were quantities, and thus prior to the heavens. For quantity was created in the beginning with bodily substance, and the heavens on the next day. And if (I thought) only five figures, among the infinite ones remaining, could be discovered that had certain properties beyond the others that were specific to the quantity and proportion of the six heavens that Copernicus established, they would fulfil my prayer. So, I approached things again. What plane figures would there be between solid spheres? Rather, solid bodies should be assumed.

Behold, reader, this discovery and the subject matter of this whole little work. For if someone mildly skilled in geometry were directed with these same words, the five regular solids, with the proportion of their circumscribed spheres to the inscribed ones, would immediately occur to him; before his eyes would immediately alight that scholium to Euclid's proposition 18 of Book 13, in which it is proved impossible that more than five regular solids could exist or be devised. It is a thing of wonder that while I had not yet worked out the preferred places of each of the solids in order, I happily hit the mark so exactly using an inartful conjecture deduced from the known distances of the planets that I would not change anything afterwards when I worked out the rules that I sought for. To help you remember, mark down for yourself this thought, conceived in words at the moment that it happened: *The*

circle of the earth is the measurer of all. Circumscribe a dodecahe-
dron around it. The circle enclosing that will be Mars. Circumscribe
a tetrahedron around Mars. The circle enclosing that will be Jupi-
ter. Circumscribe a cube around Jupiter. The circle enclosing that
will be Saturn. Now inscribe an icosahedron within the earth: the
circle inscribed in it will be Venus. Inscribe an octahedron within
Venus. The circle inscribed in it will be Mercury. You have the
reasoning for the number of the planets.

This chance event was the realization of my labour: you see
this too in the plan of this book. And indeed, the pleasure I felt
in this discovery cannot be expressed in words. The waste of
time no longer bothered me; the labour no longer wearied me;
I did not evade the bother of the calculation; I exhausted days
and nights in computation, until I could determine whether the
thought, conceived in words, would agree with the spheres of
Copernicus, or whether the winds would carry off my joy. But
if I were to discover that the matter was as I thought it to be, I
made a vow to the highest and greatest God that at the first
chance I would report this wonderful exemplar of his wisdom
openly in print among men. Since, therefore, the matter was
successful after a few days, and I discovered how closely one
body follows after another among the planets, and I rendered
the whole business into the form of this present little work, and
it was approved by Maestlin, a distinguished mathematician,
you will understand, friendly reader, that I am compelled by
my vow, and cannot follow the custom of the satirist who
directs us to hold books back for nine years.

This is the one reason for my progress: but let me remove
any bit of unfavourable suspicion that you have, and add willingly one other: I recite for you the words of Archytas from
Cicero: *If I had ascended to heaven itself and thoroughly*
observed the nature of the world and the beauty of the stars,
that wonder would be unpleasing to me if I did not have you,
a kind, attentive and eager reader, to tell about it.

You will not discover new or unknown planets here, but
rather the ones of old, displaced only very little, but
fortified – though it may seem absurd – by the interposition
of rectilinear solids, so that from now on you can respond to

the peasant asking by what hooks the sky is hung so that it doesn't fall.

Farewell.

Chapter 1: With which rules the Copernican hypotheses agree, and an explanation of the hypothesis of Copernicus

Although it is pious to see, from the start of this discussion about nature, whether anything is said that is contrary to Holy Scripture, nevertheless I believe that it is untimely to provoke that controversy here even before it is a matter of concern. In general, I promise this: that I will say nothing injurious to the Holy Scriptures, and that if Copernicus is found guilty of anything with me, then I will consider him of no value. And this was always my plan, from when I began to study the books of Copernicus's *On the Revolutions*.

Because of this, therefore, I was not prevented by any religious obligation from listening to Copernicus, if he spoke appropriately. What first made me confident was the very beautiful agreement of all the things that appear in the heavens with the claims of Copernicus, as he not only gave an account of the past motions, repeated from furthest antiquity, but also declared in advance the future ones, not with perfect certainty, but still with much more certainty than Ptolemy, Alphonso and others. Moreover, and greater still than those things which we learn to wonder at in others: Copernicus alone offers the most beautiful explanation for it, and removes the reason for wonder, which is ignorance of causes.

On this point, I could never agree with those who rely on the model of accidental demonstration, which infers something true from false premises by necessity of the syllogism – those who, that is, argued, relying on that model, that it was possible that the hypotheses that Copernicus decided upon were false, and nevertheless that true phenomena follow from them just as they do from genuine principles.

For the model does not fit. That which follows from false premises is accidental, and the nature of the falsehood is such that it asserts itself the first time it is applied to another known matter, unless you freely concede that the disputant may assume infinite other false propositions, and that he may never, in his progress, accord with his own prior claims. The matter is different for someone who places the sun at the centre. Order him to demonstrate, once the hypothesis has been specified, anything whatsoever about the things that actually appear in the heavens – to go back, to go forward, to obtain one thing from another, and to do whatever he pleases that the truth of the matter allows – he will not hesitate at all if it is genuine, and he will return from even the most complex turns of the demonstrations to the same point very consistently. You may object that the same thing can still partly be said, or could once partly be said, about the ancient tables and hypotheses, which certainly satisfy the appearances but are nevertheless rejected as false by Copernicus, and that therefore the same reasoning could be used in response to Copernicus – that although he certainly renders an excellent account of the appearances, nevertheless his hypothesis is mistaken. I answer first that the ancient hypotheses plainly offer no account of principal topics. One of these is that they disregard the causes for the number, magnitude and time of the retrogradations, and why they correspond so precisely with the position and mean motion of the sun. With respect to all these matters, since a most beautiful order is evident according to Copernicus, the cause must be found there too. The principles of Copernicus cannot be false, when so consistent an account of so very many phenomena, unknown to the ancients, is rendered by them, to the extent that it is rendered by them. The very successful Tycho Brahe, an astronomer highly celebrated by all, saw this. Although he disagreed entirely with Copernicus about the position of the earth, nevertheless he retained from him that which offers us the reasons for things previously unknown: that the sun is truly the centre of the five planets.

There were other reasons as well. For turn from astronomy to natural philosophy or cosmography: these hypotheses of Copernicus not only do not transgress against the nature of things, but rather serve it greatly. For nature loves simplicity, it loves unity. There is never anything in it that appears otiose or superfluous; rather, more often one thing is designed so that many effects follow from it. While in the common hypotheses there is no end to the fabrication of spheres, in those of Copernicus very many motions follow from very few spheres. And thus this man has not only freed nature from the onerous and useless baggage of so many immense spheres, but on top of this has also disclosed to us a boundless repository of highly divine reasoned conclusions concerning the very beautiful fitness of the whole world and all of its bodies. Nor do I hesitate to assert that whatever Copernicus inferred *a posteriori*, and demonstrated from observation, using geometrical axioms, could all be demonstrated *a priori* by Aristotle, without ambiguity.

Still, I have not embraced this path rashly and without the very important authority of my illustrious teacher of mathematics, Maestlin. For he supplied me with a different specific reason for believing thus: he recognized that the comet of the year [15]77 moved very consistently with the motion of Venus, as recorded by Copernicus, and (with a conjecture based on the superlunary altitude) that it followed its course in the sphere of Venus itself. And if anyone carefully assesses how easily the false disagrees with itself, and by contrast how consistently the true agrees with the true, a most powerful argument for the Copernican arrangement of the spheres can be commenced without fault from this alone.

Chapter 14: The primary goal of the little book, and astronomical proof that these five solids are between the spheres

Therefore let us come to the principal proposition. It is known that the paths of the planets are eccentric, and hence it is the received opinion of natural philosophers that the spheres have exactly the thickness required for demonstrating the varieties of the motions. Thus far, indeed, Copernicus agrees with our philosophers. But from now on we discern a significant difference of opinions. For the natural philosophers think that there is no empty space in the celestial spheres from the inmost surface of the lunar heaven all the way to the tenth sphere; rather, that sphere is always touched by sphere, and the inmost surface of the upper one is fully united with the highest surface of the lower one. Thus, to someone inquiring, for example, about the physical position of the heaven of Mars, they respond: the inner surface of Jupiter. And based on Ptolemy and the common description of astronomy they can perhaps support this position: because there is no opportunity there to investigate the proportion of the spheres, and no aid for doing so. For just as no one easily contradicts those who have written about the new Indies if he has not inspected those places himself, thus an astronomer cannot reject the meagre accounts of the natural philosophers about the contact of the spheres if the experience of observation and the agreement of the hypotheses has not carried him away into heaven itself and between the spheres.

In Copernicus we see that no sphere is touched by another, but rather vast intervals are left through the systems, certainly full of celestial air, but nevertheless pertaining to neither of the neighbouring systems. But since I professed at the beginning to render cause for those intervals from the five solids – that is, why such large ones were left by the best and greatest Creator between pairs of planets – by showing that specific figures produce specific intervals: let us now see how successfully this has been attempted and let us debate this cause before Astronomy as judge, and Copernicus as interpreter. I leave to the spheres themselves as much thickness as the ascent and descent of each planet requires. But if the figures are interposed, as I have said, the inmost surface of the upper sphere ought to be equal to the circumscribed sphere of the figure, and the outer of the lower sphere to the inscribed one, and, moreover, the figures to be determined in the order which I proved above rationally.

And behold how near the parallel numbers are! From this you can infer how easily it would have been noticed, and how much the inequality of the numbers would have been, if this had been attempted differently than the nature of the heavens – that is, if God himself had not looked back at these proportions in the Creation. For surely it cannot be accidental that the proportions are so near to these intervals.

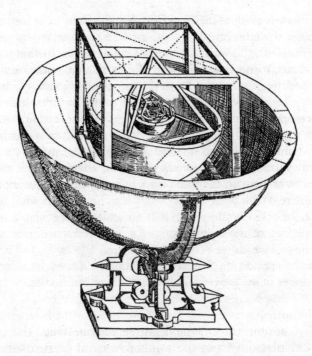

Foldout engraving of the ways in which the Platonic solids generate
the spaces between the planets, with the cube as the outermost solid,
the octahedron as the innermost and the sun at the very centre.

Chapter 16: A specific reminder about the moon, and about the material of the solids and spheres

Nor indeed should we fear that the lunar spheres are strangled
and constricted by the close proportions of the solids, if not buried and enclosed in the sphere [of the earth] itself. For it is absurd
and monstrous to position these bodies in the heavens clothed in
some material that does not allow passage to a foreign body.
Certainly there are many who do not fear to doubt whether the
heavenly spheres are entirely the adamantine kind, or whether it
is by some divine power (a comprehension of the geometrical

proportions guiding their courses) that the stars are transported through the ethereal plains and air free of those fetters of the spheres. Truly, no weight will make the steps of something moving variable and tottering, with the result that sometimes it is derailed from its circle. In fact, a point and a centre are not heavy. Rather, everything aims for the centre of that which is of the same nature as its body. Neither does a centre acquire weight from the fact that it attracts other things to itself, or is sought by them – no more than the magnet grows heavy when by its action it draws iron. Indeed, how has this earth, which we have established with Copernicus to be entirely transported, been set into its sphere – with what levers, with what chains, with what celestial adamant? Certainly with that air which all we men, around the surface of the earth, swallow (fermented and mixed with vapours) – which we penetrate with hand and body, but nevertheless do not divide or separate, although it is the vehicle for the influences of the heavens into the middle of our bodies.

Chapter 18: On the disagreement between the calculations from the solids and from Copernicus in general, and on the exactitude of astronomy

Those who want to examine everything very exactly will think that because the calculation of the solids is not consistent to the last detail with the claims of Copernicus and with his numbers, the whole of my work is simply a diversion. But I do not despair of victory, even against that.

For there are many things that Copernicus restored for us concerning the science of the motions, which had fallen into ruin: and our astronomy is purer than the memory of our fathers. But even so, if we examine the matter itself deeply, we will certainly be compelled to admit that we are almost as distant from that blessed and desired perfection as ancient astronomy is from that of today. The path on the way to the truth is long and circuitous. The ancients have shown it to us, and our ancestors advanced upon it. We move ahead, and pause one step closer, but have not yet reached the goal. I do not say this to treat astronomy with contempt – 'you can advance somewhat, even if you aren't able to go further' – but rather for this reason: lest someone rashly thinks this disagreement more serious than it is, and while attacking me and these five solids, mocks the foundations of astronomy themselves.

Indeed, the diligent reader of Copernicus will find that Copernicus himself acted as a human being in accepting whatever sort of numbers fulfilled his needs to a certain extent and established his principles. He does not reject numbers from diverse operations that, according to the power of demonstration, ought to fully agree, although they have discrepancies of some small amounts. He thus collects observations in Walter, in Ptolemy and elsewhere, so that he can use them more advantageously for building up the calculation, which is why he has no scruple in now and then disregarding or changing hours in time, quarters of degrees in angles, and more. Occasionally, as when he changes the eccentricity of Mars and Venus, he even accepts discrepancies from the truth, only because they point a finger slightly toward what he wants. He draws out from Ptolemy many things whole and untouched, though by his own admission they should have been corrected, while he changes other similar things; and he builds up the foundations of the new astronomy afterwards on all this.

Maestlin has given me very many examples of all of this, which I neglect to insert here for the sake of brevity. And indeed, Copernicus would justly seem to incur criticism, if he had not done it purposely, because he preferred to have an astronomy imperfect in certain ways rather than to be without one entirely. For these sorts of difficulties may occur while the stars hasten along. To overcome them, and to aspire without impediment to constitute a science with minimal fault, as Copernicus dared, is the mark of a courageous man, while a lazy man would sidestep it and a timid one would give up hope and abandon the whole responsibility. Further, Copernicus himself neither conceals his errors, just recounted, nor confesses them with shame. He defends himself with the example of Ptolemy and the ancients, he absolves himself with the difficulty of observing, and he sets himself everywhere before others as an example of disregarding those minute defects when establishing splendid discoveries. If that had not been done before, Ptolemy would never have disclosed to us the *Almagest*, nor Reinhold the *Prutenic Tables*.

At first I hesitated, lacking a plan, like someone who does not know how to reassemble the scattered little wheels of an automaton in order. Maestlin consoled me, or more correctly dissuaded me from this exactness of detail. 'It is not possible for us,' he said, 'to exhaust all the treasures of nature, nor must every hidden problem be well addressed. Rather, just as an injury to the human body, it must be endured and mitigated somewhat, rather than throwing the sick man into immediate danger of death via the exacting anatomist.' He mentioned to me the example of Rheticus and his similarly painstaking concern for exactitude, and Copernicus's rebuke of him. A letter of Rheticus is prefaced to his 1551 *Ephemerides*, which – because it is not available everywhere and supports this whole chapter wonderfully in many places – I will include selections from as a coda to this chapter. What follows, therefore, is – among other things – Rheticus to the reader:

However, he (Copernicus) wanted his investigations to be moderate, not excessive. Thus he purposely – not because he was lazy or weary of mental exhaustion – avoided the habit of breaking things down into tiny pieces that some have affected and others attain, like the exactness of Peurbach's tables of eclipses. You may see, moreover, certain people devoting their full attention to clearly and scrupulously scrutinizing the positions of the stars, who – while they gape at seconds, and third, fourth and fifth minutes, in the meantime disregard whole parts, and motions of phenonemona often err by hours, and sometimes even by whole days. This is, without doubt, what happened in the fable of Aesop to the man who was ordered to bring back a lost ox; while he busied himself with trying to catch some small birds, he both missed capturing these and was robbed of the ox itself. I remember when I was young and impelled by curiosity, and I wanted to arrive, as it were, at the innermost parts of the stars. I sometimes even disputed with that best and greatest man, Copernicus, concerning this exactness. But he, since he was certainly delighted by my spirit of honest desire, was in the habit of reproaching me gently, and encouraging me to learn to lift my hand from the writing tablet. 'If I,' he said, 'can approach the truth by sixths, which are ten scrupula, I will exult no less than Pythagoras was reported to have done at the discovery of the method of squares.' This bewildered me, who replied that one must strive toward more certainty. But he showed that going even as far as that was achieved with difficulty, for three reasons especially, alongside others. He said that the first of these was that he noticed that many observations of the ancients were not genuine, but rather accommodated to that theory of motions which each one of them had personally established. Therefore, singular attention and diligence were necessary in order to separate those to which the opinion of the observer had added or subtracted only very little from those which were corrupted. He said that the second reason was that the positions of the fixed stars were not known by the ancients more accurately than sixths of degrees, and it is nevertheless according to them that the positions of the planets needed to be obtained. He excepted only a small number in which the recorded declination of the star from the equator helped the matter, because from this the position of the star itself could now be established more certainly. The third reason he mentioned was this: we don't have authorities of the sort that Ptolemy had after the Babylonians and the Chaldeans, those luminaries of the art, Hipparchus, Timochares, Menelaus and the others on whose observations and precepts we are able to depend and rely. His preference indeed was to acquiesce in those whose truth he could profess, rather than to acrimoniously marshal his skill over the dubious exactitude of doubtful ones. Indeed, he was certainly not so very far off in his own values – maybe a sixth or quarter part of one degree from the truth – and this defect he not only was far from regretting, but rather greatly rejoiced that he could proceed all the way to there after a long time, great labour, maximum effort, and singular devotion and diligence. Indeed, Mercury, as if according to Greek proverb, he left behind for the general public, since he knew of no observation, either from his own study or that he had received from others, by which he would be greatly aided or which he could totally prove. Through many reminders, charges and instructions, he principally urged me to

pay attention to observing the fixed stars, most especially those
which appear in the zodiac, because the conjunction of the planets
with these could be recorded, etc.

Thus far is the part of Rheticus's letter that pertains to the matter. What do you think now, dear reader, about Copernicus? If
this work had been foretold to him, and he had understood how
near it comes to him and his plans, what do you think he would
not have tried, what labour would he not have begun, so that
the solids might be brought together with his spheres? And if
that were granted, what consensus, what perfection could not
be hoped for? In respect of which, what others might furnish –
what Maestlin himself at some time might furnish, with God's
favour – time will show. Meanwhile, I would not wish anyone
to pronounce against me rashly, and one should bear this deferral of debate with equanimity.

Galileo Galilei and Kepler, Correspondence, 1597

Kepler and Galileo had not yet heard of each other in 1597, until a copy of Kepler's recently published Cosmographic Mystery *found its way to Galileo in Italy. Galileo quickly wrote to Kepler expressing his admiration for the book and admitting that he, too, was a Copernican, though he worried about declaring this so publicly. Kepler wrote back immediately, delighted to find a fellow Copernican, and urged Galileo to openly declare his support for heliocentrism, and at the very least to write back soon. Galileo remained silent on both fronts, however, for the next thirteen years, until his telescopic discoveries of 1610.*

Galileo to Kepler, 4 August 1597

Most Learned Man,

I received your book from Paul Amberger only a few hours ago, and since Paul just told me about his return to Germany, I would certainly judge myself ungrateful of spirit if I did not give thanks to you with this letter for the gift I received. Therefore, I thank you, especially for considering me worthy to be called into your friendship. I have seen nothing of the book so far except the preface, from which nevertheless I perceived in a small way your intention, and I am extremely grateful that I have an ally in the search for truth, and indeed a friend of that same truth. For it is truly lamentable that those who study the truth are rare, and that those who do not follow a perverse method of philosophizing.

But since this is not the place to deplore the miseries of our generation, but to congratulate you on the beauty of having discovered a confirmation of the truth, let me add only this: I promise that I will thoroughly examine your book in good spirits, since I am sure that I will find beautiful things in it. I will do it, moreover, with much pleasure, since I came to the Copernican belief many years ago. From that position, moreover, I discovered the causes of many natural effects which, widely unknown, were unexplainable through the common hypothesis. I assembled many methods and refutations of arguments against the common hypothesis, but nevertheless, thus far I did not dare to carry them into the light. I was frightened because of the fate of Copernicus our teacher, who – while granted immortal fame among some people – was nevertheless put to ridicule by an infinite many (for such is the number of fools).

I would certainly dare to disclose my thoughts if there existed many like you. But since there are not, I will refrain from business of this kind. I am agitated by the scarcity of time and the eagerness to read your book, so will bring this letter to an end and maintain that I am your greatest lover and am very ready to be of service to you in all things.

Farewell and do not refrain from sending me your most delightful letters.

Most loving of your honour and name, Galileo Galilei, mathematician in the Academy of Padua

Kepler to Galileo, 13 October 1597

Kind Sir,

On 1 September, I received the letter that you wrote on 4 August, which indeed affected me with a double joy: first, on account of the friendship begun with you, an Italian; second, on account of our consensus in the Copernican cosmography. Since, therefore, you have kindly invited me at the end of your letter to write further, nor do I lack incentive to do this of my own accord, I could not do otherwise than to write to you through this present noble youth. For in fact I think that since then, if you had leisure, you would have examined my little book more deeply: therefore, I ardently desire that you send me your learned opinion. For I am accustomed to earnestly request a forthright opinion of my books from those to whom I write. And I want you to believe me – I prefer the censure of one judicious man, however harsh, to the thoughtless applause of all the public.

In truth, I wish you intended something else, gifted as you are with such intelligence! For wisely and secretly, by means of your own example, you warn that we must withdraw from the universal ignorance, nor must we rashly put ourselves before the rages of the learned crowd. In this you follow Plato and Pythagoras, our genuine teachers. Nevertheless, a great work has begun in this generation, first, by Copernicus, and then by several others, and by some of the most learned mathematicians. And henceforth, moving the earth should not be considered something new.

Therefore, be confident, Galileo, and go forth. If I conjecture well, few of the distinguished mathematicians of Europe want to separate from us: such is the force of truth. If it is less safe for you to publish in Italy, and if there are impediments there, perhaps Germany will concede this liberty to us. But enough of this. At all events, write to me privately, if you don't want to publicly, what you have discovered that is favourable to Copernicus.

Farewell, most famous sir, and respond to me in turn with a very long letter.

Brahe, *Instruments of a Restored Astronomy*, 1598

When Tycho Brahe had charted the path of the comet of 1577, he found the existing astronomical record of observations insufficiently accurate for his purposes. He subsequently undertook a lifelong project of astronomical reform, focused on the precise observation of celestial bodies using new and newly improved astronomical instruments. He did so for two decades at Uraniborg, a large observatory complex he established on the small island of Hven, granted to him by King Frederick II of Denmark. The complex included many large observing instruments, which Tycho described, along with the techniques for their use, in his Instruments of a Restored Astronomy. *Among them was the mural quadrant depicted here, used for meridian observations. In the image, one assistant on the far right sights the object with the quadrant, another calls the time, and a third, seated at a table, records the information. Tycho is depicted seated on a chair in the centre, dressed in noble garb. Behind him, some of his other instruments can be seen on the upper-level observatory; the complex's library can be seen in the middle, and the alchemical laboratory in the basement. Tycho's observations, accurate to within one minute of arc, did not reveal the parallax that a moving earth seemed to require in Copernicus's system. The alternative – a cosmos of immense size filled with unimaginably large fixed stars – was one that Tycho found absurd; his own geo-heliocentric system, with an unmoving earth, thus seemed to him a preferable alternative.*

William Gilbert, *On the Magnet*, 1600

Aristotelian physics held that the natural motion of the element earth was rectilinear, down toward the centre of the earth which was situated at the centre of the cosmos. William Gilbert (1544– 1603), an English physician, performed numerous experiments on lodestones (natural magnets), which he detailed in his 1600 On the Magnet. *Based on those experiments, the functioning of the magnetic compass needle and recent works on navigation and magnetism such as those of Robert Norman in 1581, Gilbert argued for a new kind of physics, in which the earth itself was a giant magnet that worked in exactly the same ways as a small circular lodestone. As magnetic forces often produced circular motion, this meant that the natural motion of earth might be circular, rather than rectilinear. This led Gilbert to endorse the Copernican position of the earth's diurnal rotation, though he did not take a stand on the earth's centrality or its annual revolution.*

Book I, Chapter 17: That the globe of the earth is magnetic and a magnet, and how our magnet stone contains all the primary powers of the earth, just as the earth by those same powers remains in a fixed direction

The lodestone (and all magnetics – not only the stone, but all homogenous magnetic substances) seems to contain the force of the heart of the earth and its deepest bowels, and to carry within itself most especially the earth's inner and deepest substance; and it has the ability peculiar to the globe of attracting, directing, distributing, rotating and positioning things in the world according to the rule of the whole, and thus contains and stores the dominant powers that it has.

To its body magnetic bodies converge from all sides and cling, as we see happens to the earth. It has poles, not mathematical points but natural boundaries of force, with a very powerful influence on the unity of the whole, just as there are on the earth, though our ancestors always sought them in the heavens. It has an equator, a natural differentiator between the two poles, just like the earth: for among all the lines drawn on the terrestrial globe by mathematicians, the equator is a natural boundary and not, as will be clear later, only a mathematical circle.

It maintains a constant direction toward north and south, just like the earth. It also has a circular motion toward the position of the earth, in which it arranges itself according to its rule. It observes the ascensions and declinations of the earth's poles, and conforms itself properly to them, and on its own it naturally raises its poles above the horizon according to the rule of the land or the region, or slips below it. It receives sudden properties from the earth and acquires its vertical direction from it, and iron is affected by the vertical direction of the terrestrial globe just as it is by a magnet. Magnetic things are directed and regulated by the earth, and obey the earth in all their motions.

All of its motions agree with, and are properly subject to, the form of the earth's geometry, as we will demonstrate later with very clear experiments and diagrams; and the greatest part of the visible earth is also magnetic, and has magnetic motions, although it is disfigured by infinite corruptions and mutations. Why, therefore, do we not recognize this special homogenous substance of the earth, most closely related to its internal nature and most similar to its innermost essence?

Book 6, Chapter 3: On the magnetic diurnal revolution of the globe of the earth, a probable assertion against the inveterate opinion of the Primum Mobile

Among the ancients, Heraclides of Pontus and Ecphantus, and then the Pythagoreans Nicetas of Syracuse and Aristarchus of Samos, and others (as is said), thought that the earth moves, and the stars set by the obstruction of the earth and rose when that obstruction ended. Indeed, they move the earth by turning it like a wheel on an axle, from west to east. Philolaus the Pythagorean wanted the earth to be one of the stars, and to turn around fire in an oblique circle, just as the sun and moon have courses of their own. Indeed, he was a distinguished mathematician, and a very skilled examiner of nature.

But afterwards, philosophy was discussed and disseminated by many, and opinions fashioned based on the talent of the common people or suffused with the subtleties of the Sophists took hold of most minds, and the consensus of the majority prevailed like a torrent. Thereafter many exceptional discoveries of the ancients were rejected, and sent to perish in exile; or at least, no longer being cultivated, they faded away.

Therefore Copernicus (among our contemporaries a man most worthy of literary praise) was the first to attempt to elucidate the phenomena of moving bodies by new hypotheses; and others – men very skilled in all kinds of learning – either follow or observe these demonstrations of causes in order to find a more certain symphony of motions.

Thus the assumed and imagined orbs, used by Ptolemy and others to find the times and periods of the motions, are not necessary to admit into the physical forays of philosophers. It is therefore an ancient opinion delivered to us from the antiquity of time, but one now augmented by new reflections, that the whole earth is turned in a diurnal revolution in the span of twenty-four hours. For truly, since we see the sun and moon and the other planets and the whole congregation of stars approaching and receding in the span of one natural day, either the earth itself has a diurnal motion from west to east, or the whole heavens and all remaining things in nature are rushed necessarily from east to west. But first, it is not probable that the supreme heaven and all those splendours of the visible fixed stars are urged on that extremely rapid and insane course.

Besides, who is the expert who has ever found out whether the stars that we call fixed are in one and the same sphere, or has confirmed by reason that there are any real and so to speak adamantine spheres? No one has ever demonstrated this, nor is there any doubt that the planets are at dissimilar distances from the earth. Thus those great and very numerous lights are separated from the earth by varying and very remote altitudes, and do not adhere to any sphere or firmament (as is fabricated) or vaulted body. And thus some of their intervals have been conceived on the basis of opinion rather than actuality because of their unfathomable distance. Others far exceed these and are extremely remote, and are therefore placed in heaven at varying distances, either in the very thinnest ether, or in that very subtlest quintessence, or in the void. How could they remain where they are in such a great rotation of so vast an orb of such uncertain substance?

Astronomers have observed 1,022 stars. Besides these, there are innumerable others, some of which appear to our senses very faintly; with others, our vision is dark, and they are scarcely perceived, except by those with exceptional eyesight. Nor is there anyone gifted with excellent eyes who does not sense, when the moon is silent and the air very rare, a great many lights, faint and uncertain and wavering because of the great distance. Therefore it is plausible that they are many and that they are not all caught by our eyesight. How great, then, must be the incomprehensible space all the way to the furthest fixed stars? How wide and immense must be the depth of that fantastical sphere? How far apart from the earth must be the most distantly separated stars, beyond all vision, all art and all thought? Therefore how monstrous must that motion be?

It is therefore manifest that all the stars are positioned in their designated places, press together on themselves, depend on their own centres, and that all their parts gather around them. Thus if they have motion, it will be rather each around its own centre, like the motion of the earth; or the progression of that centre on an orb, like the moon. But there would not be one circular motion of such a great unbound flock of stars. Of these stars some are positioned near the equator and would seem to be moved around very rapidly; others are nearer to the pole and would seem to be moved somewhat more gently; others are almost motionless and would seem to have very little rotation. But no difference of light, mass, or colours appears to us: for they are as bright, clear, fiery and dusky around the poles as they are near the equator and the zodiac. Those that remain in their spots and positions do not hang and are not fastened or bound to any vault.

Even more insane still is the rotation of that fictitious primum mobile, which is higher, deeper and more immense. Further, this incomprehensible primum mobile ought to be material and powerfully deep, far surpassing all inferior nature in size. For otherwise how else could it lead so many great bodies of stars, and the whole of nature all the way to the earth, from east to west? And it would be necessary to admit a universal power and perpetual tyranny ruling over the stars, and this is very trouble-some. That primum mobile carries no visible body and is not discerned at all – it is a fiction believed in and admitted by those timid people who wonder at our terrestrial mass more than at bodies so great, so incomprehensible and so distantly removed. But there can be no motion of an infinity and an infinite body, nor can there be a diurnal rotation of that very vast primum mobile.

The moon, nearest to the earth, moves around it in twenty-seven days; Mercury and Venus have motions that are somewhat slow; Mars completes a period in two years; Jupiter in twelve years; Saturn in thirty. Moreover, those who assign a motion to the fixed stars want it to be completed in 36,000 years according to Ptolemy, and 25,816 according to Copernicus's observations. Thus, the motion and consummation of the journey is always slower in the greater circles. Would there be a diurnal motion of that primum mobile, so great, so immense and deep beyond all the rest? That is indeed a superstition and philosophical fable now to be believed only by idiots, and only to be mocked by learned men, even though in earlier ages that

motion was admitted by mathematicians to the explanations
of computations and motions, at the urging of an importune
crowd of philosophers. The motions of the bodies (namely the
planets) all appear to happen in the east and following the suc-
cession of the signs.

The common mathematicians also believe that the fixed stars
proceed in the same way very slowly. And out of ignorance of
the truth they are forced to add to them a ninth sphere. But now
this first and inconceivable primum mobile, a fiction, comprehended by no judgement, marked by no visible star, but
conceived only by imagination and mathematical supposition,
and believed and admitted by philosophers wrongly, raised into
heaven and beyond all the stars, must be turned in the opposite
direction from east to west, against the inclination of the whole
rest of the world.

Ptolemy of Alexandria seems to me to be too timid and cow-
ardly when he dreads the dissolution of this lower world if the
earth were to move rotationally. Why does he not fear the ruin
of the universe, the dissolution, confusion, burning and immense
celestial and super-celestial calamities from a motion beyond all
thought, dreams, fables and poetic licences – insurmountable,
ineffable and incomprehensible? Rather, we are moved along by
a diurnal rotation of the earth (that is, by a more fitting motion),
and as a boat moves above the waters, so we turn along with the
earth, and yet we seem to stand still and be at rest. This seems
great and incredible to certain philosophers, on account of the
long-standing opinion that the earth's vast body cannot be
turned circularly in the span of twenty-four hours.

Would it be more incredible for the moon to move through
its sphere, or complete its whole course, in the span of twenty-
four hours? More so the sun and Mars, and still more Jupiter
and Saturn; and how much more astonishing would be the
speed of the fixed stars and the firmament, which are admired
as the ninth sphere, let them think as they wish. But to fashion
a primum mobile, and to attribute fictitious motion to it completed in the span of twenty-four hours, but not at the same
time to concede this too to the earth, is ridiculous.

Let theologians cast out and wipe off with sponges those old wives' fables about such a rapid rotation of the heavens, borrowed from certain thoughtless philosophers.

Aristotle fashions a philosophy for himself from simple and mixed motions: that the heavens are moved in a simple, circular motion, and the elements in a straight motion, and that the parts of the earth seeking the earth move in straight lines, falling upon its surface at right angles and pressing toward the centre, where it is always at rest. Therefore the whole earth remains immovable in its place, united and supported by its weight. That coherence of parts and congregation of matter are also in the sun, the moon, the planets and the fixed stars, and ultimately in all those round bodies whose parts cohere together and strive each to its own centre; otherwise the heaven would fall into ruin and its sublime adornment would be destroyed. Nevertheless, these celestial bodies have a circular motion. Therefore the earth can also have its own motion in the same manner; and neither is this motion unfit (as some think) for the accumulation of things or adverse to the generation of things. For since it is innate in the terrestrial globe and natural, nor is there anything external that agitates or impedes adverse movements, it turns without any ill or danger.

For these reasons, therefore, the diurnal rotation of the earth seems not only probable but evident, since nature always acts through fewer causes rather than through many; and it is more agreeable to reason that one small body of the earth goes through diurnal rotations than that the whole universe should be borne around. I pass over the reasons for the remaining motions of the earth; for I am now concerned only with the diurnal one, with which it turns around with respect to the sun and produces a natural day (which we call *nycthemeron*).

Kepler, *The Dream*,
1608 (published 1634)

While a student at Tübingen University, Kepler composed a short disputation that considered how celestial phenomena would appear to an observer on the moon. Even then, he seems to have conceived the work as a way to support Copernican theory by combatting the objection that the earth seems stationary to its inhabitants – the same, Kepler argued, would be true of the moon to its inhabitants. He revived the idea in 1608 when he drafted Somnium, *or* The Dream, *structured as a story within a story, which begins with Kepler dreaming of a mother and son in Iceland who enlist a Daemon to help them travel to the moon. The excerpt below is part of the Daemon's speech about Levania, the name he gives to the moon. In later years, Kepler wrote lengthy numbered explanatory notes that he appended to the text to clarify his arguments (some of which are included here). The book was not published until 1634, four years after his death.*

Situated fifty thousand German miles up, in the vastness of the ether, is the island of Levania. The passage to it from here or from it to this earth is rarely open. And when it is open, it is easy for those of our race, but transporting men is extremely difficult and associated with severe danger to life. Though it is very far away, the whole journey is completed in the span of four hours at most. The start of motion is very rough, for one is hurled no differently than if shot out by gunpowder across mountains and seas. [At the end of the journey], when we are ready, one descends as if from a ship onto land.

Now let me speak about the appearance of the province itself. Although the appearance of the fixed stars is the same for Levania as it is for us, it observes the motions and sizes of the planets differently than we do here, such that its science of astronomy is completely different.

Thus, just as our geographers divide the sphere of the earth into five zones according to celestial phenomena, so Levania consists of two hemispheres. One is Subvolva and the other Privolva; the former always enjoys our Volva [the earth], which for them takes the place of our moon, and the latter is eternally deprived of the sight of Volva.

To its inhabitants, while the stars hasten along, Levania seems to stand unmoved, no less than our earth does to us humans (1).

The motion of the fixed stars in their progression is very swift, as in twenty tropical years they cross the whole zodiac, which takes us 26,000 years (2).

Their Volva stands still, immobile in its place, as if it were affixed to the heavens with a nail. Above it the other stars, and even the sun itself, cross from east to west (3).

But although it seems to be unmoved from its place, nevertheless it rotates in its place, in contrast to our moon (4), and displays a remarkable variety of spots, with the spots crossing continuously from east to west. Thus, one such rotation, wherein the same spots return, is considered one hour of time by the Subvolvans; however, it equals a little bit more than one of our days plus one of our nights.

Notes

(1) Behold the hypothesis of the whole *Dream*: it is an argument for the motion of the earth, or rather a refutation of the argument against the motion of the earth that is based on the senses.
(2) Consider here the hypothesis of the little book, and learn that what are for us among the particular features of the whole world – the twelve celestial signs, the solstices, the equinoxes, the tropical years, the sidereal years, the equator, the colures, the tropics, arctic circles and celestial poles – are all tied to the

very narrow globe of the earth, and exist in the imagination of earthly inhabitants alone. Thus if we transfer the imagination to another sphere, everything needs to be conceived differently.

(3) You have here the primary hypothesis stressed in a full declaration. Without doubt we inhabitants of earth think that the surface on which we stand, and with it the balls on our towers, stand immobile, while the stars move around those balls in a course from east to west. But this does not detract from or preside over the truth at all. For the Lunarians, too, think that their lunar surface, and the ball of Volva suspended on high above it, stand still in their place, while we know for certain that the moon is one of the movable stars.

(4) The sphere of the earth also rotates once in the space of a day around its axis. This motion of the earth is evident to the eyes of the inhabitants of the moon. Nor is there any simple reason for them to suspect that the sphere of Volva does not turn on its own axis, but rather that the whole world, and with it also their own home, the moon, moves around Volva, contemplating all parts of its sphere successively – although that is the actual truth of the matter. Take as an example the appearance of the spots on the sun, which we see going around the body of the sun in the space of twenty-six days. Who could ever come to the opinion that the spots of the sun are at rest, but that this ship of ours, which is called earth, bears us in so brief a span of time around the sun that it discloses to us the separate parts of its surface and its spots successively? The Copernicans themselves, who are persuaded that they are carried around the sun in a year of time, are certain for this very reason that they do not complete this journey in twenty-six days, for these are contradictions. Therefore, vision supplies us with a very certain argument for the rotation of the sun. And, therefore, vision attests to the lunar inhabitants that their Volva rotates around its axis. Let their vision here be deceived or let it affirm something totally certain, the result is the same – whichever you choose, firm testimony is presented that the Lunarians (if there are any) are persuaded about the rotation of Volva, which was what needed to be proved. But with respect to the more secret goal of this fable, the tables are turned delightfully for us. Everyone screams that the motion of the stars around

the earth and the rest of the earth are apparent to the eyes. I
retort that to the eyes of the Lunarians the rotation of our earth,
their Volva, is apparent, as is the rest of their moon. If they were
to argue that the lunatic senses of my Lunarian people are
deceived, I retort, with equal right, that the terrestrial senses of
the inhabitants of earth lack reason.

Kepler, *A New Astronomy Based upon Causes, or Celestial Physics, Treated by Means of Commentaries on the Motions of Mars*, 1609*

When Kepler joined Tycho Brahe in Prague, where Tycho had moved to serve Emperor Rudolph II, he had hoped to use Tycho's observational data to assess and modify the theory of Platonic solids from the Cosmographic Mystery. Initially he was only given Tycho's data on Mars, and his battle with that planet's tricky orbit ultimately became the 1609 New Astronomy. Yet the book was not just about Mars, but was rather an entirely new way of doing astronomy, which he referred to in his title as 'celestial physics'. Kepler argued in the book that planetary motions needed to have physical causes and explanations; in his view, inspired in part by applying William Gilbert's theories of magnetism from the earth to the sun, a force coming from the sun might be the cause of the planetary orbits. In the course of exploring that theory with the data from Mars, Kepler discovered what became known as his first two laws. The second law (though he discovered it first) stated that the area swept out by a planet as it moves along its orbit is equal in equal time intervals; this meant that planets move faster the closer they are to the sun and slower the further from it they are. The first law (discovered second) stated that planetary orbits were shaped liked ellipses with two foci, and that the sun occupied one of those foci. His approach thus broke from tradition in two ways: the

* This translation is taken from *Selections from Kepler's* Astronomia Nova: *A Science Classics Module for Humanities Studies*, selected, translated and annotated by William H. Donahue (Green Lion Press, 2005).

insistence that astronomy was physical, and the move away
from perfect circles, which even Copernicus had embraced.

Introduction

It is extremely hard these days to write mathematical books, especially astronomical ones. For unless one maintains the truly rigorous sequence of proposition, construction, demonstration and conclusion, the book will not be mathematical; but maintaining that sequence makes the reading most tiresome. I myself, who am known as a mathematician, find my mental forces wearying when, upon rereading my own work, I recall from the diagrams the sense of the proofs, which I myself had originally introduced from my own mind into the diagrams and the text. But then when I remedy the obscurity of the subject matter by inserting explanations, it seems to me that I commit the opposite fault, of waxing verbose in a mathematical context. Furthermore, prolixity of phrases has its own obscurity, no less than terse brevity. The latter evades the mind's eye while the former distracts it; the one lacks light while the other overwhelms with superfluous glitter; the latter does not arouse the sight while the former quite dazzles it.

These considerations led me to the idea of including a kind of elucidating introduction to this work, to assist the reader's comprehension as much as possible.

The introduction to this work is aimed at those who study the physical sciences

There are many points that should be brought together here at the beginning which are presented bit by bit through the work, and are therefore not so easy to attend to in passing. Furthermore, I shall reveal, especially for the sake of those professors of the physical sciences who are irate with me, as well as with Copernicus and even with the remotest antiquity, on account of our having shaken the foundations of the sciences with the motion of the earth – I shall, I say, reveal faithfully the intent of the principal chapters which deal with this subject, and shall propose for inspection all the principles of the proofs upon which my conclusions, so repugnant to them, are based.

For when they see that this is done faithfully, they will then have the free choice either of reading through and understanding the proofs themselves with much exertion, or of trusting me, a professional mathematician, concerning the sound and geometrical method presented. Meanwhile, they, for their part will turn to the principles of the proofs thus gathered for their inspection, and will examine them thoroughly, knowing that unless they are refuted the proof erected upon them will not topple. I shall also do the same where, as is customary in the physical sciences, I mingle the probable with the necessary and draw a plausible conclusion from the mixture. For since I have mingled celestial physics with astronomy in this work, no one should be surprised at a certain amount of conjecture. This is the nature of physics, of medicine and of all the sciences which make use of other axioms besides the most certain evidence of the eyes.

On the schools of thought in astronomy

The reader should be aware that there are two schools of thought among astronomers, one distinguished by its chief, Ptolemy, and by the assent of the large majority of the ancients, and the other attributed to more recent proponents, although it is the most ancient. The former treats the individual planets separately and assigns causes to the motions of each in its own orb, while the latter relates the planets to one another, and deduces from a single common cause those characteristics that are found to be common to their motions. The latter school I again subdivide. Copernicus, with Aristarchus of remotest antiquity, ascribes to the translational motion of our home the earth the cause of the planets' appearing stationary and retrograde. Tycho Brahe, on the other hand, ascribes this cause to the sun, in whose vicinity he says the eccentric circles of all five planets are connected as if by a kind of knot (not physical, of course, but only quantitative). Further, he says that this knot, as it were, revolves about the motionless earth, along with the solar body.

For each of these three options concerning the world there are several other peculiarities which themselves also serve to distinguish the schools, but these peculiarities can each be easily altered and amended in such a way that, so far as astronomy or the celestial appearances are concerned, the three opinions are for practical purposes equivalent to a hair's breadth, and produce the same results.

The twofold aim of the work

My aim in the present work is chiefly to reform astronomical theory (especially the motion of Mars) in all three forms of hypotheses, so that what we compute from the tables may correspond to the celestial phenomena. Hitherto, it has not been possible to do this with sufficient certainty. In fact, in August of 1608, Mars was little less than four degrees beyond the position given by calculation from the *Prutenic Tables*. In August and September of 1593 this error was a little less than five degrees, while in my new calculation the error is entirely suppressed.

On the physical causes of the motions

Meanwhile, although I place this goal first and pursue it cheerfully, I also make an excursion into Aristotle's *Metaphysics*, or rather, I inquire into celestial physics and the natural causes of the motions. The eventual result of this consideration is the formulation of very clear arguments showing that only Copernicus's opinion concerning the world (with a few small changes) is true, that the other two are false, and so on.

Indeed, all things are so interconnected, involved and intertwined with one another that after trying many different approaches to the reform of astronomical calculations, some well trodden by the ancients and others constructed in emulation of them and by their example, none other could succeed than the one founded upon the motions' physical causes themselves, which I establish in this work.

The first step:
The planes of the eccentrics intersect in the sun

Now my first step in investigating the physical causes of the motions was to demonstrate that the [planes of] all the eccentrics intersect in no other place than the very centre of the solar body (not some nearby point) contrary to what Copernicus and Brahe thought. If this correction of mine is carried over into the Ptolemaic theory, Ptolemy will have to investigate not the motion of the centre of the epicycle, about which the epicycle proceeds uniformly, but the motion of some point whose distance from that centre, in proportion to the diameter, is the same as the distance of the centre of the solar orb from the earth for Ptolemy, which point is also on the same line, or one parallel to it.

Here the Braheans could have raised the objection against me that I am a rash innovator, for they, while holding to the opinion received from the ancients and placing the intersection of the [planes of the] eccentrics not in the sun but near the sun, nevertheless construct on this basis a calculation that corresponds to the heavens. And in translating the Brahean numbers into the Ptolemaic form, Ptolemy could have said to me that as long as he upheld and expressed the phenomena, he would not consider any eccentric other than the one described by the centre of the epicycle, about which the epicycle proceeds uniformly. Therefore, I have to look again and again at what I am doing, so as to avoid setting up a new method which would not do what was already done by the old method.

So to counter this objection, I have demonstrated in the first part of the work that exactly the same things can result or be presented by this new method as are presented by their ancient method.

In the second part of the work I take up the main subject, and describe the positions of Mars at apparent opposition to the sun, not worse, but indeed much better, with my method than they expressed the positions of Mars at mean opposition to the sun with the old method.

Meanwhile, throughout the entire second part (as far as concerns geometrical demonstrations from the observations) I leave in suspense the question of whose procedure is better, theirs or mine, seeing that we both match a great many observations (this is, indeed, a basic requirement for our theorizing). However, my method is in agreement with physical causes, and their old one is in disagreement, as I have partly shown in the first part, especially Chapter 6.

But finally in the fourth part of the work, in Chapter 52, I consider certain other observations, no less trustworthy than the previous ones were, which their old method could not match, but which mine matches most beautifully. I thereby demonstrate most soundly that Mars's eccentric is so situated that the centre of the solar body lies upon its line of apsides, and not any nearby point, and hence that all the [planes of] the eccentrics intersect in the sun itself.

This should, however, hold not just for the longitude but for the latitude as well. Therefore, in the fifth part I have demonstrated the same from the observed latitudes in Chapter 67.

The second step:
there is an equant in the theory of the sun

This could not have been demonstrated earlier in the work, because one of the constituents of these astronomical demonstrations is an exact knowledge of the causes of the second inequality in the planets' motion, for which some other new thing had likewise to be discovered in the third part, unknown to our predecessors and so on.

For I have demonstrated in the third part that whether the old method, which depends upon the sun's mean motion, is valid, or my new one, which uses the apparent motion, nevertheless in either case there is something from the causes of the first inequality that is mixed in with the second, which pertains to all planets in common. Thus for Ptolemy I have demonstrated that these epicycles do not have as centres those points about which their motion is uniform. Similarly, for Copernicus I have demonstrated that the circle in which the earth is moved around the sun does not have as its centre that point about which its motion is regular and uniform. Similarly, for Tycho Brahe I have demonstrated that the circle on which the common point or knot of the eccentrics mentioned above moves does not have as its centre that point about which its motion is regular and uniform. For if I concede to Brahe that the common point of the eccentrics may be different from the centre of the sun, he must grant that the circuit of that common point, which in magnitude and period exactly equals the orbit of the sun, is eccentric and tends toward Capricorn, while the sun's eccentric circuit tends toward Cancer. The same thing befalls Ptolemy's epicycles.

However, if I place the common point or knot of the eccentrics in the centre of the solar body, then the common circuit of both the knot and the sun is indeed eccentric with respect to the earth and tends toward Cancer, but only by half the eccentricity shown by the point about which the sun's motion is regular and uniform.

And in Copernicus, the earth's eccentric still tends toward Capricorn, but by only half the eccentricity of the point about which the earth's motion is uniform, also in the direction of Capricorn.

Likewise, in Ptolemy, on each of the diameters of the epicycles that run from Capricorn to Cancer, there are three points, the outer two of which are at the same distance from the middle ones; and their distances from one another have the same ratio to the diameters as the whole eccentricity of the sun has to the diameter of its circuit. And of these three points, the middle ones are the centres of their epicycles, those that lie toward Cancer are the points about which the motions on the epicycles are uniform, and finally those that lie toward Capricorn are the ones whose eccentrics (described by them) we would be tracing out *if instead of the sun's mean motion we follow the apparent motion*, just as if those were the points at which the epicycles were attached to the eccentric. The result of this is that each planetary epicycle contains the theory of the sun in its entirety, with all the properties of its motions and circles.

The earth is moved and the sun stands still.
Physico-astronomical arguments

With these things thus demonstrated by a reliable method, the previous step toward the physical causes is now confirmed, and a new step is taken toward them, most clearly in the theories of Copernicus and Brahe, and more obscurely but at least plausibly in the Ptolemaic theory.

For whether it is the earth or the sun that is moved, it has certainly been demonstrated that the body that is moved is moved in a non-uniform manner, that is, slowly when it is further from the body at rest, and more swiftly when it has approached this body.

Thus the physical difference is now immediately apparent – by way of conjecture, it is true, but yielding nothing in certainty to conjectures of doctors on physiology or to any other natural science.

First, Ptolemy is certainly condemned. For who would believe that there are as many theories of the sun (so closely resembling one another that they are in fact equal) as there are planets, when he sees that for Brahe a single solar theory suffices for the same task, and it is the most widely accepted axiom in the natural sciences that Nature makes use of the fewest possible means?

That Copernicus is better able than Brahe (of whom in all fairness, most honest and grateful mention is made, and recognition given, since I build this entire structure from the bottom up upon his work, all the materials being borrowed from him) to deal with celestial physics is proven in many ways.

1. First, although Brahe did indeed take up those five solar theories from the theories of the planets, bringing them down to the centres of the eccentrics, hiding them there and conflating them into one, he nevertheless left in the world the effects produced by those theories. For Brahe no less than for Ptolemy, besides that motion which is proper to it, each planet is still actually moved with the sun's motion, the two being mixed into one, the result being a spiral. That it results

from this that there are no solid orbs, Brahe has demonstrated most firmly. Copernicus, on the other hand, entirely removed this extrinsic motion from the five planets, assigning its cause to an exception arising from the circumstances of observation. Thus the motions are still multiplied to no purpose by Brahe, as they were before by Ptolemy.

2. Second, if there are no orbs, the conditions under which the intelligences and moving souls must operate are made very difficult, since they have to attend to so many things to introduce to the planet two intermingled motions. They would at least have to attend at one and the same time to the principles, centres and periods of the two motions. But if the earth is moved, I show that most of this can be done with physical rather than animate faculties, namely magnetic ones. But these are more general points. There follow others arising specifically from demonstrations, upon which we now begin.

3. For if the earth is moved, it has been demonstrated that the increases and decreases of its velocity are governed by its approaching toward and receding from the sun. And in fact the same happens with the rest of the planets: they are urged on or held back according to the approach toward or recession from the sun. So far, the demonstration is geometrical.

   And now, from this very reliable demonstration, the conclusion is drawn, using a physical conjecture, that the source of the five planets' motion is in the sun itself. It is therefore very likely that the source of the earth's motion is in the same place as the source of the other five planets' motion, namely in the sun as well. It is therefore likely that the earth is moved, since a likely cause of its motion is apparent.

4. That, on the other hand, the sun remains in place in the centre of the world, is most probably shown by (among other things) its being the source of motion for at least five planets. For whether you follow Copernicus or Brahe, the source of motion for five of the planets is in the sun, and in Copernicus, for a sixth as well, namely the earth. And it is more likely that the source of all motion should remain in place rather than move.

5. But if we follow Brahe's theory and say that the sun moves, this first conclusion still remains valid, that the sun moves slowly when it is more distant from the earth and swiftly when it approaches, and this not only in appearance, but in fact. For this is the effect of the circle of the equant which, by an inescapable demonstration, I have introduced into the theory of the sun.

    Upon this most valid conclusion, making use of the physical conjecture introduced above, might be based the following theorem of natural philosophy: the sun, and with it the whole huge burden (to speak coarsely) of the five eccentrics, is moved by the earth; or, the source of the motion of the sun and the five eccentrics attached to the sun is in the earth.

    Now let us consider the bodies of the sun and the earth, and decide which is better suited to being the source of motion of the other body. Does the sun, which moves the rest of the planets, move the earth, or does the earth move the sun, which moves the rest, and which is so many times greater? Unless we are to be forced to admit the absurd conclusion that the sun is moved by the earth, we must allow the sun to be fixed and the earth to move.

6. What shall I say of the motion's periodic time of 365 days, intermediate in quantity between the periodic time of Mars of 687 days and that of Venus of 225 days? Does not the nature of things cry out with a great voice that the circuit in which these 365 days are used up also occupies a place intermediate between those of Mars and Venus about the sun, and thus itself also encircles the sun, and hence, that this circuit is a circuit of the earth about the sun, and not of the sun about the earth? These points are, however, more appropriate to my *Mysterium Cosmographicum*, and arguments that are not going to be repeated in this work should not be introduced here.

7. For other metaphysical arguments that favour the sun's position in the centre of the world, derived from its dignity or its illumination, see my little book just mentioned, or look in Copernicus. There is also something in Aristotle's *On the*

Heavens, Book II, in the passage on the Pythagoreans, who used the name 'fire' to signify the sun. I have touched upon a few points in the *Astronomiae Pars Optica*, Chapter 1, p. 7, and also Chapter 6, especially p. 225.

8. But on the earth's being suited to a circular motion in some place other than the centre of the world, you will find a metaphysical argument in Chapter 9, p. 322 of that book.

Objections to the Earth's motion

I trust the reader's indulgence if I take this opportunity to present a few brief replies to a number of objections, which, capturing people's minds, use the following arguments to shed darkness. For these replies are by no means irrelevant to matters that concern the physical causes of the planets' motion, which I discuss chiefly in parts three and four of the present work [chapters 22–60].

On the motion of heavy bodies

Many are prevented by the motion of heavy bodies from believing that the earth is moved by an animate motion, or better, by a magnetic one. They should ponder the following propositions.

The theory of gravity is in error

A mathematical point, whether or not it is the centre of the world, can neither effect the motion of heavy bodies nor act as an object toward which they tend. Let the physicist prove that this force is in a point which is neither a body nor is grasped otherwise than through mere relation.

It is impossible that, in moving its body, the form of a stone seeks out a mathematical point (in this instance the centre of the world) without respect to the body in which this point is located. Let the physicists prove that natural things have a sympathy for that which is nothing.

It is likewise impossible for heavy bodies to tend toward the centre of the world simply because they are seeking to avoid its spherical extremities. For, compared with their distance from the extremities of the world, the proportional part by which they are removed from the world's centre is imperceptible and of no account. Also, what would be the cause of such antipathy? With how much force and wisdom would heavy bodies have to be endowed in order to be able to flee so precisely an enemy surrounding them on all sides? Or what ingenuity would the extremities of the world have to possess in order to pursue their enemy with such exactitude?

Nor are heavy bodies driven in toward the middle by the rapid whirling of the primum mobile, as objects in whirlpools are. That motion (if we suppose it to exist) does not carry all the way down to these lower regions. If it did, we would feel it, and would be caught up by it, along with the very earth itself; indeed, we would be carried ahead, and the earth would follow. All these absurdities are consequences of our opponents' view, and it therefore appears that the common theory of gravity is in error.

The true theory of gravity

The true theory of gravity rests upon the following axioms.

Every corporeal substance, to the extent that it is corporeal, has been so made as to be suited to rest in every place in which it is put by itself, outside the sphere of influence of a kindred body.

Gravity is a mutual corporeal disposition among kindred bodies to unite or join together; thus the earth attracts a stone much more than the stone seeks the earth. (The magnetic faculty is another example of this sort.)

Heavy bodies (most of all if we establish the earth as the centre of the world) are not drawn toward the centre of the world because it is the centre of the world, but because it is the centre of a kindred spherical body, namely, the earth. Consequently, wherever the earth be established, or whithersoever it be carried by its animate faculty, heavy bodies are drawn toward it.

If the earth were not round, heavy bodies would not everywhere be drawn in straight lines toward the middle point of the earth, but would be drawn toward different points from different sides.

If two stones were set near one another in some place in the world outside the sphere of influence of a third kindred body, these stones, like two magnetic bodies, would come together in an intermediate place, each approaching the other by an interval proportional to the bulk of the other.

If the moon and the earth were not each held back in its own circuit by an animate force or something else equivalent to it, the earth would ascend toward the moon by one fifty-fourth part of the interval, and the moon would descend toward the earth about fifty-three parts of the interval, and there they would be joined together; provided, that is, that the substance of each is of the same density.

If the earth should cease to attract its waters to itself, all these waters would be lifted up, and would flow onto the body of the moon.

Reason for the ebb and flow of the sea

The sphere of influence of the attractive power in the moon is extended all the way to the earth, and in the torrid zone calls the waters forth, particularly when it comes to be overhead in one or another of its passages. This is imperceptible in enclosed seas, but noticeable where the beds of the oceans are widest and there is much free space for the waters' reciprocation. It thus happens that the shores of the temperate latitudes are laid bare, and to some extend even in the torrid regions the neighbouring oceans diminish the size of the bays. And thus when the waters rise in the wider ocean beds, the moon being present, it can happen that in the narrower bays, if they are not too closely surrounded, the waters might even seem to be fleeing the moon, though in fact they are subsiding because a quantity of water is being carried off elsewhere.

But the moon passes the zenith swiftly, and the waters are unable to flow so swiftly. Therefore, a current arises in the ocean of the torrid zone, which, when it strikes upon the far shores, is thereby deflected. But when the moon departs, this congress of the waters, or army on the march toward the torrid zone, now abandoned by the traction that had called it forth, is dissolved. But since it has acquired impetus, it flows back (as in a water vessel) and assaults its own shore, inundating it. In the moon's absence, this impetus gives rise to another impetus until the moon returns and the impetus is restrained, moderated and carried along by the moon's motion. So all shores that are equally accessible are flooded at the same time, while those more remote are flooded later, some in different ways because of their various degrees of accessibility to the ocean.

To the objection that objects projected vertically fall back to their places

But even if the earth's power of attraction is extended very far upwards, nevertheless, if a stone were at a distance that was perceptible in relation to the earth's dimeter, it is true that, the earth being moved, such a stone would not simply follow, but its forces of resistance would mingle with the earth's forces of attraction, and it would thus detach itself somewhat from the earth's grasp. In just the same way, violent motion detaches projectiles somewhat from the earth's grasp, so that they either run on ahead if they are shot eastwards or are left behind if shot westwards, thus leaving the place from which they are shot, under the compulsion of force. Nor can the earth's revolving effect impede this violent motion all at once, as long as the violent motion is at its full strength.

But no projectile is separated from the surface of the earth by even a hundred-thousandth part of the earth's diameter, and not even the clouds themselves, or smoke, which partake of earthy matter to the very least extent, achieve an altitude of a thousandth part of the semi-diameter. Therefore, none of the clouds, smokes, or objects shot vertically upwards can make any resistance nor, I say, can the natural inclination to rest do anything to impede this grasp of the earth's, at least where this resistance is negligible in proportion to that grasp. Consequently, anything shot vertically upwards falls back to its place, the motion of the earth notwithstanding. For the earth cannot be pulled out from under it, since the earth carries with it anything sailing through the air, linked to it by the magnetic force no less firmly than if those bodies were actually in contact with it.

When these propositions have been grasped by the understanding and pondered carefully, not only do the absurdity and falsely conceived physical impossibility of the earth's motion vanish, but it also becomes clear how to reply to the physical objections, however they are framed.

The opinion of Copernicus

Copernicus preferred to think that the earth and all terrestrial bodies (even those cast away from the earth) are informed by one and the same motive soul, which, while rotating its body the earth, also rotates those particles cast away from it. He thus held it to be this soul, spread throughout the particles, that acquires force through violent motions, while I hold that it is a corporeal faculty (which we call gravity, or the magnetic faculty) that acquires the force in the same way, namely through violent motions.

Nevertheless, this corporeal faculty is sufficient for anything removed from the earth: the animate faculty is superfluous.

To objections concerning the swiftness of the earth's motion

Although many people fear the worst for themselves and for all earth's creatures on account of the extreme rapidity of this motion, they have no cause for alarm. On this point see my book, *On the New Star*, chapters 15 and 16, pp. 82 and 84.

To objections concerning the immensity of the heavens

In the same place, you will find the full-sail voyage along the world's immense orbit, which, in objection to Copernicus, is usually held to be unnatural. There it is demonstrated to be well proportioned, and that, on the contrary, the speed of the heavens would become ill-proportioned and unnatural were the earth ordered to remain quite motionless in its place.

To objections concerning the dissent of holy scripture, and its authority

There are, however, many more people who are moved by piety to withhold assent from Copernicus, fearing that falsehood might be charged against the Holy Spirit speaking in the Scriptures if we say that the earth is moved and the sun stands still.

But let them consider that since we acquire most of our information, both in quality and quantity, through the sense of sight, it is impossible for us to abstract our speech from this ocular sense. Thus, many times each day we speak in accordance with the sense of sight, although we are quite certain that the truth of the matter is otherwise. This verse of Virgil furnishes an example: *We are carried from the port, and the land and cities recede.*

Thus, when we emerge from the narrow part of some valley, we say that a great plain is opening itself out before us.

Now the Holy Scriptures too when treating common things (concerning which it is not their purpose to instruct humanity) speak with humans in the human manner, in order to be understood by them. They make use of what is generally acknowledged in order to weave in other things more lofty and divine.

No wonder, then, if Scripture speaks in accordance with human perception when the truth of things is at odds with the senses, whether or not humans are aware of this. Who is unaware that the allusion in Psalm 19 is poetical? Here, under the image of the sun, are sung the spreading of the Gospel and even the sojourn of Christ the Lord in this world on our behalf, and in the singing the sun is said to emerge from the tabernacle of the horizon like a bridegroom from his marriage bed, exuberant as a strong man for the race. Which Virgil imitates thus: 'Aurora leaving Tithonus' saffron-coloured bed.'

The psalmodist was aware that the sun does not go forth from the horizon as from a tabernacle (even though it may appear so to the eyes). On the other hand, he considered the sun to move for the precise reason that it appears so to the eyes. In either case, he expressed it so because in either case it appeared so to the eyes. He should not be judged to have spoken falsely in either case, for the perception of the eyes also has its truth, well suited to the psalmodist's more hidden aim, the adumbration of the Gospel and also of the Son of God. Likewise, Joshua makes mention of the valleys against which the sun and moon moved, because when he was at the Jordan it appeared so to him. Yet each writer was in perfect control of his meaning. Joshua meant that the sun should be held back in its place in the middle of the sky for an entire day with respect of the sense of his eyes, since for other people during the same interval of time it would remain beneath the earth.

But thoughtless persons pay attention only to the verbal contradiction 'the sun stood still' versus 'the earth stood still', not considering that this contradiction can only arise in an optical and astronomical context and does not carry over into common usage. Nor are these thoughtless ones willing to see that Joshua was simply praying that the mountains not remove the sunlight from him, which prayer he expressed in words conforming to the sense of sight, as it would be quite inappropriate to think, at that moment, of astronomy and of visual errors. For if someone had admonished him that the sun doesn't really move against the valley of Ajalon, but only appears to do so, wouldn't Joshua have exclaimed that he only asked for the day to be lengthened, however that might be done? He would therefore have replied in the same way if anyone had begun to present him with arguments for the sun's perpetual rest and the earth's motion.

Now God easily understood from Joshua's words what he meant, and responded by stopping the motion of the earth, so that the sun might appear to him to stop. For the gist of Joshua's petition comes to this, that it might appear so to him, whatever the reality might meanwhile be. Indeed, that this appearance should come about was not vain and purposeless, but quite conjoined with the desired effect.

But see Chapter 10 of the *Astronomiae Pars Optica*, where you will find reasons why, to absolutely everyone, the sun appears to move and not the earth: it is because the sun appears small and the earth large, and also because owing to its apparent slowness, the sun's motion is perceived, not by sight, but by reasoning alone, through its change of distance from the mountains over a period of time. It is therefore impossible for a previously unformed reason to imagine anything but that the earth, along with the arch of heaven set over it, is like a great house, immobile in which the sun, so small in stature, travels from one side to the other like a bird flying in the air.

What absolutely all men imagine, the first line of Holy Scripture presents. 'In the beginning,' says Moses, 'God created the heaven and the earth,' because it is these two parts that chiefly presented themselves to the sense of sight. It is as though Moses were to say to man, 'This whole worldly edifice that you see, light above and dark and widely spread out below, upon which you are standing, and by which you are roofed over, has been created by God.'

Suppose someone were to assert, from Psalm 24, that the earth is founded upon rivers in order to support the novel and absurd philosophical conclusion that the earth floats upon rivers. Would it not be correct to say to him that he should regard the Holy Spirit as a divine messenger and refrain from wantonly dragging Him into physics class? For in that passage the psalmodist intends nothing but what men already know and experience daily, namely, that the land, raised on high after the separation of the waters, has great rivers flowing through it and seas surrounding it.

If this be easily accepted, why can it not also be accepted that in other passages usually cited in opposition to the earth's motion we should likewise turn our eyes from physics to the aims of Scripture?

A generation passes away (says Ecclesiastes), and a generation comes, but the earth stands for ever. Does it seem here as if Solomon wanted to argue with the astronomers? No; rather, he wanted to warn people of their own mutability, while the earth, home of the human race, remains always the same, the motion of the sun perpetually returns to the same place, the wind blows in a circle, and returns to its starting point, rivers flow from their sources into the sea, and from the sea return, to the sources, and finally, as these people perish, others are born. Life's tale is ever the same; there is nothing new under the sun.

You do not hear any physical dogma here. The message is a moral one, concerning something self-evident and seen by all eyes but seldom pondered. Solomon therefore urges us to ponder.

It is said, however, that Psalm 104, in its entirety, is a physical discussion, since the whole of it is concerned with physical matters. And in it, God is said to have 'founded the earth upon its stability, that it not be laid low unto the ages of ages'. But in fact, nothing could be further from the psalmodist's intention than speculation about physical causes. For the whole thing is an exultation upon the greatness of God, who made all these things: the author has composed a hymn to God the Creator, in which he treats the world in order, as it appears to the eyes. It is clear that he is not writing as an astronomer here.

Advice to astronomers

I, too, implore my reader, when he departs from the temple and enters astronomical studies, not to forget the divine goodness conferred upon men, to the consideration of which the psalmodist chiefly invites. I hope that, with me, he will praise and celebrate the Creator's wisdom and greatness, which I unfold for him in the more perspicacious explanation of the world's form, the investigation of causes and the detection of errors of vision. Let him not only extol the Creator's divine beneficence in His concern for the well-being of all living things, expressed in the firmness and stability of the earth, but also acknowledge His wisdom expressed in its motion, at once so well hidden and so admirable.

Advice for idiots

But whoever is too stupid to understand astronomical science, or too weak to believe Copernicus without affecting his faith, I would advise him that, having dismissed astronomical studies and having damned whatever philosophical opinions he pleases, he mind his own business and betake himself home to scratch in his own dirt patch, abandoning this wandering about the world. He should raise his eyes (his only means of vision) to this visible heaven and with his whole heart burst forth in giving thanks and praising God the Creator. He can be sure that he worships God no less than the astronomer, to whom God has granted the more penetrating vision of the mind's eye, and an ability and desire to celebrate his God above those things he has discovered.

Commendation of the Brahean hypothesis

At this point, a modest (though not too modest) commenda-
tion to the learned should be made on behalf of Brahe's opinion
of the form of the world, since in a way it follows a middle
path. On the one hand, it frees the astronomers as much as possible from the useless apparatus of so many epicycles, and with
Copernicus's, it includes the causes of motion, unknown to
Ptolemy, giving some place to physical theory in accepting the
sun as the centre of the planetary system. And on the other
hand, it serves the mob of literalists and eliminates the motion
of the earth, so hard to believe, although many difficulties are
thereby insinuated into the theories of the planets in astronomical discussions and demonstrations, and the physics of the
heavens is no less disturbed.

To objections concerning the authority of the pious

So much for the authority of Holy Scripture. As for the opinions
of the pious on these matters, I have just one thing to say: while
in theology it is authority that carries the most weight, in philosophy it is reason. Therefore, Lactantius is pious, who denied
that the earth is round, Augustine is pious, who, though admitting the roundness, denied the antipodes, and their Inquisition
nowadays is pious, which, though allowing the earth's smallness, denies its motion. To me, however, the truth is more pious
still, and (with all due respect for the Doctors of the Church) I
prove philosophically not only that the earth is round, not only
that it is inhabited all the way around at the antipodes, not
only that it is contemptibly small, but also that it is carried
along among the stars.

But enough about the truth of the Copernican hypothesis.
Let us return to the plan I proposed at the beginning of this
introduction. I had begun to say that in this work I treat all
of astronomy by means of physical causes rather than fictitious hypotheses, and that I had taken two steps in my effort
to reach this central goal: first, that I had discovered that the
planetary eccentrics all intersect in the body of the sun, and
second, that I had understood that in the theory of the earth

there is an equant circle, and that its eccentricity is to be bisected.

The third step toward the physical explanation. The eccentricity of Mars's equant is to be precisely bisected

Now we come to the third step, namely, that it has been demonstrated with certainty, by a comparison for the conclusions of parts 2 and 4, that the eccentricity of Mars's equant is also to be precisely bisected, a fact long held in doubt by Brahe and Copernicus.

Therefore, by induction extending to all the planets (carried out in Part 3 by way of anticipation), since there are (of course) no solid orbs, as Brahe demonstrated from the paths of comets, the body of the sun is the source of the power that drives all the planets around. Moreover, I have specified the manner [in which this occurs] as follows: that the sun, although it stays in one place, rotates as if on a lathe, and out of itself sends into the space of the world an immaterial emanation of its body, analogous to the immaterial emanation of its light. This emanation itself, as a consequence of the rotation of the solar body, also rotates like a very rapid whirlpool throughout the whole breadth of the world, and carries the bodies of the planets along with itself in a year, its grasp stronger or weaker according to the greater density or rarity it acquires through the law governing its diffusion.

Once this common power was proposed, by which all the planets, each in its own circle, are driven around the sun, the next step in my argument was to give each of the planets its own mover, seated in the planet's globe (you will recall that, following Brahe's opinion, I had already rejected solid orbs). And this, too, I have accomplished in Part 3.

By this train of argument, the existence of the movers was established. The amount of work they occasioned me in Part 4 is incredible, when, in producing the planet–sun distances and the eccentric equations that were required, the results came out full of flaws and in disagreement with the observations. This is not because they should not have been introduced but because I had bound them to the millstones (as it were) of circularity, under the spell of common opinion. Restrained by such fetters, the movers could not do their work.

Fourth step to the physical explanation. The planet describes an oval path

But my exhausting task was not complete: I had a fourth step yet to make toward the physical hypotheses. By most laborious proofs and by computations upon a very large number of observations, I discovered that the course of a planet in the heavens is not a circle but an oval path, perfectly elliptical.

Geometry gave assent to this, and thought that such a path will result if we assign to the planet's own movers the task of making the planet's body reciprocate along a straight line extended toward the sun. Not only this, but also the correct eccentric equations, agreeing with the observations, resulted from such a reciprocation.

Finally, the pediment was added to the structure, and proven geometrically: that it is in the order of things for such a reciprocation to be the result of a magnetic corporeal faculty. Consequently, these movers belonging to the planets individually are shown with great probability to be nothing but properties of the planetary bodies themselves, like the magnet's property of seeking the pole and catching up iron. As a result, every detail of the celestial motions is caused and regulated by faculties of a purely corporeal nature, that is, magnetic, with the sole exception of the whirling of the solar body as it remains fixed in its space. For this, a vital faculty seems required.

Next, in Part 5, it was demonstrated that the physical hypotheses we just introduced also give a satisfactory account of the latitudes.

There are some, however, who are put off by a few extraneous and seemingly valid objections and do not wish to put such great trust in the nature of bodies. Therefore, in parts 3 and 4, some room was left for Mind, so that the planet's proper mover could attach the faculty of Reason to the animate faculty of moving its globe. These people would have to allow the mind to make use of the sun's apparent diameter as a measure of reciprocation, and to be able to sense the angles that astronomers require.

Kepler, Orbit of Mars, from *New Astronomy*, 1609

Kepler used this diagram of the orbit of Mars in the Ptolemaic and Tychonic systems to demonstrate the differences between the three world systems and the reason to prefer the Copernican. Though the three world systems were geometrically equivalent and equally capable of accounting for planetary motion, the Ptolemaic and Tychonic systems produced orbital paths that looked extremely complicated, much like this pretzel-shaped diagram. The Copernican paths were much more physically simple – not circular, but close to it – and linked to the physical presence of the sun; this, Kepler argued, was proof that the Copernican system was preferable to the others, as it was the only system that made physical sense.

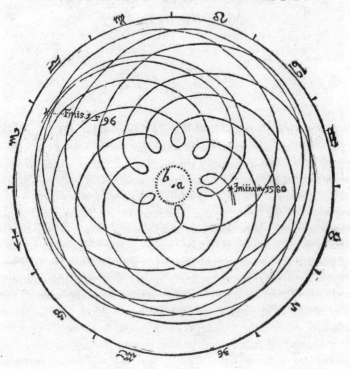

Galileo Galilei, *The Starry Messenger*, 1610

Galileo Galilei (1564–1642), Italian astronomer and natural philosopher, had been a Copernican for many years by the time he heard about a recently invented device that made distant objects appear closer. A professor of mathematics at the University of Padua at the time, in 1609 he directed his own improved telescope at the heavens and made a number of discoveries that challenged the old Aristotelian cosmological picture and removed some central objections to Copernican theory. The observations of the imperfect surface of the moon and of the light that the moon received from the earth challenged the terrestrial–celestial dichotomy that underpinned the Aristotelian world view and suggested that the moon was much like the earth, and the earth much like the other planets. Further, the observations of the four moons of Jupiter meant that the earth was not alone in having a moon, and that heavenly objects could revolve around bodies other than the earth; if this was the case, then Galileo suggested that there was no reason why the earth itself could not revolve around the sun along with its moon, just as Jupiter clearly revolved around another body with its four moons. Noteworthy here are Galileo's detailed descriptions of his slow process of discovery and interpretation of the observations of Jupiter's moons, as well as his decision to name those moons after the powerful Medici, who offered him a position at court in return.

The Starry Messenger

Disclosing great and distant remarkable spectacles,
And displaying them to the admiration of everyone,
But especially philosophers and astronomers, which have
been observed by Galileo Galilei, a Florentine gentleman
and official mathematician at the University in Padua, with
the help of a telescope recently invented by him, pertaining
to the surface of the moon, innumerable fixed stars, the
Milky Way and nebulous stars, but especially to four
planets which revolve around the planet Jupiter at disparate
distances and periods and with wonderful speed, which,
known to no one until this day, the author very recently first
apprehended, and resolved to name the Medicean stars.

To the Most Serene Cosimo II de' Medici, Fourth Grand Duke
of Tuscany:

Behold four stars, reserved for your famous name! These are not
among the rank and file and less conspicuous fixed stars, but in
the illustrious class of the wandering ones; and in their motions,
differing among themselves, around the planet Jupiter – the most
noble of all the others – as though his genuine progeny, they
complete their courses and circles with wonderful speed, while
meanwhile, in unanimous concord, simultaneously every ten
years they all realize great revolutions around the centre of the
universe, namely, the sun itself.

It pleased the Almighty God that your most serene parents
should judge me not unworthy to be engaged in teaching Your
Highness mathematics, which I fulfilled for the previous four
years, at the time of year when it is customary to take a rest
from weightier studies. And therefore because I discovered
these stars, unknown to all previous astronomers, with you as
my patron, I have very rightly decided to distinguish them with
the name of your very venerable family. And if I first investigated them, who would rightly blame me if I impose a name on
them? I will thus call them the Medicean stars, in the hope that

those stars will gain as much dignity from that appellation as other stars have conveyed to other heroes.

Accept, therefore, most merciful prince, this familial glory, reserved for you by the stars, and enjoy for a very long time those divine advantages which are conveyed to you not so much by the stars as by their creator and governor, God.

Padua, 12 March 1610

In this small treatise I relate great things for all observers of nature to inspect and contemplate. Great, I say, first on account of the excellence of the matter itself; next, on account of the novelty of something unheard of through the ages; and next, on account of the instrument whose aid made them exposed to our senses.

It is clearly a great thing to add to the great multitude of fixed stars, which can until today be observed through our natural faculties of sight, innumerable others, exposed plainly to our eyes, never seen before, and surpassing old and known ones more than tenfold.

It is most beautiful and delightful to see the body of the moon, removed from us by almost sixty [semi]diameters of the earth, looking as near as if it were distant by only two of the same, so that the diameter of the same moon appears as though thirty times greater, its surface 900 times, and its solid body nearly 27,000 times greater than while it is viewed only with the naked eye. From this, then, someone may understand with sensory certainty that the moon is not covered in a smooth and polished surface, but one that is rough and uneven, and – just as the surface of the earth itself – it is packed everywhere with vast bulges, deep cavities and ravines. It seems that it ought not to be considered of very little importance to have resolved debates about the galaxy, or the Milky Way, and to have made its essence manifest to the senses, still less the intellect. On top of this, it will be very pleasant and noble to demonstrate directly that the substance of the stars, which all astronomers until today have called nebulous, is very different than what has hitherto been believed.

But what surpasses all admiration by far and especially impelled us to acquire the notice of all astronomers and philosophers is this: namely, that we have discovered four planets, which no one has recognized or observed before us, whose periods are around a certain famous star from among those known, a counterpart of Venus and Mercury around the sun, and they sometimes precede it, and sometimes follow it, but never depart from it beyond certain limits. All of these things were recently discovered and observed a few days ago with the help of a telescope devised by me, with God's grace first having illuminated me.

Perhaps other more excellent things will be discovered daily, either by me or by others, with the help of a similar instrument. Therefore, I will first briefly recall its form and preparation, and also the occasion for devising it, and next recount a history of the observations made by me.

About ten months ago, a rumour struck my ears that a telescope had been constructed by a certain Dutchman, with the help of which visible objects, though very far from the eye of the observer, were seen distinctly, as though they were near. Some experiences of this very wonderful effect were circulated, which some people trusted, but others did not. A few days later it was confirmed to me in a letter from a French nobleman, Jacques Badovere, from Paris. This finally was the impetus for me to direct myself fully toward investigating the ideas behind it, and then devising means through which I might arrive at the invention of a similar instrument, which a little while later I achieved, supported by the doctrine of refraction. I first prepared a tube of lead, in whose ends I fitted two glass lenses, both plane on one side but on the other side one spherically convex and the other concave. Next, bringing my eye near the concave one, I saw objects sufficiently large and near: for they appeared three times closer and nine times larger than when they were observed with the naked eye alone. Afterwards I constructed another more accurate one, which displayed objects more than sixty times larger. Finally, sparing no labour or expense, I managed to construct an instrument so excellent that things appeared through it nearly one thousand times larger, and more than thirty times closer, than if observed with the natural faculty alone. It would be completely redundant to enumerate how advantageous this instrument would be both on land and at sea. But with terrestrial matters dismissed, I directed myself toward celestial investigation. First, I observed the moon as near as if it were scarcely two terrestrial [semi]diameters away. After that, I frequently observed the stars, both the fixed and the planets, with incredible pleasure of mind.

Now we shall review the observations made by us during the two months most recently past, summoning all those passionate about true philosophy to the beginnings of these great contemplations.

Let us speak first about the surface of the moon that is turned toward our gaze. In order to be more easily understood, I divide it into two parts, namely, the one brighter, and the other darker. The brighter seems to embrace and suffuse the whole hemisphere, but the darker, as if a sort of cloud, stains the surface itself and renders it spotted. Moreover, these spots, somewhat murky and quite large, are obvious to everyone, and every age has seen them. On account of this, I will call them great or ancient spots, to differentiate them from other spots, smaller in size but strewn so thickly and frequently that they sprinkle the whole lunar surface, especially the brighter part. These, however, have been observed by no one before us. Moreover, from the repeated and frequent examination of them, I have been led to the opinion and now certainly realize that the surface of the moon is not perfectly smooth, uniform and exactly spherical, as the great cohort of philosophers supposed about it and about the other celestial bodies, but rather, is uneven, rough and crammed with cavities and bulges, no differently than the surface of the earth itself, which is adorned here and there with mountain ridges and deep valleys.

This is a good place to explain another aspect of the moon that is worthy of wonder. When, both before and also after a conjunction, the moon is found not far off from the sun, it offers to our sight the spectacle of not only the part of its globe that is adorned with shining horns, but also a certain thinly glimmering periphery on the dark part (namely turned away from the sun), that seems to delineate its ambit and separate it from the darker field of the ether itself. But if we subject the matter to a

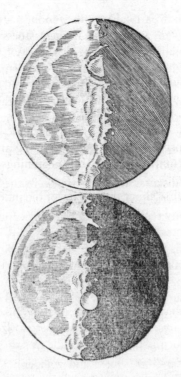

more exacting investigation, we will see not only the extreme border of the dark part shining with a certain vague brightness, but the whole surface of the moon, namely, the part that does not yet feel the brightness of the sun, made white by a certain not negligible light.

This wonderful brightness has elicited no small amount of wonder from those philosophizing, who have proffered one assertion after another in order to explain it. For certain of them have said that it is the particular and natural brilliance of the moon itself; others that it is bestowed to it by Venus; others by all the stars; and others by the sun, which passes through the deep solidity of the moon with its rays. But contentions of this sort are refuted with scant labour and defeated as false.

Therefore, because a secondary brightness of this sort is

neither innate and particular to the moon, nor borrowed from any stars or the sun, and because in the vastness of the world no other body remains except the earth alone, what, I ask, should we suppose? What should we suggest? Is it possible that the lunar body or some other dark and gloomy body is flooded with the light of the earth itself? Why marvel? It is very much so: in an equal and agreeable exchange, the earth compensates the moon with an amount of light equal to that which it receives from the moon nearly all the time in the deepest darkness of the night.

Let these few words concerning this matter suffice at present, with a broader discussion in our System of the World, where a very strong reflection of solar light from the earth is shown with a great many reasons and experiments, for those who toss around that the earth should be kept away from the dance of the stars, especially because it is devoid of motion and light. For we will confirm with hundreds of natural reasons that it moves and surpasses the moon in brilliance, and that it is not the universe's dumping ground for its filth and dregs.

Thus far I have spoken about the observations made surrounding the lunar body; now, let us briefly report what we have observed recently about the fixed stars. And first, this is worthy of note: that the stars, both fixed and wandering, when observed using a telescope, are not at all seen to be increased in magnitude in the same proportion as the remaining objects, and likewise the moon itself, are increased. For the stars the increase appears so much less that you may suppose that the telescope, which is able to multiply the remaining objects, for example, a hundred times, will scarcely increase the stars four or five times. The reason for this, however, is that when the stars are observed with the naked eye, they do not present themselves to us as they are in their simple and, so to speak, bare size, but radiating with a certain brightness, and surrounded with twinkling rays, especially when night has already spread. For this reason, they seem much bigger than they would if the plumes they have assumed were stripped from them, for the angle that they indicate is formed not by the primary body of the star, but by the brilliance that widely surrounds it.

You will also see through the telescope a crowd of other stars below the sixth magnitude, which evade the naked eye. They are so plentiful as to scarcely be believed, for you may see more than six other differences of magnitude, the greater of which – which we can call stars of the seventh magnitude, or of the first magnitude of invisible ones – appear with the aid of the telescope bigger and brighter than stars of the second magnitude with the naked eye.

And in order for you to see one or two attestations of their almost inconceivable abundance, I decided to describe two groups of stars, so that from their example you can form a judgement about the others. For the first I had determined to depict the entire constellation of Orion, but, overwhelmed by the enormous number of stars and the lack of time, I put off this undertaking for another occasion, for there stand both near and around the old stars, within the bounds of one or two degrees, more than five hundred new ones spread out. On account of this, to the three stars in the belt and six in the sword, known already for a long time, I have added eighty other recently seen adjacent ones, and I have preserved as precisely as possible the distances between them. The known (or old) stars I have depicted as bigger, for the sake of a distinction, and I have surrounded them with a double line, and the other invisible ones as smaller, with a single line. I have also preserved the differences in magnitude as much as possible.

For the second example I have depicted six stars of Taurus, called the Pleiades (I say six, moreover, since the seventh is almost never visible), enclosed in very narrow boundaries in the heavens. Adjacent to these are more than forty other invisible ones, none of which is removed from the previously mentioned six by scarcely more than half a degree. Of these I have recorded only thirty-six, and I have preserved their distances, sizes and the differences between the old and the new ones, just as with Orion.

What was observed by us third is the essence, or substance, of the Milky Way. When it is possible to observe it with the help of a telescope, all the arguments that have tormented philosophers through the ages are dissolved by the certainty of our eyes, and we are freed from wordy disputes. For the galaxy is nothing other than a mass of innumerable stars strewn in clusters everywhere. In whatever region you direct the telescope, an enormous number of stars immediately present themselves to

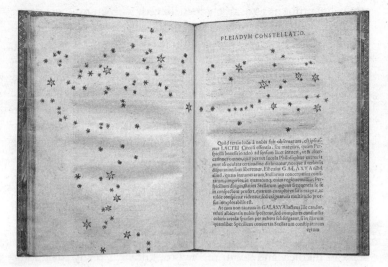

your gaze. Many of them seem relatively large and very visible, but the multitude of very small ones is completely beyond determination.

I have briefly described what I have recently observed concerning the moon, the fixed stars and the galaxy. There remains, therefore, that which it seems should be valued most greatly in the present business: that I disclose and make known four planets never seen from the very beginning of the world until our times, the circumstances in which I discovered and observed them, their positions and the observations made through the past two months of their motions and changes, summoning all astronomers to confer in examining and determining their periods, which I could not achieve until now because of too little time. I warn them again, however, lest they approach such an investigation in vain, that they will need a very accurate telescope, of the sort that I described at the beginning of this discussion.

And so, on the seventh day of January of the present year, 1610, at the first hour following night, when I observed the celestial stars through a telescope, Jupiter made an appearance. And because I had prepared quite an excellent instrument for myself (which did not happen previously because of the weakness of the other instrument) I recognized three little stars near it, small indeed but still very bright. Although I believed that they were among the fixed stars, nevertheless they provoked some degree of wonder, because they seemed to be distributed in an exact straight line parallel to the ecliptic, and brighter than the other stars, though equal in magnitude. They were arranged among themselves and Jupiter as such: namely, on the east side there were two stars, and one toward the west. The one more easterly and the western one appeared a bit larger than the remaining one. I was not concerned at all about the distance between them and Jupiter, for, as I said, I had first believed them to be fixed stars. When, however, on the eighth, led by an unknown fate, I returned to the same investigation, I discovered a very different arrangement, for there were three little stars all west of Jupiter and nearer to each other than on the previous night, all separated by equal distances, as the adjacent sketch shows. Here, although I had not turned my attention at all toward the mutual advances of the stars, nevertheless my interest was piqued by how Jupiter could be found to the east of the aforementioned fixed stars, when the day before it had been west of two of them. Hence I was afraid that perhaps Jupiter was moving forward in a straight line, contrary to the astronomical computations, and that it moved ahead of those stars due to its own proper motion. For that reason, I awaited the next night with the greatest anticipation, but my hope was frustrated, for the sky was covered completely with clouds. On the tenth of January, the stars appeared, but there were only two of them, both to the east, and the third I believed to be hidden behind Jupiter. They were arranged together just as before, in the same straight line with Jupiter, and exactly along the zodiac. When I had seen these things, and since I knew that similar changes could not in any way be due to Jupiter, and I recognized additionally that the observed stars were always the same (for there were no others, either preceding or following, within a great distance along the zodiac), I now exchanged uncertainty for wonder, I discovered that the apparent change I had recorded was due not to Jupiter but to the stars. I therefore deemed that they should be observed henceforth more conscientiously and accurately.

RECENS HABITAE.   23
dentalis proxima min. 2. ab hac vero elongabatur oc-

Ori.   * ○ *   *   Occ.

cidentalior altera min: 10. erant præcisè in eadem recta, & magnitudinis æqualis.
    Die quarta hora secunda circa Iouem quatuor sta-
bant Stellæ, orientales duæ, ac duæ occidentales in

Ori.   * *○ *   *   Occ.

eadem ad vnguem recta linea dispositæ, vt in proxima figura. Orientalior distabat à sequenti min. 3. hęc
verò à Ioue aberat min. o. sec. 40. Iuppiter à proxima
occidentali min 4. hæc ab occidentaliori min. 6. ma-
gnitudine erant ferè equales, proximior Ioui reliquis
paulo minor apparebat. Hora autem septima orientales Stellæ distabant tantum min. o. sec. 30. Iuppiter

Ori.   ** ○   * *   Occ.

ab orientali viciniori aberat min. 2. ab occidentali verò sequente min. 4. hæc verò ab occidentaliori distabat min. 3. erantque æquales omnes, & in eadem recta
secundum Eclypticam extensa.
    Die quinta Cœlum fuit nubilosum.
    Die sexta duæ solummodo apparuerunt Stellæ me-

Ori.   * ○ *   Occ.

                                        dium

Therefore, on the eleventh, I saw an arrangement of this sort:
namely, only two stars to the east, of which the middle one was
distant from Jupiter three times as much as from the more eastern one, and the more eastern one was nearly two times as big as
the other, although on the previous night that had appeared
equal. Therefore, I decided it to be established beyond all doubt
that there were three stars in the heavens moving around Jupiter,
just as Venus and Mercury move around the sun. Ultimately this
was observed as clearly as the light of day in many other subsequent investigations, as well as that there were not only three,
but four wandering stars attending to their revolutions around
Jupiter, whose changes, observed consequently more accurately,
I supply in the following account. I also measured the distances
between them with the telescope, using the method explained

above. Additionally, I have added the times of the observations, especially when many where made in the same night, for truly the speeds of the revolutions of these planets are so fast that one may frequently obtain hourly differences.

On the twelfth, on the first hour of the next night, I saw the stars arranged in this way: the star that was more to the east was bigger than the one to the west, but both were very conspicuous and bright. Each one was distant from Jupiter by two minutes. A third little star, not seen at all earlier, began to appear in the third hour. It almost touched Jupiter from the eastern side, and was very tiny. They were all in the same straight line, arranged together along the ecliptic.

On the thirteenth, for the first time I saw four little stars in this arrangement with respect to Jupiter: there were three to the west and one to the east, and they formed a nearly straight line, but the middle one of those on the west deviated a little from the straight line towards the north. The more eastern one was separated from Jupiter by two minutes, and there were gaps of only a single minute between Jupiter and the rest. All the stars seemed to be the same size, and although they were very small, nevertheless they were extremely bright, and shone far more than the fixed stars of the same magnitude.

On the fourteenth, the weather was cloudy.

On the fifteenth, in the third hour of the night, the four stars were approximately in the condition depicted here with respect to Jupiter: all were to the west and arranged approximately in the same straight line, though the third one from Jupiter was raised up a little to the north. The one closer to Jupiter was the smallest of all, and the rest consequently appeared larger. The distances between Jupiter and the other three were all equal and two minutes, but the one more to the west was separated from the one nearest to it by four minutes. They were very bright and did not twinkle at all, just as they have always appeared, both before and afterwards. But in the seventh hour there were only three stars in attendance, which looked like this along with Jupiter: they were still in exactly the same straight line. The one closer to Jupiter was very tiny, and was removed from it by three minutes; the second one was separated from that one by one minute, and the third from the second by four minutes, thirty seconds. But after another hour the middle two little stars were closer still, for they were separated by only thirty seconds from each other.

[*Galileo's descriptions continue through to 2 March, with detailed observations and images for each date.*]

These are the observations of the four Medicean planets, recently and first discovered by me. Although I was not yet able to use them to compute their periods, allow me at least to relate some things worthy of attention. First, since they sometimes follow and sometimes precede' Jupiter at similar distances, and depart from it toward the east and the west only within a very narrow spread, and they accompany it equally in retrograde and direct motion, no one can doubt that they perform their revolutions around it, while meanwhile carrying out all together a twelve-year period around the centre of the world. In addition, they revolve in unequal circles, which is clearly inferred from the fact that I could never see two planets adjoining one another while at greater distances from Jupiter, while near Jupiter, two, three and sometimes all were found crowded together at once. Further, it is understood that the revolutions of the planets describing more narrow circles around Jupiter are faster, for the stars closer to Jupiter are more often seen to the east when the day before they had been seen to the west, and vice versa. On the other hand, the planet traversing the largest circle seems, after carefully and accurately assessing its previously noted returns, to have a period of half a month.

In addition, we have an excellent and splendid argument for removing the scruples of those who bear with equanimity the revolution of the planets around the sun in the Copernican system, but are so disturbed by the motion of one moon around the earth, while meanwhile both complete an annual circuit around the sun, that they decide that this arrangement of the universe should be overthrown as impossible. For now we have not only one planet revolving around another, while both pass through a great circle around the sun, but our senses offer us four stars moving around Jupiter, just as the moon around the earth, while all at once, with Jupiter, move through a great circle of twelve years around the sun.

Kepler, *Conversation with* The Starry Messenger, 1610

Thirteen years after their brief correspondence of 1597, Galileo once again wrote to Kepler, this time via the Tuscan ambassador in Prague, to solicit Kepler's support for the recent telescopic discoveries he had described in The Starry Messenger. *Though Kepler had no telescope with which to verify Galileo's claims, he responded with a long letter of support, which he then published. In the letter he supplied a variety of reasons why Galileo's claims should be believed, including the prestige of Galileo's patrons, Galileo's own character, and the ways in which his various claims had precedents in earlier literature, both ancient and recent. In the course of the letter, Kepler reflected on the differences between theory and observation, on the implications of Galileo's claims for heliocentric theory, on the possibility of life on Jupiter and on the idea of an infinite universe (the last of which Kepler strongly opposed). Kepler also argued that though the earth was no longer centrally positioned, it was still the noblest body other than the sun, as it was perfectly positioned to observe the movements of all the other bodies in the system.*

Johannes Kepler, Imperial Mathematician

Conversation with The Starry Messenger, *recently sent to mankind by Galileo Galilei, mathematician of Padua*

Prague, 1610

Notice to the Reader:

Since many people desired my opinion about Galileo's *Starry Messenger*, I wanted to satisfy all of them with this shortcut: that I copy in print the letter I sent to Galileo (composed in great haste, among necessary occupations, within a prescribed number of days).

But after it was already printed, friends warned that the form seemed somewhat contrary to custom. One wanted the introduction removed; another wanted some words softened that might, to the improvident, seem to attribute to my opponent opinions that depart from scholarly tradition; another wanted Galileo to be praised more sparingly, so that there would be room left for the opinions of more famous men, who they hear think differently from me.

Therefore, my plan was to warn the reader that 'to each his own is beautiful', and that while most people become enraged when disputing, lightheartedness seems to me a more agreeable condiment for discussions. Others aim for dignity in relating philosophical matters through the gravity of their assertions, yet they often seem laughable despite themselves. I am rather naturally inclined to temper the labour and difficulty of a doctrine with the relaxation of mind.

Further, nothing I wrote about Galileo was coloured. I have always had the custom of praising that which I thought was said well by others, and refuting that said badly. I have never looked down on or dissembled about the knowledge of others when I lacked my own. I was never the lackey of others, nor did I ever neglect myself if, through my own battle, I discovered something better or earlier than they.

Nor do I think that Galileo, an Italian, has behaved so well toward me, a German, that in turn he must be fawned upon in advance of either the truth or my inmost sentiments. Still, no one should suppose that I, by freely agreeing with Galileo, set out to snatch away the freedom of those dissenting on their own. Everyone else's judgement is preserved, though I have praised him. In fact, if any of my own doctrines that I have undertaken to defend here is shown by a proper method to be in error by someone more learned than me, I promise to abandon it willingly, although I came to it earnestly, believing it to be true.

To the Noble and Most Excellent Lord,
Galileo Galilei, Florentine Gentleman,
Professor of Mathematics at the University of Padua:

Johannes Kepler, imperial mathematician, sends many greetings.

For a long time now I idled at home, thinking of nothing but you, most excellent Galileo, and your letters. For at the previous market day, my book, called *Commentaries on the Motions of Mars*, a labour of many years, was published and made public. And from that time, like someone who had acquired glories of war through a very difficult campaign, with some freedom introduced to my studies, I thought that among others even Galileo, the most capable of all, would discuss with me through letters the kind of new astronomy, or celestial physics, I had published, and would resume the discussion interrupted twelve years ago.

But behold, unexpectedly around 15 March, a work of my Galileo was reported swiftly in Germany: it was said that instead of reading a book of someone else's, he was occupied with his own highly groundbreaking argument, about four planets previously unknown discovered with the aid of a double-lensed telescope. When Johan Matthäus Wackher von

Wackenfels, councillor of the illustrious Holy Roman Emperor and referendary of the Sacred Imperial Council, reported this to me from his carriage in front of my house, great wonder overtook me as I considered this highly bizarre news. The excitement was so great (from an ancient quarrel between us having been unexpectedly settled) that – he from joy, I from shame, both from laughter, and confused by the novelty – he was scarcely able to talk, and I to listen.

Wackher's assertion increased my astonishment: there were very famous men, very far removed from the popular foolishness by dint of their learning, seriousness and constancy, who were reporting these things about Galileo, even while the book was still at the press, and would be available with the next run.

When I first withdrew after speaking with Wackher, the authority of Galileo, which he acquired through his talent for cleverness and his rectitude of judgement, especially moved me. Therefore, I pondered how some increase in the number of planets was possible while saving my *Cosmographic Mystery*, which I had brought to light thirteen years earlier, and in which the five Euclidean figures (which Proclus called 'cosmic', based on Pythagoras and Plato) admit no more than six planets around the sun. Moreover, it is clear from the preface of that book that then I also sought more planets around the sun, but in vain. Thinking carefully about that, this occurred to me, and I quickly reported it to Wackher: that the earth is one of the planets, and its moon is in motion around itself, which according to Copernicus makes it extraordinary. Similarly, it is certainly possible that four other very small moons, moving in very narrow circles around the bodies of Saturn, Jupiter, Mars, or Venus, have appeared to Galileo. Mercury, however, the last around the sun, is so immersed in the rays of the sun that Galileo would have as yet been able to find nothing similar there. It seemed to Wackher, by contrast, that without doubt these new planets go around some of the fixed stars (he had already presented this sort of thing to me for some time, following the speculations of Cardinal Cusa and Giordano Bruno). And if four planets had escaped notice until now there, why should we not believe that further on there are innumerable other ones right now to be discovered, after this start? Thus, either this world is itself infinite, as it pleased Melissu and the author of the magnetic philosophy, the Englishman William Gilbert, to think, or, as it seemed to Democritus and Leucippus, and more recently, Bruno and Bruce (your friend, Galileo, and mine), there are infinite other worlds (or, for Bruno, earths) similar to ours.

So it seemed to me, and so to him, while meanwhile we awaited Galileo's book, which had caused these hopes, with an extraordinary desire to read it. The first copy reached me by permission of the emperor, which I examined and flipped through swiftly. I see 'great and distant remarkable spectacles proposed to philosophers and astronomers', including to me, if I am not mistaken; I see 'summoned to the beginnings of these great contemplations all those passionate about true philosophy'. I then wished passionately to dive into these matters (called forth, as you see), and as someone who had written about the same material six years earlier, to discuss with you, the great expert Galileo, in a very delightful kind of writing, the great unexhausted treasures of Jehovah the Creator, which he discloses to us one after another. For who is permitted to be silent about such great things? Who is not filled with an abundance of love of God, pouring itself out very copiously through tongue and pen? The orders of the most august Emperor Rudolph, who asked for my judgement about this material, added to my feelings. But what should I say about Wackher? I came to him without the book, professing nevertheless to have read it. He was jealous of me and quarrelled, but finally concluded plainly that I should become very proficient in this material without delay.

While I pondered somewhat, your letter to the ambassador of the illustrious grand duke of Tuscany arrived, filled with your love for me, and bestowing upon me honour by your belief that you should write to me through such a great man, with a copy of the book and an added notice. He tendered them very kindly due to his friendship with you, and accepted me under his patronage with great benevolence. Thus, by my own inclination, because it pleases my friends, and because you yourself ask attentively, I will do it, induced by some hope that I, with this letter, may profit you (if you decide to make it known). Thus you will advance better equipped with a defender against hard-to-please critics of novelty, for whom anything unknown is unbelievable, and anything beyond the familiar boundaries of Aristotelian narrowness is impious and wicked.

I may, perhaps, seem rash to so easily accept your assertions with no experience of my own to support me. But why should I not believe a very learned mathematician, whose pen also proves the rectitude of his judgement, and who is so far removed from rendering falsehoods, and who would not say that he has seen what he hasn't in order to entice public attention? Who, because of his love of truth, does not hesitate to oppose even the most accepted opinions, and bear the public disparagement with indifference? Who writes publicly, and thus could not conceal any dishonour, if any such were committed? Should I disparage the honesty of a Florentine gentleman regarding that which he has seen? I, dim-sighted, and he, with penetrating sight? He, equipped with optical instruments, while I, une-quipped, lack these supplies? Should I not believe, when he invites all to observe the same as he, and even offers his own instrument in order to establish trust based on sight? Or would this be so inconsequential: to trifle with the family of the grand dukes of Tuscany, and set the name of the Medici in front of his own fictions, while promising true planets in the meantime? Why should I not believe, when I recognize part of the book as extremely truthful, based on my own experiments and on the affirmations of others? What reason would there be for him to have thought to deceive the world with only four planets?

In fact, the story of spots on the moon is very ancient, supported by the authority of Pythagoras and Plutarch, the greatest philosopher. I will postpone until further below, in its place, how Maestlin said this too, as did my own *Optics*, published six years ago. Thus, because others, with closely agreeing testimony, also profess these things about the body of the moon, and they are consistent with your reported very clear experiments on the same matter, there is hardly any reason why I should question the truth of the rest of your book, concerning the four planets around Jupiter. Rather, let me wish that there was a telescope already prepared for me, with which I might anticipate you in discovering two bodies around Mars (as the ratio seems to me to require), and six or eight around Saturn, and perhaps one around Venus and one around Mercury.

The first chapter of your little book focuses on the construction of the telescope. The idea of so capable a telescope seems incredible to many, but it is neither impossible nor new. Nor was it recently advanced by the Dutch; rather, it was already disclosed many years ago by Giovanni Battista Della Porta in *Natural Magic*, Book 17, Chapter 10, 'The Conditions of a Lens'.

Nor is it unbelievable that expert engravers with great industry, who use lenses for viewing fine details of an engraving, also accidentally happened upon this device while they attached convex lenses to concave ones in different ways in order to choose the combination that better served their eyes. I do not say these things in order to lessen the praise of the mechanical inventor, whoever he was. I know how far the distance is between theoretical conjectures and ocular experience; between Ptolemy's discussion about the antipodes and Columbus's disclosure of the new world; and indeed, between the double-lensed tubes circulated widely and popularly and your machine, Galileo, with which you penetrated the heavens themselves. Rather, here I strive to make the disbelieving trust in your instrument.

And now, most skilled Galileo, I commend you for your indefatigable industry, as you deserve. Suppressing all doubts, you directed yourself straight toward ocular experimentation. And now, with the sun of truth having arisen through your discovery, you have dispelled all those spectres of uncertainty along with the night, their mother; and you demonstrated what is possible.

Now with respect to the use of the instrument, you have clearly shown how to recognize to what degree things are enlarged by the instrument, and a single minute or parts of minutes can be distinguished in the heavens. With respect to this, since your industry rivals that of Tycho Brahe in the highest accuracy and certainty, it will not be too much of a departure to intersperse some remarks here.

I remember when that polymath of all sciences, Johannes Pistorius, asked me – not only once – whether Brahe's observations were so impeccable that I thought absolutely nothing was lacking in them. I strongly conceded that the height had been reached, and that there was nothing remaining for human industry, because the eyes could not achieve more exactness, and because of the trouble of refraction, which moves the positions of the stars with respect to the position of the horizon. By contrast, he very resolutely affirmed this: that in the future someone would come who would make known a more precise method with the power of glasses. I opposed him, saying that the refraction of glasses makes them ill-suited for the certainty of observations. But now, finally, I see that Pistorius was in part a true prophet.

You have proven very assuredly with experiments based on clear optical laws that many summits rise up on the lunar body through the bright part, especially lower down, the counterparts of the highest mountains on our earth. They first enjoy the light of the sun rising on the moon, and they are revealed to you using your telescope.

Another of your very pleasant observations is the twinkling form of the fixed stars, different from the circular form of the planets. What else should we deduce from this, Galileo, but that the fixed stars emit their light from within, while the planets are opaque and tinted from without – that is, to use the words of Bruno, that the former are suns, the latter moons, or earths?

Nevertheless, let him not drag us into his opinion about infinite worlds, as many as there are fixed stars, and all of them similar to our own. Your third observation comes to our relief: the innumerable multitude of fixed stars is beyond that which was known in antiquity. You do not hesitate to pronounce that there seem to be more than 10,000 stars. For the more and more crowded they are, the truer is my argument against the infinity of the world.

Let this line of argument serve as well in support. To me, with my weak vision, any of the somewhat bigger stars, like Sirius, seems a little less in magnitude than the diameter of the moon, if I take its glittering rays into account. But those with very accurate vision, using astronomical instruments which are not afflicted by those embellishments as the naked eye is, establish the magnitudes of the stars' diameters through minutes and parts of minutes. If from the fixed stars we took 1,000 only, none bigger than one minute (although most enumerated ones are bigger), those, all gathered together in one round surface, would equal (and indeed surpass) the diameter of the sun. By how much more would the little discs of 10,000 stars, brought together into one, surpass in visible magnitude the apparent disc of the sun? If this is true, and if they are suns, of the same kind as our sun, why then do these suns not all together surpass our sun in brilliance? Indeed, why do they spread only a faint light all together to the most accessible places? If you tell me they are very far from us, this does not help the cause at all. For the greater the distance, the more is each of them greater than the diameter of the sun. But perhaps the ether spread out between them obscures them? Not at all: for we distinguish them with their scintillations, and with their differences of shape and colour, which would not be if the density of an outside ether was an obstacle. Therefore it is sufficiently clear that the body of our sun is inestimably brighter than all the fixed stars, and hence that this world of ours is not one of an indiscriminate crowd of infinite others.

You have blessed astronomers and physicists with your disclosure of the essence of the galaxy, the nebulae and nebular spirals, and you confirmed to those who already professed the same thing along with you some time ago: that it is nothing other than a heap of stars, with their lights jumbled together on account of the feebleness of vision.

Finally, I turn with you to the new planets, the most marvellous topic in your little book. First, I exult that I am somewhat revived through your labours. If you had discovered planets in motion around one of the fixed stars, then I would now be chained and imprisoned within Bruno's innumerabilities, or rather, exiled to his infinity. Therefore, you have freed me for now from great fear when you say that those four planets are not moving around one of the fixed stars, but around the planet Jupiter.

What Galileo recently observed with his own eyes was both introduced many years ago as a supposition and clearly established with arguments. Greatness is clearly merited for those whose reason anticipates the senses in similar branches of philosophy. For who would not classify as more noble that astronomical doctrine – with nary a foot stepped out of Greece – revealed the properties of the frigid zone, rather than Caesar's experimentation with water clocks revealing that the nights on the shore of Britain are a little shorter than those in Rome? Or rather than the wintering of the Dutch in the north, which is very astonishing, but which would have been impossible without knowledge of doctrine? Who does not celebrate Plato's fable of Atlantis, Plutarch's of the gold-coloured islands beyond Thule, and Seneca's prophetic versions about the future discovery of a new world, after just such has been finally furnished by that Florentine Argonaut? Columbus himself keeps his reader doubting which to admire more: his talent in inferring the place of the New World from the blowing of the winds, or his fortitude in braving unknown waters and the vast ocean, and his good fortune in gaining what he desired.

Of course, in my own subject likewise the marvels will be Pythagoras, Plato and Euclid, who, rising up by the excellence of their reason, concluded that it could not be otherwise but that God had adorned the world according to the model of the five regular solids. However, they mistook the manner in which he did so. The public, by contrast, will praise Copernicus, who, though his talent was certainly above average, nevertheless offered a description of the world as it appears to the eye, bringing to light only the particulars. Far behind the ancients is Kepler, who having considered the Copernican system visually, ascends from the particulars to the causes themselves, and to the reasons that Plato revealed *a priori* so many centuries ago. He shows the reason for the five Platonic solids expressed in the Copernican system.

It is not absurd or offensive to prefer the former to the latter: the very nature of things demands it. For if the glory of the architect of this world is greater than that of its contemplator, however clever – because the former produced the reasons for its construction from within himself, while the latter scarcely recognizes, after great labour, the reasons expressed in the construction – then surely those who understand the causes of things innately before they are accessible to the senses are more similar to the architect of things than others, who, having seen the matter, reflect on its causes. Therefore, Galileo, do not begrudge our predecessors their praise here. That which you say your eyes discovered only very recently, they predicted many years before you, as is proper.

Nonetheless, your glory will be this: that Copernicus, and I, following him, demonstrated the error in the ancient approach to how the five solids were expressed in the world, and substituted the genuine and true way, while you in part correct and in part render doubtful that doctrine of Bruce, borrowed from Bruno. They thought that other bodies also are encircled by their own moons, as our earth has its. But you demonstrate that they spoke too generally. Jupiter is one of the planets, which Bruno called earths, and behold: there are four other planets around it. But Bruno did not argue this about earths, but rather about suns.

Meanwhile, I cannot refrain from adding still more seemingly absurd ideas to your discoveries. It is likely, I predict, that there are inhabitants not only on the moon, but also on Jupiter itself, or (as was pleasantly bandied about very recently at a certain meeting of philosophers) on those regions now first being discovered. And as soon as somebody teaches the art of flying, they will not lack colonists from our human race. Who would have once believed that the navigation of the very vast ocean was calmer and safer than that of the Adriatic Sea, the Baltic Sea, or the English Channel? Produce ships or sails accommodated to the heavenly breezes and there will be those who do not fear even that vastness. Therefore, as though for those who before long will attempt this journey, let us build an astronomy, I of the moon, and you, Galileo, of Jupiter.

God the Creator leads the community of men, as though it were a growing little boy, slowly maturing, successively from one level of awareness to another. Therefore, let us recognize and look back in a certain measure, at what progress there has been in the knowledge of nature, at how much still remains, and at what we may expect from men further on.

But let us return to humbler reflections, and sum up what we started: namely, there are four planets circling around the planet Jupiter with different distances and times. Wackher indeed noted very cleverly that Jupiter too, as our earth, revolves around its axis, and that the revolution of the four moons follows that rotation, just as the revolution of our moon follows that rotation of our earth in the same direction. Therefore, he finally now believes in the magnetic account through which, in my recent *Commentary on Celestial Physics*, I made clear that the cause of the planetary motions is the rotation of the sun around its axis and poles. Evidently (as you, Galileo, beautifully state) if four circulating objects accompany Jupiter before and behind as it completes its twelve-year course, why was it absurd for Copernicus to say that one moon clings to the earth while it completes its annual motion, by the same method?

But then, one might say, if there are globes in the heaven similar to our earth, are we then in competition with them over which of us possesses the better part of the universe? How, then, are all things created on account of man? How are we the masters of God's work?

I will not leave unsaid those philosophical arguments that seem to me may pertain here, through which I shall maintain not merely in general that this system of planets on which we humans live dwells in the very bosom of the world, around the heart of the world – namely, the sun – but also that in particular we humans live on the globe which clearly is destined for the primary rational and most noble of creatures.

First, consider the fixed stars, which surround this bosom of the world quite like a wall. Next, consider the brightness of our sun, beyond that of the fixed stars. Third, geometry is one and eternal, shining brightly in the mind of God; the fact that a share in it has been granted to men is among the reasons why man is the image of God. And in geometry the most perfect genus of figures, after the sphere, are the five Euclidean solids. This planetary world of ours was divided according to their pattern and archetype.

Grant, therefore, that there are infinite other worlds: they will be either dissimilar to ours, or similar to it. You will not say that they are similar to it, for what would be the purpose of an infinite number, if each one has complete perfection within it? Even Bruno himself, the defender of infinity, supposes that each world differs from the others in its type of motion. If they differ in motions, then also in distances, which produce the periods of the motions. And if in distances, then also in the order, type and perfection of the figures which generate the distances. Indeed, if you think that the worlds are similar to each other in all things, then you will also end up with similar creatures and just as many Galileos observing new stars in new worlds as there are worlds. But what would be the purpose?

On the other hand, let those infinite worlds be dissimilar to ours: then they will be adorned with something other than the five perfect figures, and therefore will be less noble than this world of ours. From this we conclude that this world of ours would be the most excellent of all of them, if there were many.

Now let us address this too: why the earth is greater than Jupiter and is worthier of being the home of the sovereign creature. Indeed, the sun is in the centre of the world and is the heart of the world: it is the source of light and heat, and the origin of life and heavenly motion. But it seems that man should unperturbedly stay away from that royal throne. Heaven is for the Lord of heaven, the Son of justice, while the earth He dedicated to the children of men. For although God has no body and needs no dwelling place, nevertheless more of the power that governs the world is in the sun (or as it says everywhere in Scripture, in the heavens) than in the other spheres. Therefore, man should recognize his poverty and the riches of God due to this difference in dwelling place. He should recognize that he is not the source and origin of the world's adornment, but that he depends on its source and true origin. Add this as well, which I said in the *Optics*: for the sake of the contemplation toward which man was made and equipped with eyes, man could not have stayed at rest in the centre. Rather, he needed to circle around on the annual voyage of this earth for the purpose of observation. This is no different from how surveyors of inaccessible things exchange station for station in order to acquire from the distances of the stations a proper foundation for triangulating measurement.

After the sun, however, there is no more noble or suitable sphere for man than the earth. For first, it is the middle in number of the primary spheres (disregarding Jupiter's satellites and that of our earth's, the moon, as is necessary). For it has above it Mars, Jupiter and Saturn, and within the embrace of its circuit run Venus, Mercury and, rotating in the middle, the sun, which stimulates all the orbits – truly Apollo, the expression Bruno frequently uses. Next, since the five solids are divided into two – the three primary ones, cube, tetrahedron and dodecahedron, and the two secondary ones, icosahedron and octahedron – the circuit of the earth separates each class like a wall, as it touches the centres of the twelve faces of the dodecahedron above and the twelve vertices of the corresponding icosahedron below. Thus by its position between the figures alone, the sphere of the earth is more remarkable than the other spheres. Third, we on earth can scarcely see Mercury, the last of the primary planets, on account of the nearby and excessive brightness of the sun. How much less will Mercury be visible on Jupiter or Saturn! Thus it seems that this sphere was allotted to man with this ultimate plan: that he be able to contemplate all the planets. Truly, who will deny that in compensation for the planets concealed from the inhabitants of Jupiter, but which we earthly inhabitants see, four others were allotted to Jupiter, based on the four lower planets, Mars, earth, Venus and Mercury, which go around the sun within the ambit of Jupiter? Based on what I have said, we humans certainly should pride ourselves on the excellent residence of our bodies, and we owe thanks to God, the Creator.

It has pleased me above to discuss philosophically with you, Galileo, these new questions which you raise through your experiments. In closing, I strongly beseech you, celebrated Galileo, to proceed vigorously with your observations, and to communicate with us at the very first opportunity what observations you have achieved.

Kepler, *Dioptrics*, 1611

In his Starry Messenger, *Galileo had not explained how the telescope worked, and some worried about whether it could be trusted to reveal what was truly out there. Continuing the support for Galileo's discoveries begun in his earlier* Conversation with The Starry Messenger, *Kepler published two further shows of support: a* Narration *in which, after he finally received a telescope, he recorded how his own observations of Jupiter agreed with those of Galileo; and a discussion of the optics of the telescope, called the* Dioptrics. *In the latter Kepler not only explained how the telescope worked but also offered suggestions for improvement, including a description of a new type of telescope using two convex lenses. In the preface to the text, Kepler included several of Galileo's more recent letters detailing his latest discoveries, including the rings of Saturn and the phases of Venus, along with Kepler's own commentary. In order to preserve his priority as first discoverer, Galileo had shared his discoveries first in anagram form before a full revelation, and Kepler offered his frustrated attempts at decoding them, along with Galileo's ultimate solutions.*

Preface

The Starry Messenger of Galileo is in everybody's hands, as is my *Dissertatio*, a conversation of sorts with it, and then my *Narratio*, which confirmed *The Starry Messenger*. The reader may therefore briefly consider the chapters of the *Messenger* and how much was revealed there with the help of the telescope, a reckoning of which I offer in this book. Vision has testified that there is a particular bright body in the heavens, which we call the moon; reasoning from optics has demonstrated that that body is round; and astronomy, with arguments established on some optical foundations, has determined that its altitude from earth is about sixty semi-diameters of the earth. Various spots have appeared in that body, and consequently there arose a vague opinion of a few philosophers that varying images of mountains and valleys, and of water and land, were visible there. But now the telescope places all these things before our eyes so precisely that whoever, while enjoying such a sight, still judges it doubtful, is clearly a coward.

Nothing is more certain than that the southern parts of the moon swarm with many vast mountains, and that the northern parts, which are lower, receive the liquid flowing down from the south in very great lakes. Now, as if a new door of the heavens has been opened, our eyes themselves are led to the sight of hidden things. And if anyone would like to explore these new observations with the force of reason – for who does not see how far it may extend its boundaries via the contemplation of nature – we may inquire of what use mountains and valleys and vast seas are on the moon, and whether some creature, more ignoble than we, might inhabit those regions.

Another question is also settled here – one that was almost born with philosophy itself, and today occupies the most noble talents: whether the earth is able to be moved (as the doctrine of planetary theory strongly requires) without the destruction of gravity or the confusion of the motion of the elements. For if the earth were exiled from the centre of the world, some fear that the waters would desert the sphere of the earth and flow into the centre of the world. But we see that there is a force of water on the moon also, occupying the low-lying cavities of that sphere: and although that sphere is led in revolutions in the same ether, beyond both the centre of the world and the centre of our earth, nevertheless the quantity of water on the moon is not impeded at all from adhering steadfastly to the sphere of the moon – that is, from directing itself toward the centre of its body. Thus optics reforms the doctrine of gravity and levity through this example

of the sphere of the moon, and confirms here my introduction in
the *Commentaries on the Motions of Mars*. The supporters of
the Samian philosophy (if I may be permitted with this epithet to
indicate its inventors Pythagoras and Aristarchus of Samos) also
have a ready aid against the immobility of the earth as apparent
to the eyes in the moon. Indeed, optics teaches us that if any of
us were on the moon, it would certainly seem to him that the
moon, his home, was totally immobile, but that our earth, and
the sun and all other things, were mobile: for thus are the reasonings of sight related to each other.

And now, with the aid of the newly invented telescope, the very
eyes of astronomers are led directly to a detailed view of the substance of the Milky Way: such that anyone who enjoys this
spectacle is forced to confess that the Milky Way is nothing other
than a mass of the most minute stars. What a nebulous star is has
been completely unknown until now: but the telescope, directed
at one such nebulous convolution (as Ptolemy calls them), again
shows that, as with the Milky Way, there are two, three, or four
very bright stars located in a very close space. And who, without
this instrument, would have believed that the number of fixed
stars is tenfold or twentyfold greater than the number in the Ptolemaic description? And how, I ask, could we find an argument
about the end or boundary of this visible world – that it is this
very sphere of fixed stars – without this telescope having detected
this very multitude of fixed stars, which is a kind of vaulting of
the mobile world? And how much would an astronomer be mistaken in determining the magnitude of the fixed stars without
surveying them anew using a telescope?

But what exceeds all in my admiration is that chapter of *The
Starry Messenger* where we are told of the discovery, with the
help of an especially perfected telescope, of almost another
world in Jupiter. The mind of the philosopher is almost stupe-
fied to consider that there is some huge globe, which is equal in
bodily mass to fourteen terrestrial globes (unless here Galileo's
telescope will shortly show us more precise proportions than
those of Brahe) around which revolve four moons, not dissimilar to our own moon, with the slowest revolving in the span of
fourteen of our days, as Galileo revealed; the next to it, but the

most visible of all of them, in the span of eight days, as I detected last April and May; and the remaining two in a much shorter time still. From this, my own reasoning in the *Commentaries on Mars*, invoked for a similar cause, persuades me that the globe of Jupiter itself also rotates very rapidly, and without doubt more rapidly than in the span of one of our days, such that the perpetual circuits of those four moons follow the rotation of that great globe around its axis.

And even this sun of ours, the common hearth of this terrestrial world and the Jovian world, which we judge to be at most thirty minutes [in magnitude], there scarcely fills six or seven minutes; and meanwhile, it is found again in the same place among the fixed stars, having passed through the zodiac, after twelve of our years. Therefore, the creatures who live on that globe of Jupiter, while they contemplate the very short course of those four moons through the fixed stars, while they observe them and the sun rising and setting daily, would swear by the stone Jupiter (for I am recently returned from those [Roman] regions) that their globe of Jupiter is at rest and immobile in one place, but that the fixed stars and the sun, bodies which are actually at rest, revolve no less than those four moons in a great variety of motions around that home of theirs. From this example, much more than from the previous example of the moon, the supporter of the Samian philosophy will learn what can be answered to someone objecting that the doctrine of the motion of the earth is absurd, and using as support the testimony of our vision. O telescope, much-knowing and more precious than any sceptre: is not one who holds you in his right hand ordained as king and lord of the works of God?

Truly, 'that which is overhead, the great orbs with their motions, you make subject to intellect'.

If anyone is somewhat sympathetic to Copernicus and the luminaries of the Samian philosophy but hesitates only because he doubts this: that it is possible for the earth to travel on a journey amid the planets through the ethereal plains with the moon clinging to it constantly as an inseparable companion revolving around the earth itself, in the manner of a faithful dog that circles its master as he wanders, running sometimes in front of him and sometimes off to the side. Let him look at Jupiter, which, as this telescope shows, clearly drags with itself not only one such companion, as the earth according to Copernicus, but fully four, which never move away from it and meanwhile press on in their individual journeys.

But enough has been said about this in the *Conversation with* The Starry Messenger. It is time to turn to those things disclosed through the use of this telescope after the publication of *The Starry Messenger*, and after my *Conversation* with it.

A year has already passed since Galileo wrote to Prague that he had discovered something new in the heavens beyond his earlier discoveries. And lest there should be someone who, out of malice, resorted to calling himself its first observer, Galileo left time for everyone to publish whatever new things he had seen. He himself, in the meantime, described his discovery in transposed letters in this way: *s m a i s m r m i l m e p o e t a l e u m i b u n n u g t t a u i r a s.*

From these letters I fashioned a somewhat uncivilized verse, which I inserted in my *Little Narration* in the month of September of last year: 'Salve umbistineum geminatum Martia proles.' 'Hail, double-circling offspring of Mars.' But I was very far off from the meaning of the letters: they had nothing to do with Mars. But I will not detain you, reader: here is the solution in the words of the author, Galileo, himself:

[*Kepler quotes the full Italian of Galileo's letter*]

Although this differs little from the Latin words, nevertheless I will translate them lest my reader is stymied. This, therefore, is what he writes:

But let me come now to another chapter, which Kepler revealed in his recent *Narration*: the letters which I sent, transposed, to Your Illustrious Lordship. And since it was signified to me that his majesty wants to learn what they mean, I send it to Your Illustrious Lordship so that you may communicate it with his majesty and with Kepler and anyone you wish.

The letters, connected properly, say this: 'Altissimum planetam tergeminum observavi.' 'I have observed the highest planet threefold.'

Indeed, I discovered with the greatest astonishment that Saturn is not one star alone, but three very close to each other, so that they seem to almost touch each other. They are absolutely immobile in relation to each other, and are arranged in this way: oOo. The middle of them is much bigger than the outer ones. They are located one at the east, and one at the west, in one straight line to a hair. However, they are not exactly along the line of the zodiac: for the one further west rises a bit towards the north perhaps parallel to the equator. If gazed upon through a glass that does not multiply very much, the stars will not appear very distinct from each other, but the star of Saturn will appear somewhat elongated, in the form of an olive, thus: O. But if you were to use a glass that multiples the surface more than a thousand times, three very distinct globes will appear, which almost touch each other, and which will seem not to be separated from each other in latitude by more than a very fine and scarcely visible thread. Thus behold: an escort was found for Jupiter, and for the decrepit old one, two servants to assist him in walking and never leave his side. I learned nothing new about the remaining planets.

That is Galileo. But if I have a choice, I will not make a funeral feast for Saturn and servants from his companion globes: but rather from those three connected together I will make a three-bodied Geryon, from Galileo I will make Hercules, and from his telescope a club, armed with which Galileo has conquered the furthest of the planets and, extracted from the deepest recesses of nature and dragged to the earth, he has revealed it to all our eyes.

Galileo seemed at the end of his letter to have finished reporting on the planets and offering new observations about them. But the ever-attentive artificial eye – the telescope, that is – in a short time disclosed many more things, concerning which read the following letter of Galileo:

[*Kepler quotes the full Italian of Galileo's letter*]
Which may be translated thus:

I eagerly await a response to my last two letters, so that I may find out what Kepler says about the marvel of Saturn's star.

Meanwhile I send to him a cipher of a new and extraordinary observation which aids in the decision of great controversies in astronomy and specifically contains within it an argument for the Pythagorean and Copernican structure of the world. In due time I will disclose the solution to the cipher and some other remarkable things. I hope that I have discovered a method for determining the periods of the four Medicean stars, which Kepler, not without very good reason, thought inexplicable, etc.

The letters, transposed, are these: 'Haec immatura a me jam frustra leguntur, o.y.'

Thus far Galileo. But if, reader, this letter has filled you with the desire to know the meaning contained in those letters, come and read the following letter of Galileo.

But first, I would like to note in passing what Galileo says about the Pythagorean and Copernican structure of the world. For he points a finger at my *Cosmographic Mystery*, published fourteen years ago, in which I took the dimensions of the planetary spheres from the astronomy of Copernicus, who makes the sun stable in the middle and the earth moving around the sun and around its own axis, showing that the distances of the orbs correspond to the five regular Pythagorean figures, which that author had already distributed among the elements of the world in a beautiful rather than successful attempt, and which were the reason for Euclid's writing this whole geometry. And thus in the *Mystery* one finds a certain combination of astronomy and Euclidean geometry, and through it a consummation and most perfect completion of both. This was the reason why I awaited so eagerly the sort of argument Galileo would produce on behalf of this Pythagorean structure of the world. Therefore, Galileo's letter about this argument follows. [*Kepler quotes the full Italian of Galileo's letter.*]

This is Galileo's letter, whose substance can be captured as follows in Latin:

It is time to reveal the method of reading the letters that I sent transposed a few weeks ago. The time is now, I say, after I have become very certain about this matter, such that there isn't even a shadow of doubt any more. You should know, therefore, that about three months ago, when the star of Venus could be seen, I began to look at it through a telescope with great care, so that I could understand with sense itself something that I held as certain in my mind. At first, therefore, Venus appeared as a perfectly circular figure, and was enclosed by an exact and clear boundary, though it was very small. Venus retained this shape for a while, until it began to approach its greatest separation from the sun, and meanwhile the apparent size of its body continued to increase. From that time it waned on the eastern side, which was turned away from the sun, and within a few days it entirely assumed the shape of a perfect semicircle. That figure lasted without even the slightest change until it began to move back toward the sun, abandoning the tangent of its epicycle. At this time it wanes from its semicircular figure more and more, and continues to diminish it until its occultation, when it will wane into a very slender crescent. With this passage made to its morning designation, it will appear to us as only sickle-shaped, with a very slender crescent facing away from the sun; after that the crescent will be filled more and more until it is at its greatest distance from the sun, where it will appear semicircular, and that shape will last for many days without any noticeable variation. Then by degrees it will wax slowly from semicircular to a full orb, and it will preserve that perfectly circular figure for well over several months. Further, at present the diameter of the body of Venus is about five times greater than it was when it first appeared as an evening star.

From this wonderful observation, a very certain demonstration, perceptible to sense itself, is supplied to us of the two most important questions which until today have been debated by the cleverest men on both sides of the issue. One is that all the planets are by nature dark bodies (so that we can understand the same thing about Mercury that we do about Venus). The other, toward which we are urged with the greatest necessity, is that we must say that Venus (and beyond it also Mercury) is carried around the sun, as are all the remaining planets: things indeed believed by Pythagoras, Copernicus and Kepler, but never proven by the senses as it is now in Venus and Mercury. Therefore, Kepler and the rest of the Copernicans may boast that they have philosophized well, and that their belief is not vain, however much it has happened to them – and may still happen to them later – that they are regarded with universal agreement by philosophers of this time, who philosophize in books, as stupid and little less than fools.

Therefore, the words which I sent in transposed letters, which said 'Haec immature a me jam frustra leguntur, o.y.' are reduced, in order, to say 'Cynthiae figuras aemulatur mater amorum', 'The mother of loves emulates the figures of Cynthia', that is, Venus imitates the shapes of the moon.

Three nights ago, when I observed an eclipse of the moon, nothing particularly noticeable occurred. Only the boundary of the shadow appeared indistinct, vague and obscured, the reason being that the shadow rises from the earth, very far from the body of the moon.

I have some other particulars, but I am prevented now from writing about them, etc.

Thus far Galileo.

What shall we make now, dear reader, of our telescope? Perhaps the staff of Mercury, with which we can cross the liquid ether sea, and, with Lucian, lead a colony to the uninhabited evening star, enticed by the pleasantness of the region? Or rather the arrow of Cupid, which, gliding through our eyes, has punctured our inmost mind, provoking in us a love for Venus? For what can I not say about the wonderful beauty of this globe, if, lacking its own light, it achieves such splendour with only the borrowed light of the sun that not even Jupiter or the moon have, though they enjoy with it an equal proximity to the sun! The latter's light, if compared to the light of Venus, will indeed appear greater due to the apparent magnitude of its body, but it will seem weak, faint and almost leaden. O truly golden Venus! Who will doubt any longer that the whole globe of Venus is exquisitely wrought out of pure gold, as its surface, when placed in the sun, sparkles with such lively brilliance? Now let me add my experiments on the alteration of Venus's light with the blinking of the eye, which I examined in my *Optical Part of Astronomy*. Reason will be able to add nothing other than this: that the star of Venus spins on its axis with a very rapid rotation, exhibiting different parts of its surface one after another, more or less accepting of the light of the sun.

But enough of my reasonings. Let us know here, in the place of an epilogue, the reasonings of Galileo, built from experiments, about all that which he produced with the telescope. Thus here he is again:

[*Kepler quotes the full Italian of Galileo's letter*]

The Latin meaning of the writing is this:

Your last letter has wonderfully delighted me, especially where you bear witness to the benevolence felt toward me by the illustrious Imperial Councillor, Wackher, which I greatly appreciate. And since it arose from this: that I have demonstrated with some necessary reason certain conclusions that he himself has long regarded as true, therefore let me make this possession of grace toward me, so dear, that much firmer. Please convey these things to him from me: that I have very certain demonstrations ready which clearly show, as he himself holds, that all the planets receive light from the sun, and that they themselves are naturally dark and opaque bodies. But the fixed stars shine with their own natural light, and do not need illumination from the rays of the sun: for God knows whether they reach the further region of the fixed stars with as much brightness as we receive from the rays of the fixed stars coming down to us.

The chief foundation of my reasoning is this: that I have clearly observed with the telescope that the planets receive greater light and reflect it back more brightly, whenever they are nearer to us and to the sun. Accordingly, Mars at perigee – namely, closest to the earth – leaves Jupiter far behind in brightness, although in terms of bodily size itself it is far behind Jupiter. Thus it is difficult to receive this radiance of Mars with the telescope, for it is so great that it impedes the sight from being able to distinguish the round boundary of the disc of Mars's body. This is not true of Jupiter, for it appears fully circular. After this Saturn, on account of its being very far away, appears with a very exact boundary, both the bigger globe in the middle and the two small balls at the sides. For it appears with a faint and subdued light, without the sort of illumination that would impede a distinct apprehension of the boundaries of its three globes. Therefore, since we see that Mars, the nearest, is very brightly illuminated by the sun; that Jupiter, more remote, has a much fainter light (although without the use of an instrument it appears bright enough, which happens on account of the size and brightness of its body); and that Saturn, the most remote, has the weakest light, and is almost watery: with what sort of light do you think the fixed stars would appear, given that they are unspeakably further from the sun than Saturn, if they were only illuminated by the sun? Altogether very weak, murky and lifeless. But we experience quite the opposite. Let us look, for example, with our eyes at the Dog Star. What we experience is a very vivid brightness, which almost pricks the eye with the very rapid vibration of its rays, so strong that compared to it the planets, like Jupiter and even Venus itself, are as demeaned and humbled as when a very cheap and impure glass is compared to the most pure and brilliant diamond. And although the disc of the Dog Star does not appear larger than a fiftieth part of Jupiter's disc, nevertheless its radiance is immense and very strong, so much so that the appearance of that disc conceals itself, as it were, within that radiation, and almost vanishes and cannot be distinguished except with some difficulty from the plumes that surround it. By contrast, Jupiter, and even more Saturn, are seen with a limit, where their light is faint and, I might say, quiet. For this reason I think we would philosophize rightly if we refer the cause of the twinkling of the fixed stars to a native and characteristic vibration of brightness, part of them innately. On the other hand, we may say that when it comes to the planets, their illumination – which is derived from whatever the sun distributes through the world – is restricted to their surfaces.

These are the scientific ideas in Galileo's letter – the rest I omit.

Therefore, attentive reader, you see how the very clever mind of Galileo, assuredly a most outstanding philosopher, uses this telescope of ours as though it were a ladder, and ascends the very furthest and highest walls of the visible world, surveys everything in his sight, and from there looks down with very clever reasoning at these little huts of ours – at these planetary globes, I mean – and compares the outermost to the innermost, and the highest to the lowest, with perfect judgement.

Now there are those who in philosophy never lack national zeal or complaints, and many here in Germany will want the testimony of Germans. Thus I give them the letter of a German concerning these matters, in which the very same thing will be evident: thus it was not a bad thing that Galileo did when he, busy with his own affairs, speedily communicated his discovery to us at Prague through a cipher.

Here, therefore, is what [Simon] Marius wrote to our common friend:

> Meanwhile I am working on another work: in which a. first I assert the immobility of the earth; b. but the arguments are only examined against the Copernican accounts, which in our time Kepler and Galileo, the Paduan mathematician, endorse and set up as true; c. I take the arguments of my assertions from the Sacred Scriptures; d. while stipulating also physical; e. and astronomical ones. Next, the opinion of those who think that the celestial bodies are monstrous in size will be refuted, and a new, more probable measure of their size will be given by me. In this matter the Belgian instrument, commonly called the telescope, has greatly aided me. Third, I will demonstrate that Venus is nothing other than illuminated by the sun, and is crescent-shaped, as I observed many times from the end of last year until the present April with the aid of a Belgian telescope when Venus was nearest to the earth to the west and to the east. Fourth, I will deal with the new Jovian planets, which are carried around Jupiter, like the rest of the planets around the sun, with an unequal interval and period. I have already investigated the periods of the two furthest ones, and have constructed tables so that at all times it can easily be known how many minutes away from Jupiter they are to the left or right. These last two chapters are entirely unheard of before now. Perhaps other things will occur while I work.

Thus far Marius.

You therefore have, reader, confirmed belief that the telescope yields observations of new celestial entities in the testimony of one German. What hinders me, therefore, from composing a geometrical panegyric to this most excellent instrument with this little book, and what hinders you, reader, with a view to honour, a willing spirit and uncommon attention of mind, from taking part while I recite it? With this work you will sharpen your mind and will become more learned in philosophy and the perception of causes, and better prepared for the invention of mechanical, useful and pleasant things, and finally, more cautious and safer from the thousand ways in which the common people tend to be led into error. Farewell and consult this prelude fairly and well.

John Donne, 'First Anniversary: An Anatomy of the World', 1611

One year after Galileo's telescopic discoveries, the English poet John Donne (1572–1631) – later to become the royal chaplain and dean of St Paul's Cathedral – considered the significance of the new astronomy in his 'First Anniversary', a poem commemorating the death of Elizabeth Drury, the daughter of Donne's patron, at the age of fourteen. In meditating on her death, Donne reflected too on the ways that the new science had brought to an end the older conception of the world. He noted that the new discoveries had called into question Aristotelian physics ('the element of fire is quite put out') and that the heliocentric model ('the sun is lost') had shattered the old cosmological order, with potentially dire consequences for the social and political hierarchy that rested upon it.

> And new philosophy calls all in doubt,
> The element of fire is quite put out,
> The sun is lost, and th'earth, and no man's wit
> Can well direct him where to look for it.
> And freely men confess that this world's spent,
> When in the planets and the firmament
> They seek so many new; they see that this
> Is crumbled out again to his atomies.
> 'Tis all in pieces, all coherence gone,
> All just supply, and all relation;
> Prince, subject, father, son, are things forgot,
> For every man alone thinks he hath got

To be a phoenix, and that then can be
None of that kind, of which he is, but he.
And learn'st thus much by our anatomy,
That this world's general sickness doth not lie
In any humour, or one certain part,
But as thou saw'st it, rotten at the heart.
Thou seest a hectic fever hath got hold
Of the whole substance, not to be controll'd;
And that thou hast but one way, not to admit
The world's infection – to be none of it.
For the world's subtlest immaterial parts
Feel this consuming wound, and age's darts;
For the world's beauty is decay'd, or gone –
Beauty, that's colour, and proportion.
We think the heavens enjoy their spherical,
Their round proportion, embracing all;
But yet their various and perplexed course,
Observed in divers ages, doth enforce
Men to find out so many eccentric parts,
Such diverse downright lines, such overthwarts,
As disproportion that pure form; it tears
The firmament in eight-and-forty shares,
And in these constellations then arise
New stars, and old do vanish from our eyes;
As though heaven suffered earthquakes, peace or war,
When new towers rise, and old demolish'd are.
They have impaled within a zodiac
The free-born sun, and keep twelve signs awake
To watch his steps; the Goat and Crab control,
And fright him back, who else to either pole,
Did not these tropics fetter him, might run.
For his course is not round, nor can the sun
Perfect a circle, or maintain his way
One inch direct; but where he rose to-day
He comes no more, but with a cozening line,
Steals by that point, and so is serpentine;
And seeming weary with his reeling thus,
He means to sleep, being now fallen nearer us.

So of the stars which boast that they do run
In circle still, none ends where he begun.
All their proportion's lame, it sinks, it swells;
For of meridians and parallels
Man hath weaved out a net, and this net thrown
Upon the heavens, and now they are his own.
Loth to go up the hill, or labour thus
To go to heaven, we make heaven come to us.
We spur, we rein the stars, and in their race
They're diversely content to obey our pace.
But keeps the earth her round proportion still?
Doth not a Teneriffe or higher hill
Rise so high like a rock, that one might think
The floating moon would shipwreck there and sink?
Seas are so deep that whales, being struck to-day,
Perchance to-morrow scarce at middle way
Of their wish'd journey's end, the bottom, die.
And men, to sound depths, so much line untie
As one might justly think that there would rise
At end thereof one of th' antipodes.
If under all a vault infernal be
– Which sure is spacious, except that we
Invent another torment, that there must
Millions into a straight hot room be thrust –
Then solidness and roundness have no place.
Are these but warts, and pockholes in the face
Of th' earth? Think so; but yet confess, in this
The world's proportion disfigured is;
That those two lees whereon it doth rely,
Reward and punishment, are bent awry.
And, O, it can no more be questioned,
That beauty's best, proportion, is dead,
Since even grief itself, which now alone
Is left us, is without proportion.

Christoph Scheiner and Galileo, *Letters on Sunspots*, 1612–1613

With the invention of the telescope, Galileo and other observers began noting spots that seemed to move like clouds across the surface of the sun. The Jesuit astronomer Christoph Scheiner (1573–1650), writing under the pseudonym Apelles, published a book in 1612 arguing that sunspots were satellites orbiting the sun, much like the moons of Jupiter. Therein ensued a series of letters between Scheiner and Galileo, written to Marcus Welser as an intermediary and published by both afterwards. In contrast to Scheiner, Galileo argued that the spots were not satellites, but were akin to clouds and connected to the surface of the sun itself. He used those observations to argue that the sun rotated on its axis. The two also discussed Galileo's observations of the phases of Venus, which he had reported earlier to Kepler in anagram form. Though the general concept of the phases of Venus was compatible with both Ptolemaic and Copernican theory, the particular phases that Galileo observed were incompatible with the former and proved that Venus orbited the sun – as Galileo noted here and Scheiner accepted. This did not necessarily prove Copernican theory, however, as those phases were consistent with Tychonic theory as well, in which Venus orbited the Sun, which then orbited a stationary Earth.

First Letter from Galileo to Marcus Welser about the solar spots, in reply to his earlier letter, 4 May 1612

These letters [from Scheiner/Apelles] contain what has been imagined until now about how to define the essence, place and motion of these spots. And first, there is no doubt that they are real things and not simple appearances or illusions of the eye or of lenses, as your friend demonstrates in his first letter. I have observed them for eighteen months now and have shown them to several close friends of mine and also, at this time last year, to many prelates and lords in Rome. As that author has established that the observed spots are not illusions of lenses or defects of the eye, he tries to determine something general about their place, showing that they are not in the air or on the solar body. As for the first, the notable lack of parallax shows that we must necessarily conclude that the spots are not in the air – that is, close to the earth, within the space that is commonly assigned to the element of air. But that they cannot be in the solar body does not seem to have been demonstrated with full necessity. Simply saying, as he does with his first reason, that it is not credible that there are dark spots in the solar body because it is very bright does not lead to his conclusion. For while we had to call it very pure and bright so long as no darkness or impurities were seen in it, when it shows itself to us as partly impure and spotted, why shouldn't we call it spotted and not pure? Names and attributes should be accommodated to the essence of things, and not essences to names.

As to what he mentions about ascertaining whether Venus or Mercury make their revolutions under or around the sun: I am somewhat amazed that he has not yet heard what I discovered almost two years ago and communicated to so many that by now it is well known: that Venus changes its shape in the same way as the moon. Now Apelles can observe it with a telescope and will see it in a perfectly circular shape and very small, although it was even smaller at its evening rising. He can then continue observing it and will see it, around its maximum digression, in the shape of a semicircle. From this shape it will pass to its horned shape, gradually tapering as it approaches the sun; around its conjunction it will appear as thin as a moon of two or three days, and the size of its visible circle will be increased in such a way that its apparent diameter at its evening rising will clearly appear to be less than the sixth part of that which it will have at evening occultation or morning rising, and consequently its disc will appear almost forty times greater in the latter position than in the former. These things leave no room for anyone to doubt what the revolution of Venus is: rather, with absolute necessity they will conclude, in accordance with the positions of the Pythagoreans and Copernicus, that it revolves around the sun, around which, as the centre of their revolutions, all the other planets are turned. Therefore, it is not necessary to await physical conjunctions to prove such a manifest conclusion, nor to produce reasons subject to some albeit weak rejoinder, to gain the approval of those whose philosophy is strangely perturbed by this new constitution of the universe. For they, when not held back, will say that Venus either shines by itself or is of a substance that can be penetrated by the sun's rays, so that it is bright not only on its surface but throughout its whole depth. And they will be able to shield themselves with this answer all the more brazenly because there has been no lack of philosophers and mathematicians who have believed this (and this must be said in contrast to the opinion of Apelles, who writes otherwise). And Copernicus himself admitted one of these positions as possible – indeed, even as necessary – having been unable to explain how Venus, when it is below the sun, does not appear horned. And nothing else could really be said before the telescope came to show us how Venus really is dark in itself, the same as the moon, and that it changes shapes like it.

However, it seems to me that Apelles, who has a free and non-servile mind and is very capable of reaching true doctrines, has begun to be moved by the strength of such novelty, and to heed and assent to true and good philosophy, especially this part that concerns the constitution of the universe. Yet he cannot completely detach himself from the fantasies already stamped on him, to which his intellect sometimes returns, having been long accustomed to agreeing with them. This can also be seen in the same passage when he tries to demonstrate that the spots are not in any of the orbs of the Moon, Venus, or Mercury, where he considers as true, real, actually distinct from each other and mobile, wholly or partially, those eccentrics, deferents, equants, epicycles, etc. These were placed by pure astronomers to facilitate their calculations, but not considered as such by philosopher astronomers, who, in addition to the job of saving the appearances in any way, try to investigate, as the greatest and most estimable problem, the true constitution of the universe, since this constitution exists and is singular, true, real and impossible to be otherwise. Because of its greatness and nobility, this is worthy of being put before any other question by speculative minds.

I do not deny circular motions around the earth and around centres other than its own, nor do I deny the other circular motions totally separated from the earth – that is, that do not go around it and enclose it within their circles. Because Mars, Jupiter and Saturn, with their approaches and departures, attest to me of the former, and Venus and Mercury and also the four Medicean planets of the latter circular motions. And consequently, I am quite sure that there are circular motions that describe eccentric circles and epicycles. But that to describe such motions nature really makes use of that mass of spheres and orbs fashioned by astronomers I do not feel it necessary to believe, but instead see it as an accommodation to ease astronomical computations. And I am of an opinion between those astronomers who admit not only the eccentric movement of the planets but also the eccentric orbs and spheres that impel them, and those philosophers who, by contrast, deny both the orbs and the motions around centres other than that of the earth. However, when it comes to the question of investigating the place of the sunspots, I would have wished that Apelles had not banished them from a real place that lies between the immense spaces in which the small bodies of the Moon, of Venus and of Mercury revolve – banished them, I say, by virtue of an imaginary supposition that such spaces are entirely occupied by eccentric, epicycle and deferent orbs that are disposed and, indeed, necessary to carry with them any other body located within them, which cannot wander by itself toward any other side other than where, with excessively hard chains, the surrounding heaven forces it.

It remains for us to consider what Apelles determines about the essence and substance of these spots, which is, in sum, that they are neither clouds nor comets, but stars that go around the sun. Regarding this determination, I confess to your lordship that I am not yet so confident as to establish and affirm any conclusion as certain, as I am very sure that the substance of the spots could be a thousand things unknown and unthought of by us, and that the accidents that we see in them – that is, their shape, opacity and motion – are very common, and can give us either no knowledge, or very little and general knowledge. Thus, I do not believe that any philosopher who confessed that he did not know and could not know what the material of the sunspots is would be worthy of blame. But if we wanted, by a certain analogy to familiar and known subjects, to suggest something of what they seem to be, I would actually be of an opinion completely contrary to Apelles. For it does not seem to me that any of the essential conditions that pertain to planets are appropriate to these spots, nor do I find in them any conditions that we do not see also in our clouds. This we will discover as follows:

Sunspots are produced and dissolved in times that are more or less brief; some of them condense together and become separated greatly from day to day; their shapes change, most of which are extremely irregular, with varying degrees of darkness; and being either very close to or on the solar body, it is necessary that they be very vast masses. They must also be very powerful, due to their varying opacity, to prevent in varying degrees the illumination of the sun; sometimes many are produced, sometimes few and sometimes none. Now very vast and immense masses, which are produced and dissolved in a short time and which sometimes last longer and sometimes more briefly, which spread and condense, which easily change shape, and which are denser and opaquer in some parts and less so in others – these are found around us only in clouds, while all other subjects are far from fulfilling these conditions. And there is no doubt that if the earth itself were luminous and not illuminated by the sun from outside it, then it would look quite similar to anyone looking at it from a very great distance. For depending on whether this or that region was crowded with clouds, it would appear scattered with dark spots, which would hinder its brightness according to the greater or lesser density of their parts, which would then appear more or less dark. We would see many, then few; they would expand, then shrink; and if the earth itself turned, then they would follow its motion; and being not very deep compared to the distance across which they commonly spread, those in the middle of the hemisphere would appear very large, and those coming toward the extremity would seem to shrink. In sum, I do not believe that anyone would see something that is not similar to what is seen in the sunspots.

From these observations and others, and from those that can be made from day to day, it is evident that there is no material among us that imitates the accidents of these spots more than the clouds, and the reasons that Apelles gives to show that they cannot be such seem to me of very little strength. When he asks, 'Who would ever imagine clouds around the sun?' I would reply, 'Anyone who saw such spots, and who wanted to say something probable about their essence: because he will not find anything known to us that resembles them more.'

A more detailed discussion of the sunspots
and the planets around Jupiter, to Marcus Welser,
by Christoph Scheiner, 1612

At the same time as Galileo observed horned Venus in various cities of Italy, other mathematicians in Rome also marvelled at it and found it in the same shapes: horned, bisected and gibbous. From this incredible phenomenon we have two unavoidable arguments. The first is that Venus, just as the moon, lacks its own light, and consequently casts a black shadow below the sun. The second, that it goes around the sun. Regarding this, since all the phenomena are in accord and all reasons thus agree, any prudent man will scarcely dare to doubt it in the future.

Galileo, *Letters on Sunspots*, Image of Sunspots, 1613

This is one of the many images of sunspots that Galileo included with the publication of his letters on the topic. Galileo's arguments about sunspots – that the spots were directly connected to the sun and that the sun rotated – challenged Aristotelian physics, which posited a perfect and unchanging sun and accepted the sun's diurnal motion around the earth and annual motion along the ecliptic, but not a rotation on its axis. The rotation of the sun also supported the parallel idea of the earth's diurnal rotation on its axis, an important feature of the Copernican model.

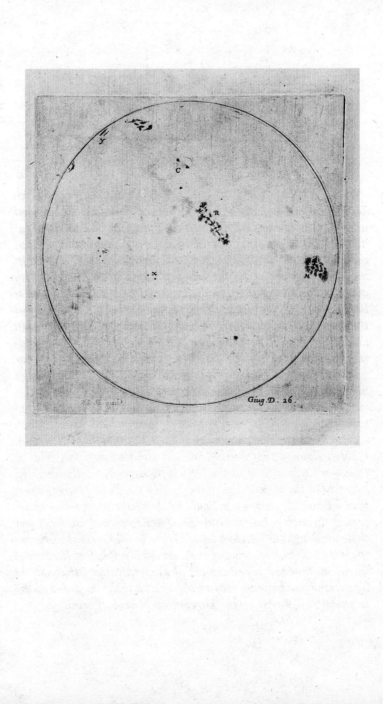

Galileo, Letter to Benedetto Castelli, 1613

Galileo helped his protégé Benedetto Castelli obtain the position of professor of mathematics at the University of Pisa in 1613. There, Castelli was invited to dine with the grand duke of Tuscany, Cosimo II de' Medici, and the Grand Duchess Christina. In the course of the meal, the grand duchess questioned Castelli about the Copernican system, and, in particular, about its relationship to Holy Scripture. When Castelli wrote to Galileo about the conversation (in which he had defended Galileo's position), Galileo responded with this letter, in which he briefly summarized his views. He argued that scriptural passages, particular ones referring to natural phenomena, should not be interpreted literally. Rather, as Scripture's primary message pertained to salvation, its references to natural phenomena were accommodated to human perceptions rather than to the truth of things, which reason and experience were better suited to investigate. Further, Galileo claimed that even if one did want to interpret biblical passages literally, one of the central passages typically heralded as proof of the immobility of the earth – God's stopping the sun in the Book of Joshua – actually made better sense when interpreted through a Copernican lens. Galileo would later expand these arguments in his 1615 letter to the Grand Duchess Christina (page 266). This earlier letter is particularly important because it set in motion the investigation of Copernicanism that would lead to the condemnation of 1616, when it was forwarded to the Inquisition in Rome by the Dominican Niccolò Lorini.

21 December 1613

Most Reverend Father and Praiseworthy Sir:

I enjoyed hearing about the arguments that you advanced at the table of the Most Serene Ladyship and in the presence of the Grand Duke and Most Serene Archduchess and some of the very excellent philosophers there. This gave me the opportunity to return to the consideration of some general questions about whether to bring Sacred Scripture into disputes about physical conclusions, and some particular questions about the passage in Joshua, which was proposed by the Grand Dowager Duchess, with some support of the Most Serene Archduchess, to contradict the motion of the earth and the stability of the sun.

As to the first general question of the Most Serene Ladyship, it seems to me that it was very prudently proposed by her, and granted and confirmed by you, that Sacred Scripture can never lie or err, and that its statements are absolutely and inviolably true. I would only have added that, while Scripture cannot err, nevertheless some of its interpreters and expounders can err sometimes in various ways. One among these is very serious and quite frequent: to always want to limit oneself to the literal meaning of the words. This would cause not only various contradictions to appear, but also grave heresies and blasphemies; it would then be necessary to give God feet, hands and eyes no less than bodily and human emotions like anger, regret, hate and sometimes even the forgetting of past things and the ignorance of future ones. Hence as there are many propositions in Scripture which appear different from the truth according to the literal sense of the words, but are conveyed in such a way to accommodate the incapacity of the common multitude, so too it is necessary for those who deserve to be separated from the multitude that wise expositors produce their true meanings, and point out the particular reasons why they have been expressed with such words. Given, therefore, that in many places Scripture is not only capable of but necessarily in need of interpretations different from the apparent meaning of the words, it seems to me that in disputes over nature it should be reserved for the last place. For Sacred Scripture and nature proceed equally from the divine Word, the former as dictation of the Holy Spirit, and the latter as the very obedient executor of God's orders. Moreover, since it behoved the Scripture to accommodate the understanding of everyone, it says many things that are different, in appearance and with respect to the meaning of the words, from the absolute truth. But nature, by contrast, is inexorable and immutable, and cares not at all whether or not her hidden causes and ways of operating are exposed to human capacity, so she never transgresses the terms of the laws imposed on her. It therefore seems that whatever either sensible experience places before our eyes or necessary demonstrations prove to us regarding natural effects should not in any way be disputed or doubted because of places in Scripture whose words have a different semblance, since not everything said in the Scripture is bound to obligations as severely as each effect of nature.

Indeed, for this very reason – because it accommodates itself to the capacity of the rough and undisciplined populace – Scripture has not refrained from obscuring its highest dogmas, attributing even to God himself conditions that are very distant from and contrary to his essence. Who, then, will want to definitively maintain that in speaking even incidentally of the earth or the sun or another creature, it has set aside this consideration and chosen to contain itself rigorously within the limited and restricted meaning of the words? This is especially so when it says things about these creatures which are very far from the primary purpose of the Sacred Scripture – indeed, things which, when said and brought forth with their truth bare and in plain view, would sooner damage its primary intention, making the populace more stiff-necked against the persuasion of the articles pertaining to salvation.

Given this, and since it is very evident that two truths can never contradict each other, it is the duty of wise interpreters to work hard to find the true meaning of scriptural passages in agreement with those natural conclusions of which we are already certain and assured because they are manifest to the senses or of demonstrated necessity. Indeed, being – as I said – that the Scripture, although dictated by the Holy Spirit, admits in many places interpretations far from the literal sense (for the reasons already discussed), and, moreover, being that we cannot assert with certainty that all its interpreters speak with divine inspiration, I would think that it would be prudent if no one were allowed to insist that scriptural passages are obliged in certain ways to support specific natural conclusions when the opposite could one day be revealed as true through the senses or through demonstrative and necessary causes. Who wants to place a limit on human capacities? Who will want to assert that we already know everything in the world that is knowable? For this reason, it would perhaps be excellent advice not to add anything unnecessary to the articles concerning salvation and the establishment of the faith, whose firmness is such that there is no danger of a valid and effective doctrine ever arising against them. And if this is the case, how much greater would be the disorder arising from adding to them at the request of people who (aside from not knowing if they speak with celestially inspired grace) we see clearly are completely lacking in the intelligence necessary to understand – I won't even say to criticize – the demonstrations with which the most exact sciences proceed to confirm some of their conclusions?

I would think that the authority of the Sacred Writings has the sole aim of persuading men of those articles and propositions which are necessary for their salvation and surpass all human reason, and so could not be made credible by any science or means other than the mouth of the Holy Spirit itself. But I do not think it necessary to believe that the same God who endowed us with sense, reason and intellect wanted us to neglect their use by giving us through other means the information we can obtain with them, especially in those sciences where only the smallest references and isolated conclusions are found in Scripture, especially astronomy, of which it contains so little that not even the names of the planets are found in it. Rather, if the first sacred writers had planned to persuade people about the arrangements and movements of the celestial bodies, they would not have treated them so briefly, which is to say almost not at all, compared to the infinity of very high and admirable conclusions contained in this science. Therefore, I hope you see, if I am not mistaken, how disorderly is the procedure of those who insist that in natural disputes scriptural passages – which do not directly concern the faith and which they often poorly understand – hold pride of place.

I now come to consider the particular passage in Joshua. Given and granted for the time being to our opponent that the words of the sacred text should be taken in exactly the way they sound – that is, that God, at the prayers of Joshua, made the sun stop and prolonged the day, so that victory ensued – I ask that the same rule should apply to me and to him, so that our opponent does not presume to bind me and leave himself free to alter or change the meaning of the words. In this case, I say that the passage clearly shows us the falsity and impossibility of the Aristotelian and Ptolemaic world system, and, by contrast, it fits very well with the Copernican one.

I would first ask our opponent whether he knows with what sorts of movements the sun is moving. If he knows, he is forced to respond that it moves with two motions, namely the annual motion from west to east, and the diurnal motion in the opposite direction, from east to west. Secondly, I would ask him if these two motions, so different and almost contrary to each other, belong to the sun and are equally its own? He must answer no, but that only one is its own and particular to it, namely the annual motion, while the other is not its own but rather of the highest heaven, that is, the Primum Mobile, which impels along with itself the sun, the other planets and the stellar sphere, forcing them to make a revolution around the earth in twenty-four hours, with a motion, as I said, almost contrary to their own natural one. I come to the third question, and would ask him with which of these two motions the sun produces day and night, that is, with its own or with that of the Primum Mobile? He must answer that day and night are effects of the motion of the Primum Mobile and that it is not day and night that depend on the sun's own motion but rather the different seasons and the year itself. Now if the day does not depend on the sun's motion but on that of the Primum Mobile, who does not see that to lengthen the day it is necessary to stop the Primum Mobile, and not the sun? Indeed, is there anyone who understands these first elements of astronomy and does not know that, if God had stopped the motion of the sun, instead of lengthening the day He would have shortened it and made it briefer? Because since the motion of the sun is contrary to the diurnal rotation, the more the sun moves toward the east, the more it would delay its course to the west, while by decreasing or cancelling the motion of the sun, it would set that much more quickly. This happens, as we see, in the moon, which completes its diurnal revolutions more slowly than those of the sun, as its own motion is faster than that of the sun. Since it is therefore absolutely impossible, in the system of Ptolemy and Aristotle, to stop the motion of the sun and lengthen the day, even while Scripture says it happened, we must therefore either not order the motions as Ptolemy wants or we must alter the meaning of the words, saying that when Scripture says that God stopped the sun, it meant that He stopped the Primum Mobile, but that, to accommodate the capacity of those who can hardly understand the rising and setting of the sun, it said the opposite of what it would have said when speaking to men of understanding. Add to this that it is not believable that God would only stop the sun, letting the other spheres move on, because this would have unnecessarily altered and exchanged the order, aspects and dispositions of the other stars with respect to the sun, and greatly disturbed the whole course of nature. But it is believable that He stopped the whole system of celestial spheres, which after that period of interposed rest could return all together to their labours without confusion or alteration. But because we have already agreed not to alter the meaning of the words of the text, we must have recourse to another arrangement of the parts of the world, and see whether the plain sense of the words conforms to that one directly and smoothly. And this is truly what we find. Having discovered and demonstrated with necessity that the globe of the sun turns on itself, making an entire rotation in about one lunar month in precisely the direction that all other celestial rotations are made, and moreover, it being very probable and reasonable that the sun, as the greatest instrument and minister of nature and almost the heart of the world, gives not only light (as it clearly does) but also motion to all the planets that revolve around it, if, according to the position of Copernicus, we attribute the diurnal motion to the earth, who does not see that in order to stop the whole system, and thus to prolong only the length of time of diurnal illumination without altering at all the rest of the mutual relations of the planets, it was enough for the sun to be stopped, exactly as the words of the Sacred text say? Here, then, is the way to lengthen the day on earth by stopping the sun, without introducing any confusion between the parts of the world and without altering the words of Scripture.

Your most affectionate servant,
Galileo Galilei

Simon Marius, *The Jovian World*, 1614

Simon Marius (1573–1625) was a German astronomer who published a book in 1614 claiming to have observed the moons of Jupiter with a telescope at exactly the same time as Galileo had in Italy, though he had waited longer than Galileo to publish his results. Much as Galileo had named them the Medicean stars after his own patrons in Italy, Marius named the moons the Brandenburg stars after his patrons, the margraves of Brandenburg. Galileo was furious over this claim to co-discovery, and in his 1623 Assayer he accused Marius of 'not blushing to make himself the author of the things I discovered'. Ever since, Marius's claim has been debated, in part because he did not publish specific observations as Galileo had done. He was, however, the first to publish tables of the motions of Jupiter's moons, and was the person who gave the moons their names based on Jupiter's 'illicit loves' – an idea he credited to Kepler.

The Jovian World, Discovered in the year 1609 with the help of the Belgian Telescope. That is, the Four planets of Jupiter, both theory and tables, founded especially on individual observations, from which their positions near Jupiter can be very readily and easily computed at any given time. Inventor and Author Simon Marius of Guntzenhausen, mathematician of the Margraves of Brandenburg in Franconia and student of higher medicine. With permission and privilege of his Sacred Imperial Majesty.

In 1608, when the Frankfurt autumn fair was being celebrated, the very noble, brave and spirited John Philip Fuchs of Bimbach in Mohr happened to be there, and happened to meet a merchant whom he already knew, who told him that right then in Frankfurt there was, at the fair, a Belgian who had devised a means by which to examine very distant objects and make it seem as though they were very near. Upon learning this, he beseeched the merchant to bring the Belgian to him, which finally happened. Whereupon that very noble man disputed at length with the Belgian inventor, doubting the truth of this new invention. At length the Belgian produced the instrument that he had brought with him, though one of its glasses had formed a crack, and insisted that he test the truth of the matter himself. He thus took the instrument into his hand and directed it toward an object, which he saw enlarged and multiplied several times. With the truth of the instrument ascertained, he asked the inventor how much money he would demand for it. When he understood that he could not have it right away, they departed without concluding their business.

Upon returning to Ansbach, the nobleman called for me and told me that an instrument had been devised with which very remote things could be seen as though they were very close. I heard this news with great wonder. He debated the matter with me at length after dinner, and ultimately concluded that such an instrument must be made of two glasses, one concave and the other convex, and with chalk sketched a drawing by hand on the table to show the sort of glasses he meant. Afterwards, we took two pieces of glass from common eyeglasses, one concave and one convex, and placed one a convenient distance behind the other, and in some way discovered the truth of the matter. But the curvature of the magnifying glass was too great. Therefore, he sent an accurate figure of the convex glass, pressed in plaster, to Nuremberg, to the craftsmen who make common eyeglasses, so that they might prepare similar glasses. But it was in vain, for they did not have suitable instruments, and he did not want to reveal the method for constructing them.

Meanwhile, no expense was spared, and several months elapsed. If we had known about the method of polishing the glasses, we would have prepared the finest telescopes immediately after our return from Frankfurt. In the interim, telescopes of this kind were being disseminated in Belgium, and one that was good enough was sent to us, delighting us greatly, in the summer of the year 1609. From that time on I inspected the heavens and the stars with that instrument at night whenever I was at the house of that nobleman I've mentioned. Sometimes he let me take it home, particularly near the end of November, when I observed the stars in my observatory as I tended to do.

Then, for the first time, I looked at Jupiter, which was opposite the sun. I detected some very small stars, some just after and some just before Jupiter, in a straight line with it. At first, I thought that they were among those fixed stars that cannot be seen elsewhere without this instrument, such as those detected by me in the Milky Way, the Pleiades, the Hyades, Orion and other places. But when Jupiter was in retrograde, I still saw these stars along with it through December. At first I was very astonished, but after that I gradually came to this belief: namely, that these stars are carried around Jupiter, just as the five solar planets, Mercury, Venus, Mars, Jupiter and Saturn, circle around the sun.

Therefore, I began to jot down observations, the first of which was on 29 December, when three stars of this kind were visible in a straight line from Jupiter to the west. At this time, I frankly confess, I believed that there were only three stars of this kind accompanying Jupiter, since several times I had seen three stars like this lined up near Jupiter. Meanwhile, two wonderfully polished glasses, convex and concave, were being sent from Venice by a very distinguished and wise man, Lord John Baptist Lenccius, who had returned to Venice from Belgium after peace was made, and to whom this instrument was already very well known. These glasses were fitted on a wooden tube and handed over to me by the nobleman I've already mentioned, so that I might test what it would reveal in the stars, especially the ones near Jupiter.

Therefore, from this time until 12 January I focused diligently on these Jovian stars, and found that there seemed to be four bodies of the same kind that looked to Jupiter for their circuit. Finally, near the end of February and the beginning of March I fully confirmed that this was the exact number of these stars. From 13 January until 8 February I was at Hal in Swabia, and I left the instrument at home, afraid that it would be damaged by the journey. After I returned home, I devoted myself to my customary observations, and in order that I might observe the Jovian stars more diligently, the nobleman gave me full use of his instrument out of a singular affection for these mathematical studies. From this time, therefore, all the way to the present, I have continued my observations with this instrument and others constructed afterwards. All this is quite true.

But I do not enumerate these things in order to diminish the reputation of Galileo, or to seize the discovery of these Jovian stars from his Italian compatriots. Rather, I would like it to be understood that these stars were not revealed to me by any mortal in any way; instead, they were discovered and observed by me in Germany, in my own investigation at nearly the same time as Galileo first saw them in Italy. The credit of first discovery of these stars among the Italians therefore deservedly is attributed to Galileo and remains his. But whether anyone among my Germans has found or seen them before me, I have hitherto been unable to ascertain, or easily believe. Indeed, I have experienced quite the contrary, namely that there were those who were not ashamed to impudently accuse both Galileo and me of errors. But I do not doubt that they have already repented for their own mistakes and are ashamed of their premature judgement of the labours of others. If, therefore, this little book of mine should come to Galileo in Florence, I ask that he receive these words from me in the spirit in which they were written. I have no desire to take away from his authority and discoveries. Rather, I want to thank him greatly for the publication of his *Starry Messenger*, which very much confirmed things for me. His observations have been especially useful to me, because they were made at about the time when I was at Hal in Swabia and had stopped my own observations. Although they do not seem to me exactly the same on all fronts, they greatly aided me.

Not so long ago, after my return from Regensburg, I procured an instrument through which not only the planets, but also all the fixed stars, are seen exquisitely rounded, especially the large and small Dog Star, the brighter stars in Orion, the Lion, Ursa Major, etc., which I had never been able to see before. I am astonished that Galileo has not seen the same thing with his quite excellent instrument. For he writes in his *Starry Messenger* that fixed stars do not terminate in a circular periphery, which some afterwards held as the greatest argument by which the Copernican world system was confirmed – namely, that because of the immense distance of the fixed stars from the earth, the round shape of the fixed stars could never be perceived at all from earth. But since it is now very certain that even the fixed stars appear circular from the earth with the telescope, this argument clearly fails and the opposite is clearly strengthened: namely, that the sphere of the fixed stars is not at all removed at such incredible distances from the earth as Copernicus's speculation makes it, but rather, the separation of the fixed stars from the earth is such that the mass of their bodies can still be seen with this instrument in a distinctly circular shape, which fits with the Tychonic and proper arrangement of the celestial spheres, as will be confirmed below in the second part of this book. But that the fixed stars shine with their own light I readily concede to Galileo, because they possess a much more excellent splendour and brightness than the planets.

Galileo, in his *Starry Messenger*, has named the Medicean Stars mainly because of this: that he himself was born and educated in Florence, under the dynasty of the great dukes of Tuscany, who for many years have descended from the illustrious family of the Medici. If I name those same Jovian satellites the Brandenburg Stars, who will blame me, since I have still more just reasons? For not only was I born under the dynasty of this most illustrious and high family, but also, from the fourteenth year of my life to the present, I was very liberally nurtured at the expense of these illustrious princes. And with this name I make them, as much as I am able, immortal, as they very much deserve. Therefore to Galileo, as their first observer in Italy, they are the Medicean stars, and to me, who first saw and observed them in Germany, they are the Brandenburg stars.

But perhaps some will not be pleased with the names listed here, and instead will demand from astronomers an individual name for each of these four Jovian stars. I think that they can be satisfied in this manner (which, however, I want to make without any superstition and with the permission of the theologians). The poets complain greatly of Jupiter's illicit loves. In particular are mentioned three young women whom Jupiter secretly captured with love: namely Io, daughter of the river god Inachus; Callisto, daughter of Lycaon; and Europa, daughter of Agenor. And still more greatly he loved Ganymede, a beautiful boy and the son of King Tros, to the extent that he assumed the form of an eagle and carried him to heaven, as the poets describe in fables, especially Ovid in Book 10.6.

There, I do not think I will have done wrong if I call the first Io, the second Europa, the third, on account of the majesty of its light, Ganymede, and the fourth, finally, Callisto. These names are included in the following couplet: 'Io, Europa, the boy Ganymede and Callisto / Too much lustful abandon was pleasing to Jove.' Lord Kepler, Imperial Mathematician, supplied to me his idea and the particular names chosen on one occasion when we were together at the Regensburg assemblies in October of 1613. Therefore, by the joke and the friendship that we entered into there, I will rightly salute him as the godfather of these four stars.

Johannes Locher, *Mathematical Disquisitions Concerning Astronomical Controversies and Novelties*, 1614

Johannes Locher (1592–1633) was a student of the Jesuit Christoph Scheiner, with whom Galileo was embroiled in a controversy over the nature of sunspots. Under Scheiner's guidance, Locher published this book arguing against the Copernican system, which Scheiner sent to Galileo and which Galileo subsequently ridiculed in his 1632 Dialogue on the Great World Systems. Yet Locher's book was not as ridiculous as Galileo later suggested. Locher acknowledged and visually depicted both the sunspots and the phases of Venus (the latter is depicted in the image below) in his book, and recognized that they suggested that the old, Ptolemaic system was wrong. Yet instead of the Copernican alternative, Locher opted for the Tychonic geo-heliocentric system, as did most seventeenth-century Jesuits. This was not only because of scriptural problems, but also because of the problematic physics of vertical falls on a spherical, rotating earth. Locher further objected to the Copernican system because, like Tycho, he found the immense distances of the stars from the earth compared to the negligible size of the earth's orbit, and the consequent immense size of those stars (required because of the lack of a detectable stellar parallax), to be absurd.

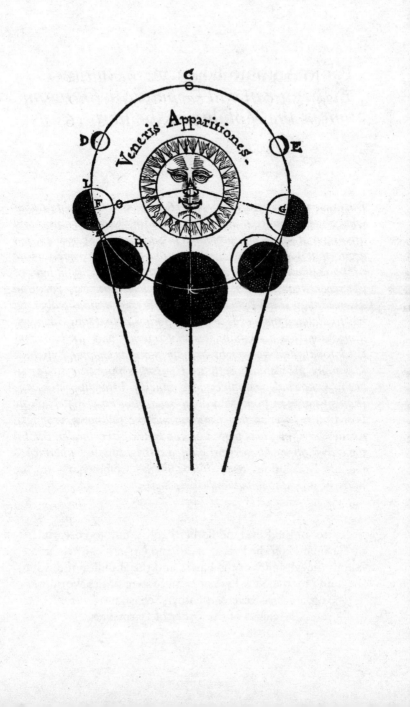

Veneris Apparitiones.

Paolo Antonio Foscarini, *Letter on the Pythagorean and Copernican Opinion about the Mobility of the Earth*, 1615

Paolo Antonio Foscarini (1565–1616) was a Carmelite father who in 1615 wrote a pamphlet defending Copernicanism and arguing that it was compatible with Scripture. Foscarini did not argue that the Copernican system was true, but merely more probable than the older geocentric system, especially in light of the recent astronomical discoveries. Given this fact, Foscarini asserted that 'if the Pythagorean view is true, undoubtedly God has so dictated the words of Sacred Scripture that they can allow a meaning that accommodates that opinion and can be reconciled to it', and he sought to offer some principles by which seemingly problematic scriptural passages might be so reconciled. Foscarini's pamphlet was quickly brought before the Inquisition. In response, Cardinal Bellarmine immediately wrote Foscarini a letter urging caution, and the following year Foscarini's pamphlet was placed on the Index of Forbidden Books, the only Copernican text prohibited outright (with Copernicus's On the Revolutions *itself, like Zúñiga's* Commentary on the Book of Job, *suspended until corrected).*

Letter of the Reverend Father Paolo Antonio Foscarini, Carmelite, on the Pythagorean and Copernican Opinion about the Mobility of the Earth, and the Stability of the Sun, and on the New Pythagorean System of the World.
To the Most Reverend Father, Sebastiano Fantone, General of the Order of Carmelites

I have decided to write in defence of the new opinion – or the opinion recently renewed and pulled into the light from the darkness of oblivion where it was buried – on the mobility of the earth and the stability of the sun, held in ancient times by Pythagoras and later by Copernicus, who derived from that hypothesis the system and constitution of the world and the position of its parts. I have previously written to you about this subject, but I wanted to briefly give you a survey of my intention, and to describe to you the foundations on which this opinion can and should rest, so that (since it is reasonable and probable for other reasons) it does not prove so repugnant and seem so opposed not only to physical causes and to principles commonly approved by all (which would be less bad) but also (which is much more important) to the great authority of the Sacred Scripture. For to those who hear of it, it seems to be one of the strangest and most monstrous paradoxes that they have ever encountered. This stems completely from old custom, confirmed through many centuries, which, for those who have made it habit, has hardened opinions that are stale, plausible and therefore commonly agreed upon by all, embraced by the learned and unlearned alike, such that they can no longer separate from them: so great is the force of custom, which is said to be another nature.

For things are ordinarily not measured or judged according to what they are, but according to the authority of those who speak of them. However, this authority, since it is no more than human, should not be deemed so important that it leads one to despise, renounce, or reject that which clearly contradicts it.

Nor should the way be closed to posterity such that we cannot dare to find anything more, or better, than that left to us by the ancients, whose genius, as their inventions, was not much superior to that of our own times. For it seems that in the perfection of discoveries they were not only equalled but also left far behind by the moderns, with our knowledge, not only in the liberal arts but also in mechanics, always improving.

What shall I say about the experiments of the moderns – have they not, in several particulars, closed the mouths of the venerable ancients and shown their very solemn and serious decrees to be vanity and lies? There were paradoxes held by

many ancients of serious and remarkable authority no less
strange than that of the mobility of the earth: the idea of the
antipodes, and that it was not possible to live in the Torrid
Zone. But the moderns have demonstrated, through much diligence and valour, that the white-bearded ancient masters erred,
and too easily believed and celebrated their false imaginings.

For the sake of brevity, I will omit here the many dreams of
Aristotle and other ancient philosophers, and I will only say
that if they had seen and observed that which the moderns have
seen and observed, and had understood their reasons, without
a doubt they would have changed their opinions and believed
in the clear truth of what was before them. So it is not necessary to attribute so much to the ancients that whatever they
affirmed should be believed and held as very certain, as if it
were revealed and descended from heaven.

What matters, therefore, in this material, is that if there is anything known to be repugnant to divine authority or to the Sacred
Scriptures dictated by the Holy Spirit, and by its inspiration
interpreted by the Holy Doctors of the Holy Church, then at
once both human reason and sense perception itself must be
abandoned, and we must judge that we are deceived, and that
what they represent to us is not true, since the knowledge which
we gain by Faith is more certain than any other that we have.

Therefore, since the opinion of Pythagoras and Copernicus has
appeared on the world stage in such strange dress, and on first
appearance seemed repugnant to different authorities of the
Sacred Scriptures (among others), it came thus (rightly, given this
presupposition) to be judged by all as mere madness. But because
the common system of the world described by Ptolemy never
fully satisfied the learned, it has always been suspected, even by
those who followed it, that some other one was truer. For though
this common system saves all the appearances that result from
the celestial bodies, nevertheless it does so with innumerable difficulties and the cobbling together of spheres (of various shapes
and sizes) of epicycles, equants, deferents and other chimeras that
are more like logical entities than something real.

Thus the followers of the common opinion confess that in
describing the system of the world they cannot guess at or teach

the true system; rather, they only investigate which is most probable and can best accommodate the celestial appearances.

Then the telescope was discovered, and with it many wonderful things were discovered in the heavens, all curious and unknown through the centuries: for example, that the moon is mountainous, that Venus and Saturn are three-formed and Jupiter four-formed, that in the Milky Way and the Pleiades and in Nebulae there are a great multitude of stars close to each other. And consequently it has revealed new fixed stars, new planets and new worlds. And with the same instrument it has been confirmed as highly probable that the courses of Venus and Mercury are not properly around the earth, but are rather around the sun, and that only that of the moon is around the earth. What, therefore, can be inferred from this, if not that the sun is in the centre, and that the earth and the other celestial spheres move around it? Therefore, from this and from many other reasons we know that the opinion of Pythagoras and Copernicus is not abhorrent to the foundations of astronomy and cosmography, but is, rather, highly likely and probable. All the more so given that among the many opinions that differed from the common system and tried to devise others, like those of Plato, Callippus, Eudoxus and, since then, Averroes, Cardano, Fracastorius and other ancients and moderns, none is more easily accommodated to the phenomena or to the calculation of the motions of celestial bodies with a determinate rule and without so many epicycles, eccentrics, deferents and rapid motion, as this one is.

And it has been supported not only by Pythagoras and then by Copernicus but also by many other famous and esteemed men, like Heraclitus and Ecphantus the Pythagoreans, and the whole school of Pythagoras, Nicetas of Syracuse, Martianus Capella and many others. And the numerous others who went (as we said above) looking for new systems because they rejected the current one and the Pythagorean one, nevertheless still, for their part, rendered it probable and came at least to indirectly confirm it when they judged the common system to be lacking and not entirely without difficult contradictions. Among these we can count the Jesuit Father Clavius, a most learned man, who, although he rejects the opinion of Pythagoras, nevertheless confesses that, in order to remove many difficulties present in the common system, astronomers must try to search for another one, as he enthusiastically urges them to do.

But what else could be found better than the Copernican system? Therefore, many moderns were induced and finally persuaded to follow it, but with some fear and remorse: because it seemed to them that the Sacred Scripture is so contrary to it that they could not be reconciled, and this repelled them. Hence this opinion remained suppressed and is considered indirectly, with a blush and a covered face. I considered all these things (because of my desire for knowledge to receive as much light and perfection as possible, and for all errors to be cleared with the pure light of truth), and I speculated as follows: Either this opinion of Pythagoras is true, or it is not. If it is not true, it is not worth speaking about or considering; if it is true, it does not matter that it contradicts all the philosophers and astronomers of the world, and that in order to follow and practise it one needs to develop a new philosophy and astronomy dependent on new principles and hypotheses. Further, what pertains to the Sacred Scripture could not harm it, for one truth cannot contradict another.

Therefore, if the Pythagorean view is true, undoubtedly God has so dictated the words of Sacred Scripture that they can allow a meaning that accommodates that opinion and can be reconciled to it. This is the reason (given the evident probability of the aforementioned opinion) that led me to consider and seek a way of harmonizing the many places in Sacred Scripture with it, and to interpret them according to theological and physical foundations such that they do not contradict it at all. That way if chance will determine expressly and with certainty that it is true (as now it seems probable), no one will find in it any impediment that prohibits it and deprives the world of the venerable and sacrosanct commerce in truth, desired by all good men.

In this undertaking I am (as far as I can imagine) the first to proceed, but I believe that scholars of those doctrines will be grateful, in particular, the most learned Galileo Galilei, mathematician of the most Serene Grand Duke of Tuscany, and Johannes Kepler, mathematician of the Sacred and Invincible Majesty of the Empire, and all the illustrious and very celebrated Academy of the Lynceans, who universally (if I am not deceived) follow this opinion. Though I have no doubt that they and other learned men would easily have found these and other similar reconciliations of scriptural passages, nevertheless I (given my profession) wanted (because of my soul's fondness for the truth) to offer in their service, and that of all the learned, this thought of mine, which I am sure will be received with the same candour of spirit in which it is offered.

To come to the point: I say that all the authorities of divine Scripture which seem to be contrary to this opinion can be reduced (in my opinion) to six classes.

The first class is those which affirm the earth to be fixed and unmoving, as in Psalms, 'the world also is established, that it cannot be moved', and also: 'Who laid the foundations of the earth, that it should not be removed for ever.' And in Ecclesiastes: 'but the earth abideth for ever' and the like.

The second is those which say that the sun moves and circles the earth, like in Psalms, 'In them hath he set a tabernacle for the sun, Which is as a bridegroom coming out of his chamber, and rejoiceth as a strong man to run a race. His going forth is from the end of the heaven, and his circuit unto the ends of it: and there is nothing hid from the heat thereof.' And so, in the book of Joshua it is posited as a miracle that Joshua made the sun stop, saying 'Sun, stand thou still upon Gibeon.' For if the sun stood still and the earth moved around it, it would not have been miracle, and to stop the progress of daylight he would not have said 'Sun, stand thou still,' but rather 'Earth, stand thou still.'

The third class is those which say that heaven is above and earth below, as in Joel, quoted by St Peter in Acts, 'And I will shew wonders in heaven above, and signs in the earth beneath,' and similar others. In this way Christ is said to have descended from heaven at the Incarnation, and to ascend to Heaven after the Resurrection. But if the earth were moving around the sun, it would be in heaven, and consequently it would be above rather than below. This is indeed the case in the opinion that places the sun in the centre, which also places Mercury above the sun, and Venus above Mercury and the earth above Venus, together with the moon, which circles around the earth. And thus the earth, together with the moon, comes to be in the third heaven. If, therefore, in spherical bodies (like the world), below is nothing other than to be the part closest to the centre, and above is that which is more toward the circumference, it follows that in order to corroborate the theological propositions about the ascension and descension of Christ, we must place the earth in the centre and the sun with the other heavens in the circumference, and not in the manner of Copernicus, which is contrary to this, according to which it does not seem that the true ascension and descension are preserved.

The fourth is those which show Hell to be in the centre of the world, as is the common opinion of the theologians. And it is confirmed by this reason: that since by denomination Hell must be in the lowest part of the world, and in a sphere there is no part lower than the centre, Hell must be in the centre of the world, which – being that it is spherical in shape – must be said to be either in the sun (because the sun would be in the centre of the world) or, if we suppose properly, that Hell is at the centre of the earth. If the earth moved around the sun, it would necessarily follow that Hell, together with the earth, was in heaven, and revolved with the earth around the sun in the third heaven – and nothing could be more monstrous and absurd than this.

The fifth is those which always contrast heaven and earth, and in turn earth and heaven, as if they had the same relation as that between the centre and the circumference and the circumference and the centre. But if the earth were in the third heaven, it would be on one side, and not in the middle, and consequently there would not be this relation between them, while in fact heaven and earth are always linked in this continual opposition, not only in the Sacred Scriptures, but also in common speech. Thus in Genesis, 'In the beginning God created the heavens and the earth,' and in the Psalms and other places repeatedly, 'He who made heaven and earth'. Therefore since these two bodies are always opposed to each other, and since heaven without doubt belongs to the circumference, the earth must, by contrast, belong to the centre of the world.

The sixth and last class (of the Church Fathers and theologians, rather than divine Scripture) is those which say that the sun, after the Judgement, must stop in the east, and the moon in the west. Which point, according to the Pythagorean opinion, should rather be said of the earth and not of the sun, because if the earth moved around the sun, it would have to stop.

These are the classes that most seriously oppose and trouble the aforementioned opinion. Nevertheless, they can be easily defended against, in my opinion, with six principles.

The first and foremost principle is this: when Sacred Scripture attributes something to God or to another creature that does not seem appropriate or proportionate to them, then it should be interpreted with one or more of the following four glosses: the first is to say that it agrees metaphorically with them – proportionally, or through similitude. The second is that it is said according to our manner of considering, apprehending, conceiving, understanding, knowing, etc. The third is that it is said according to the popular opinion or common way of speaking: and the Holy Spirit is often and with great intention accommodated to this popular and common mode. The fourth, is that it is said with respect to us.

To give some examples of all of these explanations: God does not walk, because He is infinite and immobile and has no corporeal limbs; He is a pure act, and thus has no passions of mind; and yet in Sacred Scripture we find, in Genesis, 'the Lord God walking in the garden in the cool of the day' and in Job that 'He walketh in the circuit of heaven', and elsewhere in very many places He is credited with coming, departing, waiting and hurrying, and with corporeal limbs, eyes, ears, lips, face, hands, feet, belly, clothing, arms and also many passions like anger, grief, repentance and the like. What shall we then say? Without a doubt, attributes like these are appropriate metaphorically, proportionally and by similitude.

And these things are also comparative to us, and in respect of us. Thus, God is said to be in Heaven, and to move in time, and to show and hide Himself, and to observe our steps, to seek us, to stand at the door – though He does not have a bodily place, motion, time, or act in human ways or manners. This is, rather, according to our way of understanding Him, which distinguishes these attributes in Him from one another though they are one and the same in Him, and divides His actions into several times, though they are indivisible and in one and the same instant.

Thus in Ecclesiasticus this comparison is made: 'The conversation of the godly is always wisdom, but the fool changes like the moon.' And yet in truth the moon always remains the same, as astronomers demonstrate, because one half is always bright, and the other dark; and at no time does this disposition change, except in respect of us, and according to the common opinion. Hence it is clear that here Sacred Scripture speaks according to the common mode of popular reasoning, and of according to simple folk and according to appearance, and not according to real being.

In the same way, in Genesis it is said that 'God made two great lights: the greater light to rule the day, and the lesser light to rule the night: He made the stars also,' where in both the proposition and the specification things are said that disagree with the real being of those celestial bodies. It is therefore necessary that the words of Scripture be interpreted according to the aforementioned glosses, in particular that they are to be understood according to the common sense and common way of speaking, which is the same as saying that they should be understood according to appearance and according to us, or in respect of us. For first, it is said that God made two great luminaries, meaning by this the sun and the moon; and nevertheless these are not the two greatest luminaries, according to the truth of the matter. For while the sun is one of the greatest, nevertheless the moon is not, except in respect to us: since among those that are absolutely greater, and a little less than the sun and almost equal to it, and far greater than the moon, are Saturn, or even some of the bright fixed stars of the first magnitude.

And thus coming to our principal proposition, by this reasoning, if the Pythagorean opinion is true, the authority of Sacred Scripture can be easily reconciled with it, though it seems contrary to it – particularly those of the first and second class. With this principle we may say that the Scripture speaks according to our way of understanding, and according to appearance and with respect to us, as it does with those bodies, in comparison to us, which are described by the common method of philosophizing. Thus the earth is said to be standing still and immobile, and the sun to move around it. And thus Scripture makes use of the popular and common way of speaking, because it seems according to our sight that the earth stands fixed in the centre, and the sun moves around it, rather than otherwise, just as happens to those who are carried on a boat on the sea near the shore, to whom it seems that the shore moves backwards and retreats from them, and not – as is the truth – that they actually go forward.

It is worth considering here, and not passing silently over, the question of why the Sacred Scripture so often accommodates itself to the common and popular opinions, rather than instructing the common man in the truth of the secrets of nature, since this is part of our first principle. I will say briefly that this is first because of the kind disposition of the Divine Wisdom, which accommodates Himself to all things according to their ability and nature. Thus in natural works He uses natural and necessary causes, and in liberal ones He operates liberally, and with noble men He works in elevated ways, and with the common people humbly, and with the learned wisely, and with the simple, in the common fashion. In sum, He adapts His manner with each one. Second, this is because it is not His intention to instruct us in this life about curiosities which would engage our doubt and suspicion. For He has already allowed and decreed that the world be occupied with disputes, strife and controversies, and subjected to the uncertainty of everything. Rather, His only intention here is to teach us the true path to eternal life, and this is the reason why the wisdom of God, as revealed to us in the Sacred Scripture, is called in Ecclesiastes 'saving wisdom', and not absolute wisdom: what is meant by 'saving' is that it has no other goal than to help us acquire salvation.

For God has not taught whether the original substance of the heavens is the same as the [earthly] elements; whether the continuum is composed of indivisibles or is divisible into infinity; whether the elements are formally mixed; how many celestial spheres and orbs there are; whether there are epicycles and eccentrics; nor has he taught the virtues of plants or stones, the nature of animals, the courses and influences of the planets, the order of the universe, the marvels of minerals and the whole of nature: but only things that are useful. That is, His holy law made us ultimately able to arrive at perfect cognition and a vision of the whole order and miraculous harmony, and of the sympathy and antipathy of the universe and its parts, when all these curiosities will be seen very distinctly and clearly. While in this current state He has left them to human pursuit and investigation (so far as it can attain), without meddling either directly or indirectly in their resolution. For they are of little or no use to us and might, in some cases, cause some harm: so not knowing them would not damage us and might, in some cases, be useful.

Therefore through His wonderful wisdom He made it that while all things in this world are doubtful, uncertain, equivocal, ambiguous and wavering, His Holy Faith alone is most certain. And although there are various opinions in the Church on philosophical and doctrinal matters, there is nevertheless only one truth of Faith, and of Salvation, and since that Faith is necessary to salvation, He made such that there was no doubt in it, and that it was certain and immutable, and known to all, and He gave the infallible Rule of it to the Holy Church, washed with His blood, whose visible head is the Supreme Pope.

This, therefore, is the reason why God has determined the speculative and curious questions in the Holy Scripture, as they are not useful for our edification or salvation, and why the Holy Spirit has so often conformed with the common and popular opinion without instructing us in something new, singular, or hidden. And thus consequently we see how and why no certainty can be obtained from the aforementioned authorities for resolving controversies of this sort, and why from this principle we can easily resolve and dodge the blows of the authorities of the first and second class, and any other objection drawn from Sacred Scripture, against the Pythagorean and Copernican opinion, if it comes to be known as true some other way.

And in particular the authorities of the second class can be avoided, and interpreted in another way, with the same principle already stated of common speech and our ordinary way of understanding things, as they appear to us. And in this way, it is possible to interpret the command of Joshua, 'Sun, stand thou still,' and the miracle of the sun not having any motion, saying that it was done not by actually stopping the body of the sun, but by stopping the splendour of the sun on the earth, caused not by stopping the sun, which always stands still, but by stopping the earth, which receives its splendour as it follows its usual motion and ordinary revolution toward the east, which would have made the sun go toward the west; so its stopping made the sun stop. And let this be all that is in explanation of my first principle, which required some prolixity in establishing it, due to its difficulty and importance.

My second principle is this: that all things both spiritual and corporeal, perpetual and corruptible, immovable and moveable have received from God a perpetual, immutable and inviolable law of being, and of their nature, according to the saying of Psalms: 'He hath also stablished them for ever and ever: He hath made a decree which shall not pass.' It is according to this that they always observe a perpetual tenor in their being and operations, and they come to acquire the reputation of having very certain and stable conditions. Thus Fortune (and there is nothing in the world more unstable or variable) to be constant and invariable in its continual volubility, inconstancy, vicissitude and variation, which explains the verse from Ovid 'and it is constant in its levity'.

Thus the heavens, whose motion is made to never cease by ordinary law, is said to be immobile and immutable, and the heavens to move motionlessly and earthly things to change unchangingly, because these things never cease in the first to move and in the second to change. By this principle are interpreted all the authorities of Sacred Scripture that belong to the first class, in which the earth is said to be stable and immobile: this is meant with respect to its nature. For although it includes local motion, which is threefold, according to the opinion of Copernicus (that is, diurnal, with which it turns on itself; annual, with which it moves through the twelve signs of the zodiac; and with respect to its inclination, according to which its axis always regards a certain part of the world, and the reason for the inequality of the days and the nights); and it includes other species of mutation, like generation, corruption, augmentation, diminution and alteration of various kinds – nonetheless it is always stable in all these, and never varies from the pattern began by God, moving always steadily and immutably through all the six types of motion.

The third principle is this: when a thing moves according to some of its parts, and not according to its whole, it cannot be said to be without qualification and absolutely moved, but only accidentally. For example, if you take a glass or some other portable measure of water from the sea and transport it to another place, it cannot be absolutely said that the sea is without qualification moved from one place to another, but only accidentally and according to some of its parts, because speaking absolutely it is untransferable from its place; in this way even the air is untransferable and unmoveable from its place, even if it moves and is transferred in some of its parts. This principle is clear by itself, and through it the authorities can be explained which seem to indicate the immobility of the earth. That is, the earth in essence and absolutely, according to its totality, is not changeable, given that it is not generated or corrupted, augmented or diminished, or altered according to its totality, but only according to its parts.

This is clearly the true sense of the text of Ecclesiastes, which says 'One generation passeth away, and another generation cometh: but the earth abideth for ever.' As if to say that although the earth, according to its parts, generates itself and is corrupted, and is generally subject to the vicissitudes of the generation and corruption of things, nevertheless according to its entirety it never generates itself or is corrupted, but is perpetually immutable. This is just as with a ship, which though sometimes has a board replaced with another, or a piece from its hull switched to its rudder, or some part or another removed, it nevertheless always remains the same ship. Thus the authorities do not speak here of local motion but of another sort of change, as in the substance, quantity, or quality of the earth. And if one wants to say that it does refer to local motion, then it would have to be interpreted according to this principle, that it is in respect of the natural place that it holds in the universe, as I will explain. For nothing prevents them from moving according to their parts, though that movement is then violent and not natural.

The fourth principle, therefore, is that every corporeal thing, either mobile or immobile, has its own natural and corresponding place, away from which it is moved violently, and going toward which it moves naturally. And nothing can be removed from this natural place according to its totality, because it would cause a very great disturbance and horrible disorder in the universe. Thus neither the whole earth, nor all the water, nor all the air can be moved away in their totality from their determined place, or site, or system, in respect of the order and disposition of the other bodies of the world. Thus no star can leave its place, even if it is a wandering one, and no orb or sphere can leave its, even if it is mobile in some other way. Therefore all things, even if they move, are nevertheless always said to be immobile and still in their proper place, according to the aforesaid sense, which is understood according to their totality. Therefore the earth, even if it were mobile, could be said to be stationary and immobile in this way, because it does not move in a straight motion outside the sphere given to it at its Creation, in which it always has to move circularly. Rather, it stays situated in the orb called Great, which is above Venus and below Mars, halfway between them in the heavens, where the common opinion ordinarily places the sun. And in this place it moves around the sun, and around the other two middle planets, Venus and Mercury, and it has the moon moving around it (which is another earth, but ethereal, as Macrobius said, according to the opinion of ancient philosophers). And so it never changes and never varies. Insofar as it is uniform in its assigned path, and never goes out of it, it is said to be stable and immobile, just as the heavens and all the elements can be said to be immobile in this way.

The fifth principle is little different from the one preceding. Some things are created by God in such a way that their parts are separable and divisible from each other, and from the whole; others, such that they cannot be separated, at least collectively. The former are transient, the latter perpetual. The earth, therefore, being that it is a perpetual creation, has parts

258 THE DAWN OF MODERN COSMOLOGY

that are not separable nor divisible collectively from each other, and from its centre (which is its true place), and from its entirety. In itself, in its entirety, it is always conjoined, united and coherent, nor can its parts be separated from its centre, or from each other, except accidentally and violently, after which they return naturally and immediately to their place. In this way, therefore, the earth is called immobile and immutable; and in this way not only it but also the sea, the air, the sky and everything (however otherwise mobile it is) as long as its parts are not collectively separable, can be called immobile. This principle does not differ from the one preceding except that the former concerned the parts with respect to place, and this one the parts with respect to the whole.

And another secret is derived from this speculation, because through it the formal properties of the gravity and levity of things are disclosed: something which is not so easily designated or explained without great controversy according to the common Aristotelian philosophy. According to the principles of this new opinion, gravity is therefore nothing other than a certain natural appetite and inclination of the parts to reunite with the whole. This was given by Divine Providence not only to earth and its bodies, but also to celestial bodies, including the sun, the moon and the stars. It is because of this inclination that the parts of these bodies are all amassed and joined together, such that each one does not think to be able to find rest anywhere other than the centre of the body.

What results from the same principle is the knowledge of just how false is the Aristotelian doctrine that to one simple body there is one simple motion, and these are of two kinds, straight and circular; the straight is twofold, from the middle and to the middle; the first pertains to light things, like air and fire, and the second to heavy ones, like water and earth; the circular, which is around the middle, belongs to heaven, which is neither heavy nor light. All this philosophy fails and falls to ruin, as in the new philosophy it is established that although it is true that a simple body has no more than a simple motion, nevertheless this is only the circular, and nothing else. Only according to circular motion does a simple body stay in its natural place,

and in its unity, and moves properly in its place in such a way
that the thing, though it moves, nevertheless remains united in
itself, and however it moves, it nevertheless remains as if at
continual rest. Straight motion, which is properly to a place,
applies only to those things that are out of their natural place
and find themselves far from union with each other, and unity
with the whole, but rather separated and divided from it –
which is something repugnant to the order of nature and the
form of the universe. It follows that straight motion is only
suitable for those things that do not have the perfection that is
appropriate to them by nature, so that by means of this straight
motion they try to reintegrate themselves with their whole, and
to rejoin their unity and thus return to their natural place, the
only one where they feel rest and stillness, and finally stop.

The sixth and final principle is this. Everything is referred to
simply as it is in comparison to all things, or to the many things
that make up the majority of its kind, but not in comparison to
the things that make up the minority of it. In this way a vessel
cannot be called absolutely great because it is great when compared to two or three or a few other vessels, but can be called
absolutely great if it exceeds in size all individuals of its kind,
or the majority of them. Nor can a man be called absolutely great
because he is greater than the Pygmies, nor absolutely small
because he is smaller than the Giants, but only absolutely large
and small with respect to the ordinary stature of the majority
of men.

Thus we ought not to call the earth absolutely high or low,
because it is such only with respect to some very small part of
the universe. And consequently, it ought not to be called absolutely high, because it is so only in comparison with the centre
of the world or a few parts of it that are closer to that centre,
like the sun, Mercury, or Venus. Rather, it should be referenced
in comparison with the majority of spheres and bodies in the
universe. The earth, therefore, in comparison with the whole
circuit of the eighth sphere, which includes all corporeal creatures, and in comparison with Mars, Jupiter and Saturn, and also
with the moon, and even more in comparison with other bodies
(if there are any) above the eighth sphere, and in particular with

the Empyrean Heaven, is said to be truly in the lowest place in the world, and almost in its middle, or centre, for it cannot be said to be higher than any bodies other than the sun, Mercury and Venus. Thus, it deserves the name of lowest body absolutely and simply, and not of a supreme or middle one. And so to come down from Heaven to it, or from the Empyrean, as the majority intend by the name Heaven (as Christ descended from Heaven for his Holy Incarnation), and to go from it to Heaven (as Christ went to Heaven for his glorious Ascension) are properly and truly to descend from the circumference to the centre, and to ascend from the parts closest to the centre of the world to its furthest circumference. It is possible therefore to verify the theological propositions very well in this way. And this principle is especially confirmed insofar as I have observed that nearly all of the authorities of Sacred Scripture that contrast Heaven in the singular to Earth are very conveniently understood, and with a very suitable interpretation, of the Empyrean Heaven in particular (which is the highest of all, and spiritual with respect to its end) and not of the lower and intermediate heavens, which are corporeal, and were fabricated for corporeal creatures. And with this principle in particular, the authorities and arguments of the third, fourth and fifth classes are resolved.

From these principles and their assertions it is clear that the Pythagorean and Copernican opinion is so probable that it may be more so than the common one of Ptolemy, since we can deduce from it a very well-ordered system and mysterious structure of the world more founded in reason and experience than that of the common one; and since we have reason to doubt that the authority of Sacred Scripture and theological propositions oppose it; and since with it one can not only save the phenomena and the understanding of all the celestial bodies, but also discover many natural causes which would be difficult to understand while on the other path; and since, finally, it makes it easier to bring astronomy and philosophy together, removing all superfluous and imaginary things developed only because no one knew how to reduce to some kind of cause the rule for the great variety of celestial motions.

This is what I wanted to say now, speaking as a theologian, about the not-improbable opinion of the mobility of the earth and the stability of the sun. I offer it to you, Most Reverend Father, and I do not doubt that you will accept it gratefully, because of your great inclination toward virtue and good learning.

Your Most Humble Servant,
Paolo Antonio Foscarini

Robert Bellarmine, Letter to Foscarini, 1615

Cardinal Robert Bellarmine (1542–1621) was an Italian Jesuit, rector of the Collegio Romano (the Jesuit University in Rome) and a member of the Congregation of the Holy Office, charged with protecting the Church from heresy. When Foscarini wrote his letter attempting to reconcile Copernicanism and Scripture, Bellarmine responded directly with a letter of his own. He began by endorsing the position articulated in the 'To the Reader' written by Osiander at the start of Copernicus's On the Revolutions: *that astronomical models were best understood as mathematical tools that could be used for calculation, not as a real, physical description of the cosmos. He then took issue with Foscarini's contrast between the spiritual and moral lessons of the Bible, which Foscarini had deemed matters of faith, and its physical statements, which he had not. Bellarmine argued that anything on which the Church Fathers agreed was a matter of faith; this applied to the earth's centrality and the sun's motion. Finally, Bellarmine conceded that if someone were to demonstrate unequivocally that the heliocentric model were true, then biblical passages would need to be reinterpreted accordingly. But, as he correctly noted, such a demonstration was not yet available; thus, he insisted, adherence to the traditional and accepted view was required.*

Rome, 12 April 1615

I have gladly read the Italian letter and the Latin text that you
sent to me; I thank you for both and confess that they are full
of intelligence and learning. But since you ask for my opinion,
I will confer it briefly, since you now have little time to read
and I to write.

First, I say that I believe that you and Mr Galileo are prudent
in contenting yourselves with speaking suppositionally and not
absolutely, as I have always believed Copernicus spoke. Because
it is very learned to say that if one supposes that the earth
moves and the sun stands still then one saves all the appearances better than by placing eccentrics and epicycles, and there
is no danger to anyone; and that is enough for the mathematician. But it is very dangerous to want to affirm that the sun
really is in the centre of the world and only turns on itself without moving from east to west, and that the earth is in the third
heaven and revolves with great speed around the sun; this will
not only irritate all scholastic philosophers and theologians,
but will also harm the Holy Faith by rendering the Holy Scripture false. For while you have well demonstrated many ways
of interpreting Holy Scripture, you have not applied them specifically; no doubt you would have found it very difficult if you
had wanted to interpret all the passages that you yourself
mentioned.

Second, I say that as you know, the Council [of Trent] prohibits interpreting Scripture against the common consensus of
the Holy Fathers; and if you want to read not only the Holy
Fathers but also the modern commentaries on Genesis, on
Psalms, on Ecclesiastes and on Joshua, you will find that all
agree to interpret literally that the sun is in heaven and goes
around the earth with very great speed, and that the earth is
very far from heaven and is in the centre of the world, immobile. Consider now, prudently, whether the Church can support
giving a meaning to Scripture that is contrary to the Holy
Fathers and all the Greek and Latin interpreters. Nor can it be
answered that it is not a matter of faith, since if it is not a matter of faith with respect to the topic, it is a matter of faith with

respect to the speaker. And so it would be as heretical to say that Abraham did not have two children and Jacob twelve as it would be to say that Christ was not born of a virgin, because the one and the other were said by the Holy Spirit through the mouths of the prophets and apostles.

Third, I say that if there were a true demonstration that the sun is at the centre of the world and the earth in the third heaven, and that the sun does not circle the earth but the earth circles the sun, we would have to proceed with great consideration in interpreting the Scriptures that seem contrary, and rather say that we do not understand them than say that something false is demonstrated. But I will not believe that there is such a demonstration, until it is shown to me. Nor is it the same to demonstrate that supposing that the sun stands at the centre and the earth in heaven, the appearances are saved, and to demonstrate that in truth the sun is at the centre and the earth in the heaven: because I believe that the first demonstration is possible, but I have great doubts about the second. And in the case of doubt one should not reject the Holy Scripture interpreted by the Holy Fathers.

I would add that the one who wrote 'The sun rises and the sun goes down, and hurries to the place where it rises' was Solomon, who not only spoke inspired by God, but was a light above all others, very wise and very learned in human science and in the knowledge of created things, all of which wisdom he got from God. Hence it is not likely that he affirmed something that was contrary to demonstrated truth, or to what was capable of demonstration. And if you say that Solomon speaks according to appearances, since it seems to us that the sun revolves, while actually the earth does, much as to someone who departs from the shore it seems that the shore departs from his ship, I will reply that for someone who departs from the shore, though it does seem to him that the shore departs from him, nevertheless he knows that this is an error and corrects it, seeing clearly that the ship is moving and not the shore. But as for the sun and the earth, there is no wise man who needs to correct the error, because he clearly experiences that the earth is still and that the eye is not deceived

when it judges that the moon or the stars are moving. This is
enough for now.

With this I greet you and pray that you gain all contentment
from God.

From home, 12 April 1615

As a Brother,
Cardinal Bellarmine

Galileo, Letter to the Grand Duchess Christina, 1615

Galileo expanded upon his letter to Benedetto Castelli with this letter written directly to the Grand Duchess Christina, which included many of the original points about scriptural interpretation as well as the section on reinterpreting the passage in the book of Joshua wherein the sun is told to stand still (not reproduced below). Galileo argued in the letter that the attacks against him were unfair, motivated by personal animus, and that the theological arguments of his opponents were invoked only to falsely prop up their shaky philosophical positions. He cited St Augustine to show that scriptural passages often needed to be interpreted in non-literal ways, and that one should be modest in one's claims about the complicated physical world. The letter circulated in manuscript form and was first printed, in both Italian and Latin, only after Galileo's condemnation in 1636.

To the Most Serene Madam, The Grand Duchess Mother, From Galileo Galilei:

I discovered a few years ago, as Your Serene Highness well knows, many particulars in the heavens that were unseen until this age. Their novelty, and some of the consequences that followed from them and contradicted the natural doctrines commonly held by the schools of philosophy, agitated no small number of professors against me, as if I had placed such things newly in the heavens by my own hand, to muddy nature and the sciences – with them forgetting that the proliferation of truth contributes to the investigation, growth and establishment of the disciplines, and not to their diminution and destruction. At that time, showing more affection for their own opinions than for the truth, they rushed to deny and overturn these novelties, about which their own senses, had they wanted to look carefully, would have convinced them. And thus they produced various items, and published some writings full of vain speeches, and – which was their most serious error – scattered these with sayings from the Holy Scripture, taken from places they did not well understand and which were distant from their purpose. They would perhaps not have made this error if they had attended to a very useful doctrine of St Augustine, about how we ought to take care in definitively determining the obscure and difficult meanings of things based on words alone. Speaking of certain natural conclusions about the celestial bodies, he writes thus: 'Always preserving the moderation of a pious and serious person, we should believe nothing rashly about an obscure matter, lest if later the truth happens to be revealed, although it may not be in any way opposed to the sacred books of the Old or New Testament, we may nevertheless despise it on account of our love of our own error.'

As it happened, time has revealed to everyone the truths that I first indicated, and along with the truth of the matter, another thing has become clear: the difference between those who had straightforwardly, and with no other malice, not accepted the discoveries, and those who had added some other motivating emotion to their disbelief. Hence, most of the experts of astronomical and natural science were persuaded by my first notice; and others, who had denied or doubted only because of the unexpected novelty, and who had not had the opportunity to experience the matter with their own senses, were calmed by degrees. But those who, besides attachment to their original error, have some imagined interest of their own in opposing not so much the things themselves as their author, but unable to deny them anymore, now cover themselves in continuous silence, and are diverted to other fantasies: bothered even more by the things that have soothed and calmed others, they try to harm me in other ways.

I would really give them no more regard than I do others who have contradicted me (about which I have always laughed, certain in the outcome of the business), if I did not see that their new slanders and persecutions do not revolve around some large or small matter of learning, but extend to me, as they try to attack me with offences that must be, and are, more abhorred by me than death. I cannot be satisfied that they are known to be unjust only by those who know me and them, but not to others.

Persisting in their original plan to destroy me and mine in every imaginable way, they are aware of my studies of astronomy and philosophy. They know that I maintain, with regard to the constitution of the parts of the world, that the sun, without changing its place, remains located at the centre of the revolution of the celestial orbs, and that the earth turns on itself and moves around it. And more, that I confirm this position not only by refuting the arguments of Ptolemy and Aristotle, but by producing many arguments to the contrary: in particular some relating to physical effects whose causes can perhaps not be ascribed in any other way, and other astronomical ones, depending on the evidence of the new celestial discoveries, that clearly refute the Ptolemaic system and admirably agree and confirm with the other position. Perhaps they are bewildered by the known truth of other propositions that I affirm, which differ from the common ones, and thus they are doubtful about defending themselves while they remain in the philosophical field. Therefore, they have resolved to shield the fallacies of their arguments with the mantle of simulated religion and with the authority of the Sacred Scriptures, and have applied them, with little intelligence, to the refutation of arguments they do not understand nor have even heard.

First, they tried to spread around the opinion that such propositions are contrary to the Sacred Scriptures, and consequently damnable and heretical. Afterwards, seeing how human nature is more inclined to embrace those enterprises in which a neighbour is oppressed, although unjustly, than those in which he is justly uplifted, it was not difficult for them to find people who preached that it was damned and heretical from the pulpits, which they did with insolent confidence, little pity and still less consideration for the aggravation caused not only to this doctrine and those who follow it, but to all mathematics and mathematicians. Therefore, having become more confident, and hoping in vain that the seed which first took root in their insincere minds would spread its branches and rise toward heaven, they went murmuring among the people that it would shortly be declared heretical by the supreme authority. They know that such a declaration would not only suppress these two conclusions, but would also make damnable all other astronomical and physical observations and statements that have any necessary connection with them.

To make the business easier, they try, as much as they can, to make this opinion (at least among the common people) appear as a new thing of mine, hiding the fact that Nicolaus Copernicus was its author, or rather, its restorer and confirmer, and that he was not only Catholic, but a priest, a canon, and so esteemed that he was called to Rome from the furthest parts of Germany for the reform of the ecclesiastical calendar in the Lateran Council under Leo X, and that he dedicated his book *On the Revolutions* to Paul III, which was printed and received by the Holy Church and read and studied across the world without anyone ever showing the slightest shadow of concern about this doctrine.

Now as for these false accusations that they so unjustly try to cast on me, I have deemed it necessary to justify myself to everyone whose judgement in matters of religion and reputation I must hold in high esteem. I will speak about the details that they produce in order to make this opinion detested and have it abolished as not only false but heretical, always shielding themselves with a simulated zeal for religion. They want to involve the Sacred Scriptures and have them minister to their insincere plans, against the intention of the text and of the Holy Fathers, if I am not mistaken. They want to extend (not to say abuse) the authority of these passages to purely physical matters, and not matters of faith, such that we must fully abandon our senses and our reason for some scriptural passage, though beneath the surface the words may contain a different meaning. I hope to show that I proceed with more piety and religious zeal than they do, when I propose not that this book should be condemned, but rather that it should not be condemned without understanding it, considering it, or even reading it.

I declare (and I hope that my sincerity will be clear) that I intend not only to submit myself freely to abandoning any errors that I incur in this writing through some ignorance in matters relating to religion, but also that I do not want to engage in quarrels with anyone on these matters, even on disputable points. My goal is nothing but this: that though there may be some errors in these considerations, so remote from my own profession, there may still be something in it useful to the Holy Church in making a determination about the Copernican system. If not, let my writing be torn and burned, since I do not intend or pretend to gain anything from it that is not pious and Catholic.

The reason that they produce for condemning the opinion of the mobility of the earth and the stability of the sun is that in many places in the Sacred Scriptures, one reads that the sun moves and that the earth is still. Since the Scriptures can never lie or err, it follows as a necessary consequence that anyone who asserts that the sun is immobile and the earth mobile holds a position that is erroneous and damnable.

With respect to this argument, it seems to me first that it is quite pious to say and prudent to establish that the Sacred Scripture cannot lie, whenever its true meaning has been understood. But I do not think it can be denied that that meaning is often hidden, and very different from what the simple meaning of the words suggests. It follows that if anyone, in expounding it, stopped at the bare grammatical sense, he could err, by making it appear that the Scriptures contain not only contradictions and propositions remote from the truth, but also grave heresies and blasphemies, since it would be necessary to give God feet and hands and eyes, and also bodily and human affections like anger, repentance and hatred, and even sometimes the forgetfulness of past things and the ignorance of the future. These propositions, dictated by the Holy Spirit, were communicated in this way by the sacred scribes to accommodate themselves to the capacities of the very crude and undisciplined common people. For those who deserve to be separated from the common crowd, it is necessary that wise interpreters produce the true meanings, and point out the particular reasons why they were communicated with such words. This doctrine is so commonplace and clear among all theologians that it would be superfluous to produce examples of it.

If, to remove this [Copernican] opinion and doctrine from the world, it was enough to close the mouth of one man only, then it would be very easy to do – as those men perhaps persuade themselves who measure the judgements of others by their own, and think it impossible that such an opinion should be able to subsist and find followers. But the situation is otherwise. To make that happen it would be necessary not only to ban the book of Copernicus and the writings of other authors who follow the same doctrine, but also to ban the whole science of astronomy. Still more, it would be necessary to prevent men from looking toward the heavens, so that they would not see that Mars and Venus are sometimes close to the earth and sometimes very remote, with such great variation that Venus is sometimes forty times and Mars sometimes sixty times greater than at other times; or so that they would not see that Venus is sometimes round and at other times forked, with very thin horns; or many other sensible observations, which cannot in any way be accommodated to the Ptolemaic system, but which are very solid arguments for the Copernican. And to ban Copernicus now, as many new observations and the applications of his reading by many other writers daily reveal his ideas to be more true, and after it has been allowed for so many years, when the doctrine was less followed and confirmed, would seem, in my opinion, to be a contravention of the truth, and an attempt to conceal and suppress it the more it proves evident and clear.

To ban the whole science would be to criticize a hundred passages in the Sacred Scriptures that teach us how the glory and greatness of the Supreme God are marvellously seen in all His works, and are read divinely in the open book of the heavens. Do not believe that reading the very lofty concepts in that book ends only in seeing the splendour of the sun and the stars, and their rising and setting (which is as far as the eyes of brutes and commoners can penetrate). Rather, within it there are such profound mysteries and sublime concepts that the vigils, efforts and studies of hundreds and hundreds of the brightest geniuses have not fully penetrated them, with ongoing investigations for thousands of years. That which the eyes of an idiot understand when looking at the external appearance of a human body is almost nothing compared with the admirable artifice of a skilled and diligent anatomist and philosopher, who investigates in that same body the use of so many muscles, tendons, nerves and bones; or examines the functions of the heart and other principal organs and seeks out the sources of the vital faculties, studies and observes the marvellous structures of the sense organs, and – without ever ceasing his amazement or admiration – contemplates the receptacles of imagination, memory and reason. Thus what pure sight represents is as nothing in proportion to the lofty wonders that the ingenuity of intelligent men see in the heavens, through long and accurate observations.

Galileo and the Inquisition, 1615–1616*

Foscarini's letter led the Catholic Church to take a closer look at the heliocentric theory, especially given its focus on scriptural reconciliation, newly problematic (as Cardinal Bellarmine's letter suggested) after the Council of Trent. At approximately the same time, the Dominican Niccolò Lorini filed a written complaint against Galileo with the Inquisition and included his letter to Castelli as evidence. The consultants who reviewed the letter for the Inquisition concluded that though it contained some problematic language, it was free of heresy. But given that the real question was Copernicanism itself, the Inquisition convened a committee of eleven expert consultants to review the Copernican doctrines. In February of 1616, that committee unanimously declared that the doctrine of the sun's centrality was 'foolish and absurd in philosophy and formally heretical', and the doctrine of the earth's motion 'at least erroneous in faith'. Galileo was subsequently warned by Bellarmine to avoid promulgating the theory. The next month, On the Revolutions *and Diego de Zúñiga's* Commentary on the Book of Job *were placed on the Index of Forbidden Books until certain problematic passages were removed, and Foscarini's letter banned entirely.*

* Translations taken from *The Galileo Affair: A Documentary History*, translated by Maurice A. Finocchiaro (University of California Press, 1989).

Lorini's Complaint, 7 February 1615

Most Illustrious and Most Reverend Lord:

Besides the common duty of every good Christian, there is a limitless obligation that binds all Dominican friars, since they were designated by the Holy Father the black and white hounds of the Holy Office. This applies in particular to all theologians and preachers, and hence to me, lowest of all and most devoted to Your Most Illustrious Lordship. I have come across a letter that is passing through everybody's hands here, originating among those known as 'Galileists', who, following the views of Copernicus, affirm that the earth moves and the heavens stand still. In the judgement of all our Fathers at this very religious convent of St Mark, it contains many propositions which to us seem either suspect or rash: for example, that certain ways of speaking in the Holy Scripture are inappropriate; that in disputes about natural effects the same Scripture holds the last place; that its expositors are often wrong in their interpretations; that the same Scripture must not meddle with anything else but articles concerning faith; and that, in questions about natural phenomena, philosophical or astronomical argument has more force than the sacred and the divine one. Your Most Illustrious Lordship can see these propositions underlined by me in the above-mentioned letter, of which I send you a faithful copy. Finally, it claims that when Joshua ordered the sun to stop one must understand that the order was given to the Primum Mobile and not to the sun itself. Besides this letter passing through everybody's hands, without being stopped by any of the authorities, it seems to me that some want to expound Holy Scripture in their own way and against the common exposition of the Holy Fathers and to defend an opinion apparently wholly contrary to Holy Scripture. Moreover, I hear that they speak disrespectfully of the ancient Holy Fathers and St Thomas; that they trample underfoot all of Aristotle's philosophy, which is so useful to scholastic theology; and that to appear clever they utter and spread a thousand impertinences around our whole city, kept so Catholic by its own good nature and by the vigilance of our Most Serene Princes.

For these reasons I resolved, as I said, to send it to Your Most Illustrious Lordship, who is filled with the most holy zeal and who, for the position that you occupy, is responsible, together with your most illustrious colleagues, for keeping your eyes open in such matters; thus if it seems to you that there is any need for correction, you may find those remedies that you judge necessary, in order that a small error at the beginning does not become great at the end. Though perhaps I could have sent you a copy of some notes on the said letter made at this convent, nevertheless, out of modesty I refrained since I was writing to you who know so much and to Rome where, as St Bernard said, the Holy Faith has lynx eyes. I declare that I regard all those who are called Galileists as men of goodwill and good Christians, but a little conceited and fixed in their opinions; similarly, I state that in taking this action I am moved by nothing but zeal. I also beg Your Most Illustrious Lordship that this letter of mine (I am not referring to the other letter mentioned above) be kept secret by you, as I am sure you will, and that it be regarded not as a judicial deposition but only as a friendly notice between you and me, as between a servant and a special patron. And I also inform you that the occasion of my writing was one or two public sermons given in our church of Santa Maria Novella by Father Tommaso Caccini, commenting on the book of Joshua and Chapter 10 of the said book.

Consultants' Report on the Letter to Castelli, 1615

In the composition shown me today, except for these three following items, I found nothing else to question. On the first page, where it says 'that in the Holy Scripture one finds many propositions which are false if one goes by the literal meaning of the words', etc., granted that this sentence can be taken in a benign sense, nevertheless at first impression it sounds bad. Certainly it is not right to use the word falsehood, in whatever manner it be attributed to Holy Scripture; for it is the infallible truth in every way. As for the other item, on the second page where it says that 'Holy Scripture has not abstained from perverting its most basic dogmas', etc., since the verbs 'to abstain' and 'to pervert' are always employed with negative connotations (we abstain from evil, and one perverts when something just is made unjust), they sound bad when they are attributed to Holy Scripture. Bad-sounding, too, are those words on the fourth page that read, 'Let us then assume and concede', etc., for in this statement one sees only a desire to concede the story of the sun stopped by Joshua as truly in the text of Holy Scripture, although from what follows those words can be understood in the right sense. For the rest, though it sometimes uses improper words, it does not diverge from the pathways of Catholic expression.

Consultants' Report on Copernicanism, 24 February 1616

Propositions to be assessed:

(1) The sun is the centre of the world and completely devoid of local motion.

Assessment: All said that this proposition is foolish and absurd in philosophy, and formally heretical since it explicitly contradicts in many places the sense of Holy Scripture, according to the literal meaning of the words and according to the common interpretation and understanding of the Holy Fathers and the doctors of theology.

(2) The earth is not the centre of the world, nor motionless, but it moves as a whole and also with diurnal motion.

Assessment: All said that this proposition receives the same judgement in philosophy and that in regard to theological truth it is at least erroneous in faith.

Inquisition Minutes, Thursday, 25 February 1616

The Most Illustrious Lord Cardinal Millini notified the Reverend Fathers Lord Assessor and Lord Commissary of the Holy Office that, after the reporting of the judgement by the Father Theologians against the propositions of the mathematician Galileo (to the effect that the sun stands still at the centre of the world and the earth moves even with the diurnal motion), His Holiness ordered the Most Illustrious Lord Cardinal Bellarmine to call Galileo before himself and warn him to abandon these opinions; and if he should refuse to obey, the Father Commissary, in the presence of a notary and witnesses, is to issue him an injunction to abstain completely from teaching or defending this doctrine and opinion or from discussing it; and further, if he should not acquiesce, he is to be imprisoned.

Special Injunction, 26 February 1616

At the palace of the usual residence of the said Most Illustrious Lord Cardinal Bellarmine and in the chambers of His Most Illustrious Lordship, and fully in the presence of the Reverend Father Michelangelo Segizzi of Lodi, O.P. and Commissary General of the Holy Office, having summoned the above-mentioned Galileo before himself, the same Most Illustrious Lord Cardinal warned Galileo that the above-mentioned opinion was erroneous and that he should abandon it; and thereafter, indeed immediately, before me and witnesses, the Most Illustrious Lord Cardinal himself being also present still, the aforesaid Father Commissary, in the name of His Holiness the Pope and the whole Congregation of the Holy Office, ordered and enjoined the said Galileo, who was himself still present, to abandon completely the above-mentioned opinion that the sun stands still at the centre of the world and the earth moves, and henceforth not to hold, teach, or defend it in any way whatever, either orally or in writing; otherwise the Holy Office would start proceedings against him. The same Galileo acquiesced in this injunction and promised to obey.

Decree of the Index, 5 March 1616

This Holy Congregation has learned about the spreading and acceptance by many of the false Pythagorean doctrine, altogether contrary to the Holy Scripture, that the earth moves and the sun is motionless, which is also taught by Nicolaus Copernicus's *On the Revolutions of the Heavenly Spheres* and by Diego de Zúñiga's *On Job*. This may be seen from a certain letter published by a certain Carmelite father, whose title is *Letter of the Reverend Father Paolo Antonio Foscarini, on the Pythagorean and Copernican Opinion of the Earth's Motion and Sun's Rest and on the New Pythagorean World System* (Naples: Lazzaro Scoriggio, 1615), in which the said father tries to show that the above-mentioned doctrine of the sun's rest at the centre of the world and the earth's motion is consonant with the truth and does not contradict Holy Scripture. Therefore, in order that this opinion may not creep any further to the prejudice of Catholic truth, the Congregation has decided that the books by Nicolaus Copernicus (*On the Revolutions of Spheres*) and Diego de Zúñiga (*On Job*) be suspended until corrected; but that the book of the Carmelite father Paolo Antonio Foscarini be completely prohibited and condemned; and that all other books which teach the same be likewise prohibited, according to whether with the present decree it prohibits, condemns and suspends them respectively.

Cardinal Bellarmine's Certificate, 26 May 1616

We, Robert Cardinal Bellarmine, have heard that Mr Galileo Galilei is being slandered or alleged to have abjured in our hands and also to have been given salutary penances for this. Having been sought about the truth of the matter, we say that the above-mentioned Galileo has not abjured in our hands, or in the hands of others here in Rome, or anywhere else that we know, any opinion or doctrine of his; nor has he received any penances, salutary or otherwise. On the contrary, he has only been notified of the declaration made by the Holy Father and published by the Sacred Congregation of the Index, whose content is that the doctrine attributed to Copernicus (that the earth moves around the sun and the sun stands at the centre of the world without moving from east to west) is contrary to Holy Scripture and therefore cannot be defended or held.

Kepler, *Epitome of Copernican Astronomy*, Book One, 1618

Kepler's Epitome of Copernican Astronomy *was published in three separate volumes between 1618 and 1621. Written in question-and-answer format, the first volume was intended as a textbook for beginners or for those wishing for a large-scale survey of astronomy. Yet it differed from past epitomes or textbooks in two crucial ways. Firstly, it was explicitly Copernican; secondly, it insisted unequivocally that astronomy was physical, and not just mathematical, and that it could not be understood without examining the physical causes of astronomical motions.*

Preface

Immediately after I published my *Commentary on the Motions of the Planet Mars*, at the urging of my astronomer friends I began to prepare an edition of the new astronomy, restored under Emperor Rudolph, adapted for children at the school-bench. For since this science is productively learned only if each child who wants to secure his fruit as an adult has nurtured the seeds as a boy, I wanted to help them with a model that was easily understood, low in price and appropriately rich.

Reiterating the doctrine of the sphere should not be considered a useless or idle thing, as though there were no reasons why this doctrine should be summarized yet again after the ancient conceptions of Euclid, Aratus, Cleomedes, Geminus, Proclus and Theon, and after the more recent ones, of Sacrobosco especially, and the endless commentaries on him (among them the highly learned and copious one of Christoph Clavius). For first, even if nothing new were added to the doctrines of the ancients, it would still be expedient for the same thing to be related by various authors. This is because there are different aptitudes for teaching, and the same teacher is not appropriate for every student, nor the same pen for all those learning. For this reason, I mixed both familiar and necessary definitions and also certain higher speculations, as seemed suited to the method. For both, I presented things in a question-and-response form, so as not to neglect beginners, since it is easier for them to understand, and also so adults and those of mature judgement are able to blend the tedium of common and well-known definitions with some recreation. Also, this form of speaking makes it easier to examine more difficult speculations. Therefore, I think that experts will judge this method that I follow to be more convenient on all fronts, given that I have only omitted some unnecessary accessory material, and in turn have added things which others had not yet related. In particular, I have related the methods of astronomical calculation of the very recent Tycho Brahe through several examples, with a careful and faithful imitation of the model using the relevant numbers, so that almost nothing is lacking when it comes to practise.

But the type of philosophy that I follow supplied me with more, and more urgent, reasons for refining the doctrine of the sphere. For when you consider the unique diurnal motion of the earth, a great many other motions are removed, motions of unspeakably incredible speed and contrary to the genuine and particular motions of the planets in their vicinity. And when you consider the unique annual motion of the earth itself, it casts out all the epicycles of the ancients, blindly tethered to the motion of the sun, and still further all of their eccentrics upon eccentrics, all the inclinations, deviations and reflected circles. Thus I will speak further on the doctrines of this theory. And though many things seem absurd initially and difficult to believe when it comes to this axiom of the earth's motion, it happens that if you embrace that subject, and first admit the motion of the earth, it is much easier to fashion a universal astronomy.

This hypothesis is so great, therefore, that it is entirely worthy of serving as a model for all of today's astronomers, and astronomy is made easier to understand in all its parts. Let each man consider it to the extent that he wants. Certainly I understand that I owe it this duty: that those things which I inwardly accept in my mind as true, and whose beauty I admire and enjoy with incredible pleasure, I will defend outwardly to my readers with all the force of my talent.

If there is anyone who wants to accuse me of loving novelty, he should know that that itself is not a crime in philosophy, since all philosophy is novelty compared to ancient ignorance. What matters is this alone: whether you innovate for glory or for truth. But if I was seeking glory, I probably would not have neglected the chance to devise something on my own. However, in this philosophy, and in most others to which I adhere, I follow the work of others. Not slavishly, but rather with a judgement fashioned from many things, so that it should be clear to anyone that I uphold and am on the side of truth.

It is certainly my pleasure to accord with the masses, as long as they do not err; and because of that I make it my job in this matter to persuade as many people as possible. By this plan, I will enjoy agreeing with the majority with the greatest pleasure. Meanwhile, because a good leader is in the habit of desiring peace before all other things, but – if it cannot be achieved – desiring victory, I am moved by the same idea, beset by the passing troubles of a dissenting crowd. For though the fog of common beliefs may be long-lasting, nevertheless in the end the clear brightness of truth emerges. And since I fight, as often as I can, on the side of truth, ultimately victory is mine. But this victory would not exist had it not been ceaselessly fought for with great labour and danger.

In the end, I understand that I am ordained as a priest of God the Creator with respect to the Book of Nature: therefore, I composed this sacred hymn to God the Creator (Galen also referred to Him by this title in his book *On the Use of Parts*): a new type of song, but one adjusted to the ancient, or rather youthful, Samian philosophy.

What is astronomy?

It is the science that relates the causes of the things in the heavens and in the stars, and the changes they display over time, as they appear for those of us who dwell on and rotate with the earth. With these things understood, we can predict the future appearance of the heavens – that is, the apparent celestial motions – and assign reliable times to those of the past.

Why is it called astronomy?

From the law or rule of the stars [*astra/astrorum*] – that is, of the motions by which the stars are moved, just as Economy is based on the rules governing domestic affairs, and Paedonomus is based on the rules governing children.

What is the relationship between this science and the other sciences?

1) It is a part of physics, because it investigates the causes of natural things and events: and because the motions of celestial bodies are among its subjects: and because its singular goal is to discover the arrangement of the structure of the world and its parts.
2) Astronomy is the soul of geography and hydrography, that is, of matters nautical. For the diverse things that happen in the heavens, [seen] from different places and regions on land and at sea, are adjudicated only by astronomy.
3) Chronology is subordinate to it, because celestial motions prescribe the seasons and political years, and date histories.
4) Meteorology is subordinate to it. For the stars move and influence sublunary nature, and people themselves in a certain way.
5) It embraces a large part of Optics, because it has something in common with that subject – the light of heavenly bodies – and because it reveals many deceits of vision surrounding the form of the world and of its movements.
6) However, it is subordinate to the class of mathematical disciplines, and it uses geometry and arithmetic as its two wings: considering quantities and figures of heavenly bodies

and motions, and enumerating times, and thus developing its demonstrations: and drawing speculation fully toward use, or praxis.

Of how many parts, then, is the task and office of the astronomer?

The astronomical office principally has five parts: historical, concerning observations; optical, concerning hypotheses; physical, concerning the causes of hypotheses; arithmetical, concerning tables and calculation; and mechanical, concerning instruments.

How do they differ from each other?

None of them are able to do without geometrical demonstrations, which build theory, or numbers, which build praxis, since there are some numbers that are practically the language of the geometricians. Nevertheless, the first three parts pertain more to theory, while last two more to practice.

Describe for me the first of these, the historical part.

The historical part initially records how the world appears to us; what changes in its appearance daily, annually, or over longer periods of time; what things appear different in different places on land and at sea, and what appear the same. From historical records it culls especially rare or very notable events, like solar or lunar eclipses and significant conjunctions. From the books of authors worthy of trust, like Hipparchus, Ptolemy, Albategnius, Azracheles and others cited by these, it collects subtler observations of singular stars and brings them together, adding also those things which the present time has observed. In this task, Tycho Brahe exceeds all others in incredible diligence, having left us over thirty-eight years most copious and trustworthy observations, made personally and almost continuously.

Therefore, observations of this kind must be compared with each other skilfully, and arranged into fixed classes, through fixed periods of time, so that similar things are joined with similar things: in about the same way that Aristotle proceeded to explain the nature of animals: first he fashioned a most skilful history of animals, enumerating in summary for all the species arranged in the same genus that which was common to them.

Describe for me also the second part of the astronomical task.

Having closely considered these varieties of observations, in which there is a consonance of diverse things, the second part, the optical, seeks to penetrate to the causes of why it happens that sights appear to human eyes that are extremely different from the truth – sights that astronomers call appearances, or *phenomena* in Greek. So that whenever someone has powerful skill, he can save and produce most of the diversity of appearances through some single form of movement or figure of bodies, always similar to itself, accommodating his entire method to either geometrical or optical laws. Such a method is subordinated to geometry, and it therefore happens that by devising forms of these kinds of motions, that individual comes closer than another to the nature of things itself.

In this way, although on this difficult and blind hunt for causes he reaches the plans of nature only to wander from the truth in some rooms of his opinions, nevertheless others, less skilled than him, may save the celestial appearances through his work. The prevailing custom is to call 'hypotheses' the opinion of any of the more famous authors by which he explains the causes of the celestial appearances. This is because the astronomer is accustomed to say 'with this or that set up or supposed (*hypotithentos*)' – which he himself asserts about the world – it necessarily follows by geometrical demonstrations that those things which are in the preceding historical record have appeared in just that many ways and at that time.

Thus today people speak of three forms of hypotheses: that of Ptolemy, of Copernicus, and of Tycho Brahe. Moreover, the contemplation of nature and the properties of light, or the practice of the doctrine of refractions, also pertains in general to these two first parts.

What, then, is the third part of the astronomical task?

The third part, pertaining to the physical, is not generally deemed necessary for the astronomer, although, to the contrary, it is extremely relevant to the goal of this part of philosophy, nor can it be secured except by an astronomer. For there ought not be unchecked licence on the part of astronomers to fashion whatever they want without reason; in fact, it is appropriate that you are able to render probable causes for your hypotheses, which you offer up as true causes of the appearances, and thus establish the principles of your astronomy first in a higher science, pure physics, or metaphysics, without nevertheless being shut off from those geometrical, physical, or metaphysical arguments that are supplied to you by the accounts proper to the discipline itself, which pertain to those higher disciplines, provided that you do not beg the question.

Through this bargain, it happens that the astronomer – who has gained mastery over his practice to the extent that he devises causes for the motions in consonance with reason and suited to producing all those things present in the history of observations – now brings together in one view that which he had decided bit by bit previously. And although the goal he stated until now was a pretense (namely, the demonstration of phenomena and utility for daily life flowing from it), he aims himself at a higher goal. With the great rejoicing of philosophers, he directs all his pleas, by geometrical and physical arguments, toward that goal: to set before human eyes the genuine form and arrangement, or adornment, of the whole world. And indeed, this is the Book of Nature itself, in which God the Creator, in a certain kind of writing, partly made manifest and depicted His essence and His will toward man.

What does the astronomer do in the fourth place?

The fourth and fifth parts are recalled to that lower goal, namely to utility in daily life. For the fourth, which I have called arithmetic, connects the discovered causes of motion to numbers, teaching the method by which the visible appearance of the heavens and the configuration of the stars is calculated at whatever time you please, whether past or future. From this, astronomical tables are born, by which the method already mentioned is made easier and shorter. This is why the Greeks called them *Kanonas Procheirous*, whose model are the *Ptolemaic*

Tables, which were corrected four hundred years ago by the *Alphonsine Tables*, and eighty years ago by the *Copernican Tables*, and which Reinhold made more exact and copious and called the *Prutenic Tables*. However, since there are errors in all the preceding tales, the *Rudolphine Tables* promise to set the crowning piece, pushed forward and influenced by Tycho Brahe and now awaiting the light. Therefore, this part supplies chronologers, astrologers, meteorologists, doctors, sailors and farmers with the principles necessary for their own art.

Explain also the fifth part of the astronomical task.

The fifth part, the mechanical, generally supplies instruments, as the fourth does numbers, and has many aspects. For on the one hand it attends to history, because in order for observations, which are the foundation of astronomy, to be excellent and sufficiently exact, instruments are used to aid the eyes, both to direct them more reliably and to calculate on a smaller scale and without deception. In this type of endeavour, the book of Tycho Brahe stands out, called *Instruments of a Restored Astronomy*, as it furnishes a large abundance of excellent instruments. These instruments commend the highest faith in the observations acquired through them and recorded by Brahe.

As for that which supports the subsequent parts: either models made of wood, metal, paper and the like, with which we portray the astronomical hypotheses and place them before our eyes. These models teach the non-expert or beginner and alleviate the labour of computation, which are of help to the matter of the sphere and to theories. Or we prepare sources of pleasure for important or wealthy people, like celestial automata, which imitate to a certain extent the heavens itself through an artificial motion placed within them. Beyond pure delight, utility sometimes also follows from these, especially when the sky is cloudy. Or in turn we build observational instruments, which are at first designated for private use but later come to serve for the popular and common use. Thus are born an infinite variety of instruments, with mechanics competing in manual skill with the most ingenious demonstrations of the geometers. Yet they nevertheless agree in this: that although the motion and celestial appearances are spherical and curvilinear, to us, however, inhabitants of the earth, things appear spread out flat beneath the horizon, with walls erected perpendicular to it. Thus the bodies that we work with by hand are demarcated by planes alone, or at least by a mixture of the straight and the curved. Therefore, the instruments that are easier for us to use are those in which the curved is transformed into the straight. Of this sort are those whose rules are carved out from the primary divisions of the circle, called geometric squares, or astrolabes, and, because of the flat walls of the structures, sundials, instruments whose very broad use on land and at sea is clear, and which human life could scarcely do without any longer. This utility mainly serves geography and nautical matters, because geographic charts have celestial circles projected on a plane.

To what part chiefly does this book refer?

This Epitome touches on all five of these parts. For in its principal method it runs through the variety of celestial appearances, and in order that their causes can be rendered, it explains the hypotheses, especially the Copernican but also that of Tycho Brahe, through figures and suitable instruments, as relevant to the Sphere and Theories; it weaves in physical and metaphysical debates about their truth; it places the inner Idea of the entire universe before the eyes, and draws and defines its boundaries into limits – circles, arcs, lines and angles – which must be used in astronomy; it relates the doctrine of eclipses of the sun and moon, and of the configurations of the planets between them and with the luminaries (in which function it stops at the inferior planets); it teaches the form of calculation for both the *Rudolphine Tables* and the positions of the planets themselves without tables; and finally the laws for constructing instruments, either that have been calculated or that can be calculated.

Is the region of the fixed stars infinite?

Astronomy reports nothing here: for at such an altitude the sense of our eyes is abandoned. Astronomy teaches this alone: how far the stars are perceived, even the smallest, assuming space to be finite.

If the world is finite, then what figure determines its outer boundary?

It is nothing other than a sphere.

What support do you have for this argument?

There are hardly any astronomical ones, but there are two principal metaphysical ones: the first follows from the world itself, and the second from its archetype.

State the first.

We debate about the figure that encloses the outer boundary of the world. All things, therefore, are within that figure, and nothing outside. If it actually encompasses all things, it is extremely likely that it is the most capacious form. And a round figure is most capacious.

State the second.

The Archetype of the world is God Himself, to whom no figure is more similar (if we may speak of similarity) than a spherical surface. For as God is the Being of Beings, preceding all things, unbegotten, simplest, most perfect, immovable, most sufficient to all creatures, creating and sustaining all things, one in essence, three in persons, thus the sphere alone has these properties among other figures.

Show how the image of the venerable Trinity is within the sphere.

There are three parts of the sphere: the centre, the surface, and the intermediate space, so that with one denied, the rest collapse, and they are distinct among themselves so that one is not the other.

What makes it seem to us that the stars rise up every day from one part of the horizon and set in the opposite part after a certain number of hours – the motion of the heavens or the earth?

Copernican astronomy teaches that our vision is deceived with respect to this first motion: for the stars do not really rise above the mountains and climb toward our zenith, but rather the mountains, which surround us, and stand on the surface of the globe of the earth, are indeed part of it, and are turned with the whole globe around its axis from the west to the east, and by this rotation the immobile stars of the east are revealed to us, some after others, and the western stars hidden; thus it is not the stars which cross overhead, but the vertical point that moves through the fixed stars, which is the first motion.

You say that all the apparent motions are able to be sufficiently explained through this marvellous position of the first motion, and the whole doctrine of the sphere related?

Completely and exactly, and that is precisely the goal of this singular little book, namely, to prove the very matter that has been promised in words.

When, at the start of this little book, you discussed hypotheses, you directed this law to the astronomer: that he not do anything that he desires, but rather that he confirm that his positions accord with the plans of nature. Therefore, I ask you: do you really hope to be able to prove this absurd position, and by what arguments?

That the first motion is produced by the continual rotation of the earth around its axis, with the heavenly bodies at rest (as much as pertains to this first motion) can be proven by seven types of principal arguments. These are: 1) from the subject of the motion, 2) from the speed of the motion, 3) from the uniformity of the motion, 4) from the causes of the motion, or the moving faculty, 5) from the instruments of the motion, that is from the axis and the poles, 6) from the goal of the first motion, and 7) from the signs it leaves, or effects.

Prove it from the subject of the motion.

Nature does not work in difficult, roundabout ways, but always choses simpler ones when it can. But through the rotation of the earth, a very tiny little body, around its axis toward the east, nature can clearly effect the same thing as is accomplished by the rotation of the very great world around its axis (stretched from one end of the world to the other) toward the west. Just as it is more plausible that a man turns his head in the auditorium, rather than that the auditorium turns around the head of the immobile man, so it is more plausible too that the earth rotates from west to east, than that the rest of the machine of the world rotates from east to west, since the same thing follows from either one.

If the first motion is in the heavenly spheres, then there are two motions to which they are subject, one common to all the spheres and the other particular to each of the spheres. But it is much more likely that the first motion and the second motions are distinct with respect to their subjects, such that the second ones, which are many, reside in each sphere, but the first, which is single, resides in the single body of the earth, and belongs to it alone.

This argument rests only on likelihood. Demonstrate necessity.

Our eyes bear witness without error to the fact that some motion occurs. If the earth is at rest, then the whole remaining machine of the world rotates: there is no third extra option debated. But the whole machine of the world cannot be moved with a diurnal motion, with only the earth at rest. Thus it is necessary for the earth to be moved with a diurnal motion.

Why can't the whole machine of the world be moved?

The world is either infinite or finite. Assume it is the first, according to the opinion of William Gilbert, who thinks that the omnipotence of God is illustrated if the world stretches out infinitely, so that the infinite power of the creator would be recognized from the infinite quantity of created things. Although this has been refuted by metaphysical arguments, nothing can be brought to refute it from astronomy, in which men trust more in the evidence of the senses than in forms of reasoning that are very far from the senses. But if we posit that the world is infinitely extended, Aristotle has demonstrated that the whole world cannot be moved in a rotation, insofar as it is whole.

But assume that the world is finite: in this case it follows that there is nothing beyond the world, which furnishes a place for the world but which itself is at rest. But where there is nothing that is at rest, there is no motion. For 1) motion is the separation of something moving, insofar as it is moving, from its place and its transfer to some other place. 2) The appearance of motion of a machine around an axis and resting poles cannot be understood if there is nothing in respect to which the poles are understood to be at rest. For in rotating the sphere, its pole remains immobile at the meridian. But there is nothing beyond the machine of the world with respect to which the meridian can be understood to be at rest. And what has no place in our understanding cannot come to be, even in these matters pertaining to geometry. 3) Maestlin asks this – and he is not wrong: how could it be that, with the whole system of the world shaken about, with none of its spheres, not the sphere of fire (whatever it may be), not the higher region of air excepted, only this little globe alone, whose diameter is twenty million parts less than the diameter of the world, is not dragged around?

How do you think one should respond to the authorities of all the ages, and of all the Orders, sacred and secular, who accept the contrary, without controversy – that the earth rests in the first motion, and the heavens are moved?

Copernicus responds thus: 1) Since the public is in charge of speaking, it expresses the sense of its eyes in speech. A philosopher expresses the truth, and inquires about that which is beneath the appearances of things: thus it is not absurd that the thoughts of a philosopher are remote from the judgement of the public. 2) Some scriptural passages are badly distorted into astronomical propositions: and those who are in the habit of this should have their reckless judgements scorned however possible; for it will appear plain that the Holy Fathers of the Church sometimes spoke childishly about astronomical matters, which they had not learned by profession, and searched for a defence of their own errors in the Scriptures. Thus Lactantius, who believed that the earth could not be round, distorted the Book of Job beyond the plan of God's speech toward support of philosophical speculation. This response can be extended in many ways.

For astronomy discloses the causes of the nature of things, and investigates by profession the deceptions of vision: the holy books, relating more sublime things, use the speech of men, so that they can be understood. 3) With respect to the authority of philosophers, Copernicus shows that among the principles of the new astronomy there is no lack of those who stated that the earth is moved from west to east. But this abundance is unnecessary: for even if there were no witness in earlier times to this truth, this philosophy would still need to be embraced just as much. For although in Christian theology it is preposterous for someone to first search for a judgement based on reason and only afterwards to finally consider the authorities, in philosophy it is no less foolish to first weigh the authorities and only afterwards to pass over to reasoning.

Kepler, *The Harmony of the World*, 1619

Kepler's 1619 Harmony of the World *was the book he considered his masterpiece, as it linked music, mathematics, astrology and astronomy to the harmonic archetypes that he now believed underpinned all reality. God was still a geometer, as Kepler had first argued in the* Cosmographic Mystery, *but he had relied on harmonic ratios to construct the cosmos as well, and those ratios were articulated not in the distances between planets but rather in their varying speeds; thus, both the maximum and minimum speeds of each planet and the maximum and minimum speeds of adjacent planets produced harmonic ratios. Kepler argued that each planet produced its own song as it moved from perihelion to aphelion, and that together the planets produced polyphonic harmonies. He also used this idea to develop his third law, otherwise known as the harmonic law, which established that the ratio between a planet's orbital period and the size of the semi-major axis of its orbit was sesquialterate, or* 3:2. *Also noteworthy here is Kepler's description of the equivalence of the Tychonic and Copernican models (via the metaphor of a writing tablet on a turning table), as well as his suggestion that his theory of celestial harmony might provide a model for earthly peace.*

Book V

Introduction

That which I foretold twenty-two years ago when I first discovered the five solid figures between the celestial spheres; that
which I firmly persuaded myself, before I had seen the *Harmony*
of Ptolemy; that which I promised to friends when I named this
book, before I was certain of the matter itself; that which I
stressed must be pursued sixteen years ago in my public writing;
that for the sake of which I devoted the best part of my life to
astronomical contemplation, visited Tycho Brahe, chose Prague
as my home; finally – through God the Best and Greatest, who
had inspired my mind and awoken a great desire while prolong-
ing my life and the strength of my talents, and with the remainder
supplied by two Emperors and the generosity of the Princes of
this province of Upper Austria, with my previous astronomical
duties sufficiently completed – finally, I say, I have brought it
forward into the light, and have discovered so great a truth that
it is beyond what I could ever have hoped: that the whole nature
of harmony, however great it is, with all of its parts (explained
in Book III), is discovered among the celestial motions. And not
even in the way that I had imagined (and this aspect delights me
no less), but in a completely separate way, though at the same
time also one that is most excellent and perfect.

It happened in this intermediate time, in which the very
laborious restoration of the motions kept me in a state of anxious uncertainty, that my interests and plans were especially
stimulated by the reading of Ptolemy's *Harmony*, where I discovered, unexpectedly and with the greatest astonishment, that
nearly the whole of his third book was devoted to the same
contemplation of celestial harmony, 1,500 years before.

The lack of sophistication of ancient astronomy, combined
with the exact and precise agreement of this meditation with it,
separated by fifteen centuries, strongly encouraged me to press
on toward my objective. For what need is there to say more?
The nature of things itself came forth and appeared to men,

through interpreters separated by the distance of centuries. It was the finger of God, as the Hebrews say, that this same conception about the constitution of the world was in the minds of two men who had given themselves over completely to the contemplation of nature, although neither of them had led the other to advance on this path.

Now, with the first light eighteen months ago, the day of truth three months ago, and only a very few days ago when the pure sun of most admirable contemplation began to dawn, nothing holds me back. It is a pleasure to give way to the sacred frenzy, and it is a pleasure to laugh at mortals with the frank confession that I am stealing the golden vessels of the Egyptians in order to build from them a tabernacle for my God very far from the boundaries of Egypt. If you forgive me, I will rejoice; if you are angry with me, I will endure it. Behold, I throw the die, and I write the book. Whether it must be read by those of the present or those of the future matters not at all. May it await its reader for a hundred years, if God himself stood ready for six thousand years for someone to contemplate him.

Chapter 3: Summary of Astronomical Doctrine, Necessary for the Contemplation of the Heavenly Harmonies

At the start, readers should understand this: that the ancient astronomical hypotheses of Ptolemy, as they were explained in the *Theoricae* of Peurbach and by other writers of Epitomes, should be completely removed from consideration here, and cast out of mind: for they do not relate the truth about both the arrangement of the heavenly bodies and the state of their motions.

In place of them, though, I can do no other than substitute especially the opinion of Copernicus about the world, and if it were possible, convince everybody. But since the matter is still new to interested common readers, and it is a doctrine most absurd for many to hear – that the earth is one of the planets and is carried among the stars around the immobile sun – know, therefore, those who are offended by the strangeness of this opinion, that these harmonic speculations also have a place in the hypotheses of Tycho Brahe. For that author has in common with Copernicus all other things which pertain to the arrangement of bodies and the proper combination of their motions. It is only the annual Copernican motion of the earth that he transfers to the whole system of planetary spheres and to the sun, which occupies the middle of it by the agreement of both authors. For through this translation of motion, it nevertheless happens that the earth, if not in that very vast and immense space of the fixed stars, then at least in the system of the planetary world, occupies the same place, at any time whatsoever, according to Brahe as it does according to Copernicus. Indeed, just as when one person draws a circle on a piece of paper by moving the writing foot of the compass around, and someone else fastens his paper or writing tablet to a turning lathe while keeping his compass leg or pen immobile above the turning tablet, still both mark out the same circle; so too here. For Copernicus the earth marks out a circle by a real motion of its body, advancing between the circles of Mars on the outside and Venus on the inside; but for Tycho Brahe, the whole system of the planets (in which among the others are also the circles of Mars and Venus) turns, just as the tablet on the lathe, applying to the immobile earth as if to the pen of the lathe-turner the gap between the circles of Mars and Venus. And as a result of this motion of the system, the earth marks the same circle around the sun between Mars and Venus, itself staying immobile, as it marks according to Copernicus by a true motion of its body, with the system at rest.

Since, therefore, harmonic contemplation considers the motions of the planets to be eccentric, as if seen from the sun, it is easily apparent that if an observer were on the sun, even if it were mobile, it would nevertheless seem to him that the earth, though at rest (let us concede this for now to Brahe), were travelling an annual circle between the planets and also in an intermediate amount of time. Therefore, should there be someone so weak in confidence that he cannot accept the motion of the earth among the stars, he will nevertheless be able to rejoice in the most excellent contemplation of this divine machine, if he applies whatever he hears concerning the diurnal motion of the earth on its eccentric to their appearance on the sun, which is the very appearance that Tycho Brahe demonstrates with the earth at rest. Nor will true supporters of the Samian philosophy have good cause to begrudge such people this fellowship in a most delightful speculation, as their joy will be far more perfect – certainly with respect to the complete perfection of the speculations – if they also accept the immobility of the sun and certainly the motion of the earth.

First, therefore, readers should understand this: that today according to all astronomers it is absolutely certain that all the planets go around the sun, except the moon, which alone has the earth as its centre (and whose sphere of course is not of sufficient quantity for it to be delineated in the proper proportion to the rest in this account). The earth, therefore, is added to the remaining five as a sixth, which either by its own motion, with the sun at rest, or with itself unmoved, and the whole system of the planets turning, itself marks a sixth circle around the sun.

Second, this is also certain: that all the planets are eccentric: that is, they change their distances from the sun, such that in one part of the circle they are very far from the sun, and on the opposite part they come closest to the sun.

Third, let the reader recall from my *Cosmographic Mystery*, which I published twenty-two years ago, that the number of the planets, or of courses around the sun, was chosen by the wisest Creator from the five regular solid figures, about which Euclid, so many centuries ago, wrote a book called *The Elements*.

Fourth, regarding that which pertains to the ratio of the planetary orbits: the ratio between neighbouring pairs of orbits is indeed always one wherein it is evident that every one of them is approximately the unique ratio of the spheres of one of the five solid figures, that is, of the figure of the circumscribed sphere to the inscribed sphere. But it is nevertheless not completely equal, as I had once dared to promise about astronomy when finally completed. In short, the cube and the octahedron pair penetrate their planetary spheres to some degree; the dodecahedron and icosahedron pair do not completely follow theirs; and the tetrahedron precisely touches both of them. The first has not enough of an interval between the planets; the second too much; and the third is just right.

From this it is clear that the proportions of the planetary intervals from the sun themselves were not chosen from the regular figures alone. For the Creator does not deviate from His Archetype, as He is the very source of geometry and, as Plato wrote, the operator of an eternal geometry. And this could certainly be gathered from the fact that all the planets change their intervals over fixed periods of time, such that each of them has two notable distances from the sun, a maximum and a minimum. The comparison of distances from the sun between pairs of planets is possible in four ways: either at greatest distance from the sun, at the least, or at greatest distance between the two, or least. Thus there are twenty comparisons of pairs and pairs of neighbouring planets, while by contrast there are only five solid figures. But it is certainly appropriate that the Creator, if He was concerned with the ratio of the spheres in general, was also concerned with the ratio between the varying distances of each one in particular, and that this concern was the same for both, and that the one was connected to the other. Upon weighing this carefully, we will clearly arrive at this: that for constituting the diameters and eccentricities of the spheres together, there are need of more principles beyond the five regular figures.

Fifth, let us come to the motions: when it comes to that which the harmonies are constituted between, I again drive home to my reader that it was demonstrated by me in the *Commentaries on Mars*, from the very reliable observations of Brahe, that the daily arcs on one and the same eccentric are not traversed at equal speeds. Rather, these different *durations of time on equal parts of the eccentric observe the ratio of their distances from the sun*, the source of motion. And in turn, supposing equal times – suppose one natural day for each of them – then agreeing with them, *the true daily arcs of a single eccentric orbit have a ratio between themselves that is the inverse of the ratio of the two distances from the sun*. However, at the same time, it was demonstrated by me that *the orbit of a planet is elliptical, and the sun*, the source of motion, *is at one of the foci of that ellipse*. Thus it happens that the planet, having completed out of its whole circuit *a quadrant from its aphelion, is precisely at mean distance from the sun*, between its greatest at aphelion and its least at perihelion.

From these two axioms it is rightly determined that the daily *mean motion* of the planet on its eccentric *is the same as the true* daily arc of its eccentric, at those moments when the planet is *at the end of the quadrant of the eccentric calculated from aphelion*, although that true quadrant still appears smaller than a regular quadrant. Further, it follows that *any two very true daily arcs of the eccentric, which are truly equally distant, one from aphelion, and the other from perihelion, equal two mean daily arcs together*, and consequently, because the ratio of the circles is the same as that of their diameters, it follows that *the ratio of one mean daily arc to all of the mean daily arcs* (which are equal to each other) *added together* in the whole circuit *is the same as that of the mean daily arc to that of all the true* eccentric arcs, *added together*, equal in number but unequal to each other. And these things concerning the true eccentric daily arcs and the true motions must be recognized first so that now from them we can understand the apparent motions, supposing the eye at the sun.

Sixth, as far as pertains to the apparent arcs (as if) from the sun, it has been known even from ancient astronomy that of the true motions – even the ones equal to each other – one that has retreated further from the centre of the world (like *one in aphelion*) *appears to be smaller*, to an observer at that centre; one that is nearer (like *one in perihelion*) *also in the same way appears greater*. Therefore, in addition the true daily arcs are also greater when near, on account of their faster motion, but smaller when far at aphelion, on account of their slower motion: hence I have demonstrated in the *Commentaries on Mars* that *the ratio of the apparent daily arcs on a single eccentric is near precisely the square of the inverse ratio of their distance from the sun*.

Seventh, if by chance the diurnal motions occur to anyone (not the ones that appear as if from the sun, but to observers on the earth), concerning which Book VI of the *Epitome of Copernican Astronomy* deals: let him know that there is no account of them at all here, nor, truly, ought there to be, because the earth is not the source of their motions, nor can it be, as with respect to deceptive appearance those motions not only deteriorate into bare rest or apparent pause, but into clear retrogradation. For this reason, the entire infinity of ratios is allotted to all the planets at the same time and equally. In order, therefore, to have it be certain what sort of ratios are set up particularly for the diurnal motion of the true individual eccentric orbits (granted that they themselves are still apparent, as it were, to an observer on the sun, the source of motion), this fantasy, common to all five, of an annual extrinsic motion must be removed from those particular motions, whether it originates from the motion of the earth itself, or whether from the annual motion of the whole system, according to Tycho Brahe. The motions particular to each planet must be brought out straightforwardly into view.

Eighth, thus far we have attended to the diverse times or arcs of one and the same planet. Now we must attend to those of all the pairs of planets compared with each other. Accordingly note the definition of the terms that will be necessary for us. Let us designate as *nearest apsides* of the two planets the perihelion of the upper one and the aphelion of the lower one, notwithstanding the fact that they incline not to the same region of the world, but to different and perhaps opposite regions. *Extreme motions* understand as the slowest and fastest of the whole planetary circuit. *Converging* or *approaching extremes*, as those which are at the nearest apsides of the two – that is, in perihelion of the upper and aphelion of the lower. *Diverging or retreating*, as those which are in opposite apsides – that is, in aphelion of the upper and perihelion of the lower. Again at this point, therefore, some part of my *Cosmographic Mystery*, suspended twenty years ago because it was not yet clear, must be settled and brought in here. For with the true distances of the orbs discovered, through the observations of Brahe, with continuous labour over a great deal of time, at last, at last, the genuine proportion of the periodic times to the ratio of the spheres was revealed.

And if you ask at what moment of time it was: it was conceived in my mind on 8 March of this year, 1618, but called to calculation without success, and thus rejected as false. When it was finally returned to on 15 May, with a new plan for approaching it, it broke through the darkness of my mind. The agreement of my labour of seventeen years with the observations of Brahe and this meditation was so much in accord that I first believed that I was dreaming, and begging the question from the start. But it is extremely certain and accurate that the *ratio between the periodic times of any two planets is precisely the sesquialterate* [3:2] *ratio* of their mean distances, that is, *of the spheres* themselves; attending nevertheless to the fact that *the arithmetic mean between both diameters of an elliptical orbit is a little less than the longer diameter.* Therefore, if someone took up one third of the ratio *from the period*, of, for example, the earth (which is one year), and *from the period* of Saturn (which is thirty years) – that is, the *cube roots* – and doubles that ratio, *squaring* the roots, *he has in the ensuing numbers the exactly correct ratio of the mean distances from the sun* of the earth and Saturn. For the cube root of 1 is 1, and the square of that is 1; and the cube root of 30 is greater than 3; therefore, the square of it is greater than 9. Also Saturn's mean distance from the sun is a little higher than nine times the mean distance of the earth from the sun. The use of this theorem will be necessary in Chapter 9 in order to demonstrate the eccentricities.

Chapter 4: In what properties relating to the motions of the planets have the harmonic ratios been expressed by the Creator, and how?

Therefore, with the fantasy of stations and retrogradations removed and the particular motions of the planets clearly explained in their genuine eccentric orbits, there still remain these distinctions among the planets: 1. Distances from the sun. 2. Periodic times. 3. Daily eccentric arcs. 4. Daily times on their arcs. 5. Angles at the sun, or apparent daily arcs, as though to observers from the sun. Finally, all these are features of either a single planet at different times, or of different planets; therefore, supposing an infinite amount of time, all the dispositions of the circuit of a single planet can coincide and be compared with all of the dispositions of the circuit of another planet.

However, because God has established everything with geometrical beauty unless it is bound to another prior necessary law, we easily deduce that the durations of the periodic times, and even the mass of the moving bodies, come from something which is prior, in the Archetype. These seemingly disproportionate masses and periods are adapted to that measure, in order to portray it. But I have said that the periods are obtained from the times, very long, medium and very slow. It is therefore necessary that the geometrical elegance be discovered either in these times, or in something which is prior to them in the mind of the Craftsman. We know that the ratios of the distances are bound to the ratios of the daily arcs, because the ratio of the arcs is the inverse of the ratio of the times. In turn, we have said that both the ratios of the times and the distances of any single planet are the same. Therefore, with respect to the individual planets, examination of these three – the arcs; the times on equal arcs; and the separation of the arcs from the sun, or the distances – will be one and the same. And now, because all of these are variable for the planets, there can be no doubt but that if some geometrical beauty had been arranged for them by the fixed plan of the highest Craftsman, they have received it at their extremes – thus at their distances at aphelion and perihelion – and not at their mean separations. Once the ratios of the extreme distances are given, then there is no need for the plan to be adapted to the intermediate ratios by a determined number: for they follow unaided from the necessity of the planetary motion from one extreme, through all the intermediates, to the other extreme.

But if you compare the extreme distances of the different planets with each other, then some light of harmony begins to dawn. For the divergent extremes of Saturn and Jupiter make a little more than a diapason [an octave], and their convergent extremes make the mean between major and minor sixths. So the divergent extremes of Jupiter and Mars encompass nearly a double diapason, and their convergent extremes almost a diapente [a fifth] with a diapason. But the divergent extremes of the earth and Mars have encompassed somewhat more than a major sixth, and their convergent extremes an excessive diatesseron [perfect fourth]. In the following couple of the earth and Venus again there is the same excessive diatesseron between their convergent extremes, but between their divergent extremes we are left without a harmonic ratio: for it is less than a hemidiapason (if we may call it that); that is, less than a semi-double.

Finally, the divergent path of Venus and Mercury is a little less than the compound of a diapason and a minor third; between their convergent path is an excessive diapente plus a little.

Therefore, although one distance was a bit far off from the theory of harmonies, nevertheless this success was an invitation to progressing further. This was my reasoning. First, these distances, in as much as they are lengths without motion, are not suitable to be examined with respect to harmony, because harmonies are rather part of a group whose subject is motion, because of speed or slowness. Next, with respect to these distances, in as much as they are diameters of spheres, it is plausible that the plan of the five regular bodies should be preferentially kept, by analogy. Because the rule for the geometrical solid bodies with respect to the celestial spheres, either enclosed from all around by celestial matter, as antiquity prefers, or having to be enclosed by an agglomeration of many successive rotations, is the same as the rule of the plane figures inscribed in a circle (which are the figures that beget the harmonies) with respect to the celestial circles of the motions, and to the other spaces in which the motions take place. Therefore, if we seek the harmonies, we seek them not in these latter distances (the semi-diameters of the spheres) but in the former ones (the measure of the motions) – that is, we seek them rather in the motions themselves. Truly no other distances can be accepted for the semi-diameters of the spheres but the means; but we deal here with the extreme intervals. Therefore, we do not deal with the distances with respect to their spheres, but rather with respect to their motions.

And certainly if we consider the matter more carefully, it will be apparent that it is not very reasonable for the very wise Creator to attend to the harmonies chiefly between the planetary paths themselves. For if the ratios of the paths are harmonic, everything else that pertains to the planets will be restricted and bound by those paths, with no place for attending to the harmonies elsewhere. But who will benefit from harmonies between the paths, and who will perceive these harmonies? There are two things that reveal to us harmonies in natural properties: either light or sound. The former is received through the eyes, or hidden senses analogous to the eyes; the latter is

received through the ears. The mind, capturing these emanations, judges the pleasing from the jarring either by instinct (see
Book IV concerning this) or by astronomical or harmonic reasoning. Besides, there are no sounds in the heavens, nor is there
so much turbulent motion that a buzzing would be elicited
from the friction of the celestial air. Light remains: if it should
teach us something about the planetary paths, it will teach us
that either the eyes, or a sensorium analogous to them, are situated in a certain place, and that if light itself is to teach us
something immediately in and of itself, it seems that it is necessary for it to be right before that sensorium. Therefore, there
must be a sensorium throughout the whole world – that is,
such that it is present simultaneously in one and the same way
for all the planetary motions. For the way – crossed over from
observations, through very long meanderings of geometry and
arithmetic, through ratios of the sphere and other things that
need to be learned first, all the way to these paths themselves –
is much too long for it to have seemed appropriate to introduce
the harmonies in order to stimulate some natural instinct.

Therefore, gathering all these things into sight at once, I have
rightly concluded that with the true paths of the planets through
the ethereal air dismissed, it is fitting that we turn our eyes to
the apparent daily arcs – and indeed, all are apparent from one
particular and notable place in the world, namely from the body
of the sun itself, the source of motion for all the planets. Further,
we need to consider not how high any of the planets is from the
sun, nor through what space it travels in one day – for that is
rational and astronomical, not instinctual – but rather how
much of an angle the daily motion of each one subtends at the
body of the sun itself, or how much of an arc on a common circle marked out from the sun, such as the ecliptic, it seems to
traverse in a day. Thus this appearance, borne with the aid of
light to the body of the sun, can flow with the light itself straight
to living beings, who participate in this instinct (just as in Book
IV we said that the form of the heavens flows to a foetus through
the aid of the rays).

Therefore, Tychonic astronomy (removing from the particular motion of a planet the parallaxes of the annual orbit, which

dress them with the appearance of stations and retrogradations) teaches that the daily motions of the planets in their orbits (appearing as though to spectators on the sun) are these:

Harmonia binorum		Apparentes diurni diurni. Prim-Sec.		Harmoniæ singulorum propria Prim. Sec.	
Diver.	Conv.				
a 1	b 1	♄ Aphelius 1.46. a. Perihelius 2.15. b.	Inter 1.48 & 2.15.	est $\frac{4}{5}$	Tertia major.
d 3 c 8	c 2 d 1	♃ Aphelius 4.30. c. Perihelius 5.30. d.	Inter 4.35. & 5.30.	est $\frac{5}{6}$	Tertia minor.
f 1 e 5	e 5 f 2	♂ Aphelius 26.14. e. Perihelius 38.1. f.	Inter 25.21. & 38.1.	est $\frac{2}{3}$	Diapente
h 12 g 3	g 3 h 5	Tel. Aphelius 57.3. g Perihelius 61.18. h	Inter 57.28. & 61.18.	est $\frac{15}{16}$	Semitoniu
k 5 i 1	i 3 k 3	♀ Aphelius 94.50. i. Perihelius 97.37. k.	Inter 94.50. & 98.47.	est $\frac{24}{25}$	Diesis
m 4	l 5	☿ Aphelius 164.0. l. Perihelius 384.0. m.	Inter 164. 0. & 394. 0.	est $\frac{5}{12}$	Diapason. cum tertia minore

I could therefore anticipate inwardly from the ratios of the daily eccentric arcs, related above, that there would be harmonies and consonant intervals between the apparent extreme motions of single planets, because I saw there that halves of harmonic ratios ruled everywhere, while I knew that the ratio of the apparent motions was double the ratios of the eccentric motions. And what is asserted here is possible to prove with experience itself, or without reasoning.

Hence these are the harmonies of the planets, distributed between them; nor is there any particular comparison (of convergent and divergent extreme motions) that does not come truly close to some harmony. Thus even if musical strings were tuned in such a way, the ears would not easily discern an imperfection, except that excess of the one between Jupiter and Mars.

Moreover, it follows that we will not deviate much from

the harmonies if we compare the motions on the same side either.

Therefore, perfect harmonies are discovered between the convergent extreme motions of Saturn and Jupiter (a diapason); between the convergent extreme motions of Jupiter and Mars (a diapason with almost a soft third); between the convergent extreme motions of Mars and the earth (a diapente), and between the two of them at perihelion (a soft sixth); between the earth and Venus at aphelion (a hard sixth) and between them at perihelion (a soft sixth); between the converging extremes of Venus and Mercury (a hard sixth) and between either their diverging extreme motions or even between them at perihelion (a double diapason).

You will observe, however, that where there is not a perfect major harmony, as between Jupiter and Mars, there alone I have discovered a near perfect interpolation of a solid figure, because the distance of Jupiter at perihelion is nearly triple that of Mars at aphelion. Thus this pair aims in its distances at the perfect harmony that it does not have in its motions.

Chapter 7: Universal harmonies of all six planets, as if in common counterpoint, are given in four parts

Now, Urania, we need a greater sound, while I climb through the harmonic ladder of the celestial motions to higher things where the genuine Archetype of the fabric of the world is kept hidden. Come along, modern musicians, and judge the matter with your arts, unknown to antiquity: in these final centuries, after being watched over for two thousand years, Nature – ever bountiful – has finally brought out you, the first genuine models of the universe. Through your concord of various voices, and through your ears, she has whispered to the human mind – most beloved daughter of God the Creator – about herself, and about what she is like in her innermost bosom.

Therefore, the motions of the heavens are nothing but a certain perpetual concord (rational, not audible) through dissonant tunings, as if certain syncopations or cadences (by which men imitate those natural dissonances), tending toward fixed and prescribed resolutions specific to these six terms (just as with voices), and marking and distinguishing with those notes the immensity of time. It is therefore no longer surprising that a method of singing together was finally discovered by man, aping his Creator, that was unknown to the ancients: namely, so that he could play the perpetuity of all the time of the world in a few short parts of an hour, through the artificial symphony of many voices. He can thus taste, to an extent, the pleasure of God the Creator in His works, by the most agreeable sense of pleasure he feels from this imitator of God, Music.

Chapter 9: The source of the eccentricities in the individual planets from the management of harmonies among their motions

Therefore, we see that universal harmonies of all six planets cannot come about blindly, especially with respect to the extreme motions, which we have seen all coincide with universal harmonies (except two, which coincided with the nearest universal ones).

It also follows that the Creator, the source of all wisdom, the perpetual supporter of order, the eternal and transcendent wellspring of geometry and harmony – the Craftsman of the heavens Himself – I say, has harnessed the harmonic proportions arising from the five regular plane figures to the five regular solid figures. From both classes, He fashioned the most perfect Archetype of the heavens, from which shone, just as through the five solid figures, the ideas of the spheres on which the six stars are carried. Thus indeed the measures of the eccentricities of the single spheres for the proportioning of the motions of bodies were comprised through the offspring of the plane figures, the harmonies (determined from them in Book III).

May the Author of the heavens Himself bring it about that we imitate the perfection of His work with the sanctity of life, toward which He has chosen His Church in the lands, and cleansed it of sins with the blood of His Son and the power of His Holy Spirit, holding far off all dissonances of animosity, all contention, rivalry, anger, disputes, dissension, divisions, jealousy, provocation, verbal incitement and other works of the flesh. All those who have the spirit of Christ will not only desire these things with me, but also strive to express them in actions and to make this their calling, scorning all corrupt habits of all the parties (veiled and painted with the excuse of zeal, or love of truth, or singular erudition, or obedience to contentious teachers, or any other specious pretext). Holy Father, preserve us in the consonance of mutual delight, so that we may be one, as You are one with Your Son, our Lord, and the Holy Spirit, and as You have made all of Your works one, through the most agreeable bonds of consonances, and so that from the restored concord of Your people, the body of Your Church may be built on these lands, just as You have built up heaven itself from the harmonies.

Kepler, *Epitome of Copernican Astronomy*, Book Four, 1620

Book Four of Kepler's Epitome of Copernican Astronomy, *published in the* Epitome's *second volume after his* Harmony of the World, *was far more complex than the three books in his first volume, geared more toward experts than the beginner. Its focus was strictly on physical astronomy, so much so, he wrote, 'that you might doubt whether you were doing a part of physics rather than astronomy'. More than just a survey of astronomy in Copernican terms, it was a Keplerian update of Copernican astronomy, complete with the innovations from Kepler's previous books, including both his theory of Platonic solids from the* Cosmographic Mystery *and what would later become known as his three laws: elliptical orbits and the area law from his* New Astronomy, *and the harmonic law relating planetary distances and periods from* The Harmony of the World.

Preface

This is the eleventh year since I published my *Commentary on the Motions of the Planet Mars*. Since just a few copies were made, and the book had nearly concealed the doctrine of celestial causes among a thicket of numbers and the remaining astronomical trappings, and since the price of the book deterred those with thinner pockets, it seemed to my friends that it would be appropriate for me to fulfil my duties by writing an *Epitome*, in which a summary of both the physical doctrine concerning the heavens and the astronomical one would be related, with the tedium of the demonstrations shortened, in simple and plain speech.

I did that many years ago, but meanwhile various delays interrupted the edition: not only was the little book itself repaired several times and, if I am not mistaken, improved in its content, but the plan of the book itself began to waver. For it seemed to certain people that in the doctrine of the sphere – published three years earlier – I was more long-winded in the debate about the diurnal motion or rest of the earth than served the plan of an epitome. I therefore thought that if readers did not stomach that part well – even though no epitome of astronomy lacks it – how much more contrary to custom would this fourth part be, which brandishes many new and unexpected things concerning the whole nature of the heavens? So much so that you might doubt whether you were doing a part of physics rather than astronomy, unless you knew that speculative astronomy itself is a whole part of physics.

Therefore it seemed it would be very appropriate for this book, which was written as much with physics as with astronomy, to be produced separately, so that based on the choice of the astronomer-buyer it could be either omitted or inserted into the rest of the *Epitome*. You have, kind reader, the reasons for this edition, which I hope will satisfy you.

My attempt here ought not to have been held back by modesty. It needed to be brought forward into the light with strength of mind, and certainly with the highest faith in the visible works of God (if someone is free to examine them), or with the encouragement of Aristotle himself, who believed that neither probable things, nor certain or very well-investigated things, should be silenced or suppressed in these questions.

What holds back the academies is that they were introduced in order to shape the studies of those who are learning, and it is in their interest that the laws of teaching not be changed frequently. There, because attention is on the progress of those learning, it frequently happens that things have to be chosen not because they are especially true, but because they are especially easy. Thus, if some intricate and difficult-to-understand topics should not be placed before beginners, or if they should not be placed before the accepted and necessary topics, it does not follow that those things ought neither to be written nor read privately.

Mention must be added concerning boundaries. For the boundaries to investigation must not be fixed in the small minds of a few men. The world is a trifling thing, unless all of the world can discover in it that which they seek, says Seneca. For the true boundaries of speculation are the structure of the world itself, not the false barriers thrown up by the Christian religion against overturning things, lest someone be led precipitously astray who otherwise would remain unharmed. Antiquity has taught us by example how man fixes boundaries in vain where God has not established them.

My doctrines can attest to whether I love glory or truth, for most of the ones I accept are taken from others: I build my whole astronomy from the hypotheses of Copernicus about the world, from the observations of Tycho Brahe, and, finally, from the Englishman William Gilbert's magnetic philosophy. If I delighted in novelty, perhaps I could have invented something similar to the Fracastorian or Patrician conceptions. As someone who delights in his work, and stops for a game of dice or brigands only rarely for the sake of a companion, and never for his own sake, I am engaged only in the true doctrines of others, or in correcting those which aren't fully well composed, but my talent never idles to play with new things, contrary to truth and invented by me alone. That which I declare outwardly, I believe inwardly: no cross is harder for me to bear than – I don't say to speak contrary to my mind – to be unable to assert my inmost feelings.

What are the hypotheses or principles by which Copernican astronomy saves the appearances in the individual motions of the planets?

They are chiefly: 1) that the sun is located in the centre (nearly so) of the sphere of the fixed stars, and is immobile in place; 2) that the single planets are, in reality, moved around the sun in single systems, which are constructed from many perfect circles, revolved in a perfectly uniform motion; 3) that the earth is one of the planets, so that it marks out a sphere between Mars and

Venus in an annual motion around the mean sun; 4) that the ratio of this sphere, compared to the diameter of the sphere of the fixed stars, is imperceptible, and thus as though it were immeasurable; 5) that the sphere of the moon is arranged around the earth, so that its centre, and thus its annual motion around the sun (and from one place into another) are all common to the sphere of the moon with the earth.

Do you think that those principles should be retained in this Epitome?

Since astronomy has two goals, to save the appearances and to contemplate the genuine form of the structure of the world, there is no need for all these principles for the first goal – rather, some can be changed, and some omitted, and the second one must necessarily be corrected; and although many of these are necessary for the second goal, they are nevertheless not yet sufficient.

Which of these principles can be changed or omitted and the appearances still saved?

Tycho Brahe explained the appearances while having changed the first and the third: for he, with the ancients, located the earth immobile in the centre of the world, but the sun, which is also for him the centre of the spheres of the five planets, he made circle the earth in a common annual motion, with its own system of all the spheres, while meanwhile each planet completes its own individual motion in this common system. Also he omits completely the fourth principle, and presents the sphere of the fixed stars as not much greater than the sphere of Saturn.

What, in turn, do you substitute in place of the second principle, and what, in addition, do you add to the genuine form of this worldly abode, or to things pertaining to the nature of the heavens?

Although the true individual motions must be left to the individual planets, nevertheless they are not moved by themselves to these motions, nor through the rotation of the spheres, which are not solid. Rather, the sun, in the centre of the world, rotating around its axis and the centre of its body, causes the circling of the individual planets through its rotation. Further, although the planets in truth are moved eccentrically from the centre of the sun, nevertheless there aren't any smaller circles, called epicycles, which, through their revolution, vary these intervals of the planet and the sun, but the bodies of the planets themselves, through an innate force, supply the occasion for this variation.

What, therefore, will the material of Book Four be?

This Book Four will contain especially celestial physics itself, or rather the form and plans of the composition of the world and the genuine causes of the motions. And this will be the distinguished duty of an astronomer: the demonstration of his hypotheses.

What do you suppose is the arrangement of the principal parts of the world?

The philosophy of Copernicus classifies the principal parts of the world into distinct regions of the figure of the world. For in the sphere – the image of God the creator and the archetype of the world (as was proved in Book One) – there are three regions, symbols of the three persons of the Holy Trinity: the centre, of the Father; the surface, of the Son; and the intermediate space, of the Holy Spirit. Thus likewise just as many principal parts of the world have been made, each one a different region of the sphere: the sun in the centre, the sphere of the fixed stars on the surface, and finally the system of the planets in the intermediate region between the sun and the fixed stars.

I thought that the principal parts of the world should be classified as heaven and earth?

Certainly our vision cannot show us, dwelling on earth, other more notable parts, because we trample upon the one with our feet, and are covered by the heavens, and both seem to be connected and stuck together at the common border of the horizon, like a palace in which the stars, clouds, birds, man and various types of terrestrial animals are enclosed. But since we are engaged in a discipline which uncovers the causes of things, shatters the deceptions of vision and carries the mind higher and further from the boundaries of vision, no one should be amazed that vision learns from reason, as a student learns from his master, something new, which it did not know before. Namely, that the earth, considered alone and by itself, should not be considered among the primary parts of the great world, but should be added to one of the primary parts, that is, the region of the planets, or the movable world, and that there it maintains a certain plan; and that the sun, in turn, is separate from the number of stars, and should be established as one of the principal parts of the whole universe. However, I speak now about the earth, inasmuch as it is a part of the structure of the world, but not of the dignity of the ruling creature who inhabits it.

By what properties are these three members of the great world distinguished from one another?

The perfection of the world consists in light, heat, motion and harmony of motions. These are analogous to faculties of the soul: light to the sensitive, heat to the vital and the natural, motion to the animal, and harmony to the rational. Indeed, the decoration of the world consists in light; its life and vegetation in heat; a certain type of its action in motion; and its contemplation in harmony, in which Aristotle places fulfilment.

Therefore the sun is made the principal body of the whole world in all these ways. For as pertains to light, the sun itself is not only the most beautiful and is even, as it were, the eye of the world, but it also illuminates, paints and adorns the bodies of the rest of the world, as a source of light or a very bright torch. With respect to heat, the sun is the hearth of the world. The sun is fire, as the Pythagoreans said, or a glowing rock or mass, as Democritus said. With respect to motion: the sun is the first cause of the motions of the planets and the first motor of the universe, by the very reason of its body. Finally, with respect to the harmony of motions, the sun occupies that place in which alone the motions of the planets acquire the appearance of quantities moderated harmonically.

Are there solid spheres in which the planets are carried,
and are the spaces all the way between the spheres empty?

Tycho Brahe refuted the solid spheres with three reasons. The
first is based on the motion of comets, the second on the fact that
light is not refracted, and the third on the ratio of the spheres.
For if the spheres were solid, comets would not be observed
crossing from one sphere into another, for they would be impeded
by the solidity; but they do cross from one into another, as Brahe
has demonstrated. Further, from light: since the spheres are eccentric and the earth and its surface – where our eyes are – are not
situated at the very centre of each sphere, therefore if the spheres
were solid, they would doubtless be denser than the very limpid
ethereal air, and the rays of the stars would arrive refracted to
our air, as optics teaches, and thus the planets would appear
irregularly and as if in places far from those which could be predicted by an astronomer. The third reason was adapted from the
principles of Brahe himself: they testify, as the Copernican principles also do, that Mars sometimes is found closer to the earth
than the sun is. But Brahe could not believe that this exchange
was possible, if the spheres were solid, because the sphere of
Mars would have to intersect the sphere of the sun.

What, then, is in that region of the planets besides the
planets?

Nothing except the ethereal air, common both to the spheres
and the intervals; that air is very limpid and yielding to mobile
things just as freely as it yields to the light of the sun and stars
so that it can descend to us.

I beseech you: do you really hope that causes can be
imputed for the number of the planets?

This concern has been surmounted, with the help of God, and
not badly. Geometrical reasons are coeternal with God; in them
first is the distinction between the curved and the straight. Above,
in the first book, it was said that the curved bears a certain resemblance to God, and the straight represents created things. And in
the embellishment of the world, first the outermost region of the

fixed stars was made spherical, toward that geometrical resemblance with God, because it – as some corporeal God (worshipped by the gentiles under the name Jupiter) – needed to contain all remaining things in itself. Straight quantities, therefore, pertained to the innermost contents of the outermost sphere; the first and most beautiful, to the primary. But among straight things, the first, the most perfect, the most beautiful and the simplest are those which are called the five regular bodies, which already two thousand years ago the Pythagoreans had said were the figures of the world, believing that the four elements and the heavens (the fifth essence) corresponded to their archetype.

But the truer reason is that these five figures correspond to just as many intervals of spheres, including one another mutually. If, therefore, there are five spherical intervals, it is necessary that there are six spheres, just as for four linear intervals it is necessary that there be five digits.

What are these five regular figures?

The cube, tetrahedron, dodecahedron, icosahedron and octahedron.

If the intervals agree with the ratios of the figures so closely, then why does there remain some discrepancy?

Because the archetype of the movable world corresponds not only to the five regular figures, by which the paths of the planets and the number of courses were assigned, but also to the harmonic proportions, through which the courses themselves have been attuned to a certain idea of celestial music or harmonic concord in six voices. Moreover, because that ornate music required a difference in the motion of each individual planet, from the slowest to the fastest – a difference brought about by the variation of the interval between the planet and the sun – and because the quantity or ratio of this variation was required to be different in different planets, therefore it was necessary for some very little bit to be taken away from the intervals determined by those figures, which are produced by the figures uniformly, without variation, so that freedom should be left to the harmonist to represent the harmonies of the motions.

In what way is the ratio of the periodic times that you have assigned to the mobile planets related to the already discussed proportion of the spheres in which the planets are carried?

The ratio of the times is not equal to the ratio of the spheres, but is greater than it, and indeed in the primary planets it is precisely in a ratio of 3:2. That is, if you calculate the cube root for the thirty years of Saturn and the twelve of Jupiter, and you increase it by the square, the genuine ratio of the spheres of Saturn and Jupiter will be in these squared numbers.

By what causes are you induced to make the sun the moving cause, or the source of the motions of the planets?

1. Because it is evident that by as much as any planet stands further from the sun than others, it advances that much more slowly, so that the ratio of the periodic motions is 3:2 powers of the ratio of the distances from the sun. From this, therefore, we infer that the sun is the source of motion.

2. Below we will hear the same thing come into use for the single planets, so that by as much as any one planet approaches the sun at any time, by that much faster it is carried in an exquisitely squared ratio.

3. Nor does the dignity or aptitude of the body of the sun fall short of this, because it is most beautiful and absolutely round, and is very great, and the source of light and heat, from which all life gushes into things capable of growth: such that heat and light can be thought of as instruments suitable for the sun to bring motion to the planets.

4. But what especially satisfies all the tallies of probability is the rotation of the sun in its own space around its immobile axis in the same course as that which all the planets follow. And indeed, Mercury, the swiftest of all the planets, has a period shorter than the planet nearest to it.

In fact, today it is revealed by the telescope and seen daily that the body of the sun swarms with spots, which cross the disc of the sun, or its lower hemisphere, within twelve, thirteen, or fourteen days, slowly at the beginning and end, and more quickly in the middle (which is an argument for the fact that they adhere to the surface of the sun and rotate with it). It is clearly necessary that this happens, which was proven by me in my *Commentary on Mars*, Chapter 34, by reasons deduced from that very same motion of the planets, long before it was made evident by the spots on the sun.

What do you think should be maintained concerning the body of the sun, and the force by which it is turned around its axis?

It was said in the first book that this body, and any which is rotated around its axis, was not only driven in a circuit at the very beginning of things by the omnipotent creator, but also seems to continue this motion with the help of a motor soul. For even if there were some other reason extended there that allowed this motion to be continued, nevertheless the long and perpetual duration of this motion, in which the life of the whole world consists, is more rightly obtained with the help of a soul.

Well, does the sun, by the rotation of its body, carry the planets around? And how can it do this when the sun lacks hands by which it might grasp the planet, physically so far away, and, whirling together, conduct it around with itself?

In place of hands, it has the power of its body, sent out in straight lines through the whole extent of the world, which itself, because it is an emanation of the body, rotates together with the body of the sun, resembling a very rapid vortex; it extends through the whole breadth of the circuit, however far it is, with equal speed, and the sun rotates in the narrowest space around its centre.

Could you show this by some example?

Of course: the sympathy of a magnet and an iron pointer, imbued by the rubbing of the magnet so that it absorbs its force, comes to our aid. Rotate the magnet near the pointer: the pointer will be turned at the same time. Although the grasp is of a different form, you nevertheless see that no contact of bodies intercedes here.

*If this concerned the body of the sun itself, I could concede
to it this natural power of moving. But you draw out this
material power from the body, and set it in the very wide
ether, without a subject: this seems absurd.*

It is evident from the example of the magnet, to which this same
thing could be objected, that it should not seem absurd. But in
neither of the two cases is the force without a proportional subject. For in the source itself, the body of the sun (or the fibres
stretched out from its centre to its circumference) is the subject
for the natural faculty. But in that which comes out of the
source, I think we must use reason to make a distinction between
the immaterial emanation of the body of the sun, flowing all the
way to the planets and beyond, and its force or energy, which
closely grasps and moves the planets. Thus let us grant that it is
not the body that is its subject, but the immaterial emanation of
the body.

Can you give an example of this thing?

The light and heat of the sun is a genuine example. There is no
doubt that just as the whole sun is luminous, so too it is entirely
aflame, and because of the density of the material it should be
compared to a fully glowing mass of gold, or something denser.
Now from that light of the sun an emanation comes out and
descends toward us that is incorporeal and immaterial, which we
call *lumen*, or rays of the sun, and which nevertheless accepts
quantity and accident, as it flows out in straight lines, can be
made denser or thinner, and is fully capable of being cut by a mirror or glass, namely, by reflection and refraction, as we are taught
in Optics. And this emanation of the light of the sun also carries
its heat, and in proportion to the strength, whether greater or
lesser, with which it falls on illuminable bodies, it also heats them
more or less. Thus just as that emanation, or *lumen*, which we
know certainly flows from the light of the sun, is subject to the
heating faculty, which similarly emerges from the sun through
the emanation: so too an immaterial emanation of the body of
the sun, descending all the way to the planets, has as a companion the emanation of that energetic virtue in the body of the sun

which struggles to unite similar things and to repel dissimilar
things.

*State what the ancients thought about the causes of the
inequalities [in planetary motion].*

The ancients wanted this to be the duty of the astronomer: that
he should convey the sort of apparent causes of this inequality
that would supply evidence of the true motion of the planet or
sphere itself, which is most regular, most uniform and most
constant, and of the simplest figure – namely, a perfect circle.
And they believed that someone who thought that there was
some true inequality in the real motions of the heavenly bodies
should not be listened to.

Do you believe that this axiom should be retained?

I respond in three ways: 1) That the motions of the planets are
regular – that is, ordered and described according to a fixed and
immutable law – is beyond controversy. For if this were not so,
there would be no astronomy, nor could the heavenly motions be
predicted. 2) Therefore, it follows that there is some agreement
between the whole periods. For the law about which I spoke is
one and never-ending; the turns or measures of the celestial course
are innumerable. But if they all have the same law and rule, then
all the turns are similar to each other, and the flow of time uniform. 3) But it was not yet conceded that even in the diverse parts
of each one of its circuits the motion is truly uniform. For astronomy bears witness that if we remove from our mind all the
deceptions of vision from the apparent confusion of the planetary
motions, what is left to the planet is the sort of circuit in which
the diverse parts are truly uniform, but the speed of the planet is
not uniform, just as in the angles to the sun, uniform with respect
to time, there is an apparent inequality. And Ptolemy himself,
having established different centres based on the rule of motion
of eccentrics and epicycles, made it so that those circles of his
were moved more quickly at one time and more slowly at another.
Finally, astronomy testifies concerning this, handled with appropriate subtlety: the journeys or single circuits of the planets are
not arranged exactly in a perfect circle, but happen elliptically.

Tommaso Campanella, *Apologia for Galileo*, 1622

Tommaso Campanella (1568–1639) was a Dominican friar and philosopher who opposed Aristotelian philosophy. He wrote this defence of Galileo in 1616, while imprisoned in Naples for political reasons, although it was not published until 1622. Campanella did not insist that Galileo's views were necessarily true (as some of those views, in fact, contradicted his own), but rather that they should not be prohibited by the Church, and that they were more probable than the Aristotelian-Ptolemaic alternative, which, Campanella argued, was often accepted out of deference to Aristotle rather than to Christian tradition.

Here I am invited to address the controversy of those who detest the method of philosophizing extolled by Galileo of Florence, because it seems to them that it establishes teachings contrary to the Sacred Scriptures. I will be pleased to take up these matters.

The question is *whether the method of philosophizing that Galileo prefers supports the Sacred Scriptures, or is opposed to them.*

I will complete the task in five chapters. In the first I will bring up the arguments against Galileo. In the second, I will submit the arguments in his defence. In the third, I will prepare some hypotheses for the subsequent double settlement of the issue. In the fourth, I will reply to the arguments against Galileo. In the fifth, I will say how reasoning defending him should be considered.

Chapter 2: The arguments for Galileo

1. In favour of Galileo there is the authority of the theologians who decided that the books of Nicolaus Copernicus's *On the Revolutions of the Spheres*, based on observations made by him from 1525, should be printed, because they contained nothing contrary to the Catholic faith. In those books he discussed the motion of the earth and the fixity of the firmament – that is, of the starry heavens – and the centrality of the sun. And Galileo does not offer anything new or unknown beyond that system. Therefore, if the books of Copernicus do not hinder the Catholic faith, then neither does Galileo.

2. Pope Paul III Farnese, to whom Copernicus dedicated those books, approved them, as did certain cardinals (who, before they were published, endeavoured to transcribe them at their own expense, as is apparent in the prefatory letters). During the time of Paul III, the most remarkable intellects flourished in the Church. Hence it would be quite surprising if those men were blind about Copernicus, while our contemporaries, of far less reputation, have more eyes than Argus and more certain observations than Galileo.

3. After Copernicus, Erasmus Reinhold, Johannes Statius, Michael Maestlin, Christoph Rothmann and many others wrote defending the same opinion. Indeed, more recent mathematicians do not believe that they can build correct ephemerides without the Copernican calculations, nor that they can speak correctly about the celestial motions without the ruin of certain mathematical principles and the testimony of the senses and of all nations, except using the theses of Copernicus. Nor are these only recent: Francisco Maria of Ferrara, before Copernicus himself, taught that astronomy needed to forge a new method of observing appearances in the heavens, which is what his disciple Copernicus did.

4. The most learned Cardinal Cusa embraced this view, and recognized that there are other suns and planets revolving in

the firmament of stars. And a certain Nolan [Giordano Bruno], and other heretics whom it is not permitted to name, defend this opinion. But this is not why they were condemned as heretics, nor were Catholics prohibited from publishing their books. Among others Johannes Kepler stands out, who supports this opinion in his *Conversation with Galileo's Starry Messenger*; and William Gilbert, a very learned Englishman, in his book *On the Magnetic Philosophy*, and many other Englishmen, whom I do discuss. Also Giovanni Antonio Magini, a Paduan mathematician, who from the year 1581 until the present year of 1616 testified in his *Ephemerides* that he embraces the calculations of Copernicus and Reinhold, and grumbles about other opinions.

5.   The Jesuit Father Clavius, in the last edition of his works, when he judged that Mercury and Venus wander around the Sun – although along with the followers of Aristotle he had previously believed otherwise – urged astronomers to make provisions for another system of the heavens. With this instruction in mind, a recent mathematician, the fictitious Apelles, is pushed toward the opinion of Galileo and Copernicus in his observations of sunspots.

6.   This opinion of Galileo – including the motion of the earth, the position of the sun at the centre, and the systems of the stars and waters and other elements – is very old, and indeed originated with Moses himself. Pythagoras, a Jew by birth, although born in a Greek city, as Ambrose attests, brought it into Italy and Greece, and taught it at Croton in Calabria. It was attacked by Aristotle for empty reasons, and without a mathematical demonstration, but rather with moral and shoddy conjectures, just as he rejected the books of Moses because he could not comprehend with his logic their depth and concealed methods and mysteries. But Galileo vindicates our ancestors from this injury of the Greeks. On account of this, those who are hostile toward Galileo's doctrines and method of philosophizing and prefer Aristotle to the Pythagoreans seem to do harm to Italy and Moses, when the buried truth now begins to shine forth. But our ancestors did not err in this way, since there

was not yet a new earth, a new celestial system and newly revealed phenomena, nor yet a harmony of the Scriptures with this philosophy.

7. Theologians, from the time of Casella and Francisco Maria of Ferrara until us, not only did not condemn this astronomy, but also ordered that it should be printed. Nor are there fewer in recent times. So it seems that those who oppose Galileo rise against him not out of zeal for Christian doctrine, but rather out of either envy or ignorance.

8. In the Sacred Scriptures the starry heavens are called the firmament, because they stand still. Therefore the earth moves, and therefore the sun is at the centre. For thus all the phenomena and principles of mathematics are saved, as Copernicus and his followers prove. Indeed, the followers of Ptolemy admit the same thing.

9. The sunspots and new stars in the starry heavens, and comets above the moon, clearly indicate that the stars follow this system.

10. We will prove below from the most Holy Doctors that the text of Moses cannot be correctly explained unless the stars are world systems.

Chapter 5: Where the arguments for Galileo from Chapter 2 are considered

I think that all the propositions listed could only be overturned now with difficulty. For many years I believed that the heavens were made of fire, that the heavens themselves were the source of all fire, just as Augustine, Basil and other Fathers did, and recently our Telesio. I tried in my *Questions* and *Metaphysics* to overturn all the reasoning of Copernicus and the Pythagoreans. But after the observations of Tycho and Galileo, in which they demonstrate a new star in the starry heavens, and comets there, both above the realm of the moon, and spots around the sun, I doubted that all the stars were made of fire. This doubt was supported by the waxing and waning of the moon and of Venus, and the great spot on the moon and Jupiter. Moreover, the

Medicean and Saturnian stars around Jupiter and Saturn do not seem to allow a single sun and a single centre of attraction, that is, the sun, and another of repulsion, that is, the earth, as we said in our *Physiologia*. But the idea that the appearance of the fixed stars is the same as that of the other planets creates doubt about the opinion of Galileo and others concerning suns.

Therefore, I suspend my judgement and respond to the arguments for Galileo. I am prepared to obey the commands of the Church and the better judgement of others.

To the first, second, third, fourth, fifth and seventh

To 1, 2, 3, 4, 5 and 7 the answer is the same: that the opinion of Copernicus and Galileo, approved by so many theologians, is probable, but not necessarily true. This is because this determination was not made in a General Council, nor by the individual motion of the Supreme Pope Paul III, assisted by the Holy Spirit, who only allowed for the book to be printed, maintaining this opinion as not opposed to faith. For when a Pope approves a doctrine he does not approve it fully, with respect to faith, but only as useful and worthy of reading. I therefore say that it is probable, and that there is nothing contrary to the Scriptures in these doctrines, given that the authority of the Pope and the previously mentioned theologians permitted it, but nevertheless not that it is necessarily true. For modern theologians might see what they did not, especially if they observe the Scriptures and the heavens more painstakingly and cleverly, as I propose here they should do, or if they have a new revelation. I nevertheless confess that I do not see any harm coming to the Sacred Scriptures from the doctrines of Galileo; on the contrary, I believe that they benefit them, as is clear from my words.

To the eighth, ninth and tenth

To 8, 9 and 10, I do not know whether they are demonstrations in favour of Galileo, for theologians avoid the mystical sense and celestial equivocations. But they are nevertheless without doubt demonstrations against Aristotle. Moreover, in the responses in favour of Galileo we have examined all that the theologians have said about it, and from this we understand the Scriptures to favour him no less than those who speak for the doctrines of other philosophers. I am prepared to yield to someone with a superior understanding, but let the Church judge whether Galileo should be allowed to write and discuss these things.

To the sixth

The most ancient and modern expositions of the Sacred Scriptures agree with the observation of Galileo more than with other philosophers. And the Pythagoreans acquired their doctrines from the Jews, and Galileo now agrees very well with them, not moved by only a casual opinion but by sensory observations.

I believe that the study of Galileo cannot be prohibited without some danger either of causing mockery to the Scriptures, or rather, the suspicion that we ourselves are opposed to the Scriptures along with the heathens, or of creating the suspicion that we envy those with great intellects. I believe that should his writings be suppressed, this would cause our enemies to embrace and celebrate them all the more eagerly.

However, in the things I have said and written and will write, I always submit myself to the censure of the Holy Mother Church, and to the judgement of those better than I.

Galileo, *The Assayer*, 1623

This book, written in Italian, was part of a debate between Galileo and the Jesuit mathematician Orazio Grassi (whose published pseudonym was Lotario Sarsi) over the nature of comets: Grassi believed that comets were actual celestial bodies, while Galileo was uncertain. The book is noteworthy less for the particular question about comets and more for Galileo's articulation of his larger approach to questions of science. Galileo heralded natural philosophy as a mathematical discipline and argued that it should be based on reason and expertise, rather than traditional authority or untutored experience. He dedicated the book to Cardinal Maffeo Barberini, who became Pope Urban VIII in 1623. Barberini particularly enjoyed the fable of sound included here, which seemed to suggest that explanations of nature were often only hypothetical and provisional (as Cardinal Bellarmine had argued was true of Copernican theory itself). With respect to the question of world systems, however (unlike cometary theory), Galileo was convinced that the Copernican approach was undeniably superior to its Ptolemaic alternative, as he suggested throughout his later Dialogue on the Great World Systems.

I perceive in Sarsi a firm belief that in philosophizing it is necessary to rely on the opinions of some famous author, and for our minds to remain completely sterile and barren when not wedded to someone else's discourse. Perhaps he believes that philosophy is a book of some man's fantasy, like the *Iliad* or *Orlando Furioso*, books in which the least important thing is whether what is written is true. But Sarsi, that is not the case.

Philosophy is written in this great book (the universe), which is continually open before our eyes. But it cannot be understood unless you first learn to understand the language and recognize the characters in which it is written. It is written in the language of mathematics, and the characters are triangles, circles and other geometric figures, without which it is humanly impossible to understand a word of it; without these one only wanders futilely through a dark labyrinth.

But given that it seems to Sarsi that our intellect must be yoked to the intellect of another man, and that in the contemplations of the celestial motions one must adhere to someone else, I do not see why he selected Tycho as opposed to Ptolemy and Nicolaus Copernicus, both of whom have constructed and completed entire systems of the world with supreme skill – which I do not think that Tycho has done, unless it is enough for Sarsi that he denied the other two and promised another one, though he did not accomplish it well. I would have liked if, instead of convicting the other two of falsehood, he had recognized some in Tycho's, because as for that of Ptolemy's, neither Tycho nor other astronomers, nor even Copernicus himself, could clearly contradict it, being that a principal reason taken from the movements of Mars and of Venus always opposed their approach. That is, since the disc of Venus, in its two conjunctions and separations from the sun, appears to have very little difference in its magnitude, and that of Mars appears hardly three or four times greater at perigee than at apogee, one would never be persuaded that the former really shows itself to be forty and the latter sixty times greater in one place than in the other, as is necessary if their revolutions are around the sun, as according to the Copernican system. However, I have demonstrated that this is true, and made it manifest to the senses with a perfect telescope that allows anyone who wants to see and access it through their own hand. As for the Copernican hypothesis, had we Catholics not, for our own benefit, been removed from error by the most sovereign wisdom and had our blindness illuminated, I do not believe that this grace and benefit could have been obtained from the reasons and experiences established by Tycho. Therefore, since the two systems are certainly false and that of Tycho is null, Sarsi ought not to reprimand me if, with Seneca, I desire the true constitution of the universe. But instead, in a very important and difficult question, Sarsi wants to persuade me, not as a joke, that there is a particular celestial sphere for comets in nature, and while Tycho could not drag himself out from his explanation of the differences in the apparent motion of his comet, he wants my mind to be calmed and stay satisfied with a poetic little flower that is not followed by any fruit at all. This is what S. M. [Mario Guiducci] rejected when he said, truly and with reason, that nature does not delight in poetry: a very true proposition, though Sarsi shows that he does not believe it, and pretends not to know nature or poetry. He does not understand that poetry needs fables and fictions and cannot exist without them, while untruths of this sort are so abhorrent to nature that it is as impossible to find one of them there as to find darkness in light.

Perhaps Sarsi believes that the full company of good philosophers can be enclosed within some set of walls. I, Sarsi, believe that they fly like eagles, and not like starlings. It is certainly true that the former, because they are rare, are little seen and less heard, and the latter, who fly in flocks, fill the sky with screams wherever they pause, and, with an uproar, spread filth around the world. But if, Sarsi, philosophers are like eagles, and not like the phoenix, then infinite is the crowd of fools – that is, of those who know nothing; many are those who know very little of philosophy; few are those who know some small part of things; very few who know a portion; and there is only one God who knows it all. Indeed, to say what I mean, dealing with science as a means of demonstration and with the reasons that people can obtain from it, I firmly believe that the more it participates in perfection, the fewer will be the number of conclusions it promises to teach; the fewer still will be the number of them that it will demonstrate; and consequently the fewer people it will entice and the smaller will be the number of its followers.

By contrast, magnificent titles and grand and numerous promises attract the natural curiosity of men and keep them perpetually ensnared in fallacies and chimeras, without ever letting them enjoy the sharpness of a single demonstration that might awaken their taste and allow them to recognize the insipidity of the usual fare. This will occupy an infinite number, and very fortunate will be the one who, by some extraordinary natural light, is led from the shadowy and confused labyrinth in which he would have gone perpetually around with everyone else and become still more ensnared.

Therefore, I deem it ill-advised to judge someone's opinion in philosophical matters based on the number of his followers. But although I think that the number of followers of the best philosophy may be very small, I do not conclude conversely that those opinions or doctrines which have few followers are necessarily perfect, since I am very well aware that some people are able to hold opinions so erroneous that they are abandoned by everyone else.

Sarsi thinks much of the sense of sight, such that he considers it impossible to be deceived if one were to place a fake object next to a real object. I confess that I do not have such a perfect capacity to distinguish. Rather I am like the monkey who firmly believes he sees another monkey in the mirror, and does not recognize its mistake until, after four or six times, it runs behind the mirror to attack it – so alive and genuine did the image seem.

And suppose that those whom Sarsi sees in the mirror are not genuine and real men but empty images, like those that we others see, I have a great desire to know what the visual differences are through which he so promptly distinguishes the genuine from the fake. As for me, I have found myself a thousand times in some room with shuttered windows, and through some small hole have seen a small reflection of the sun on the opposite wall – as far as sight is concerned a star no less bright than the Dog Star or Venus. Walking in the countryside when the sun is out, you will see the reflection of the sun appearing like the brightest stars in the many thousands of little straws and pebbles that are smooth or wet. Sarsi has only to spit on the ground and undoubtedly from there to where the sun's rays reflect he will see the appearance of a very natural-looking star. Will not any object placed at a great distance, when touched by the sun, appear as a star, if it is so high that it can be seen at night, when other stars are seen? And who would distinguish the moon, seen during the day, from a cloud touched by the sun, if not for the differences between their shape and size? Definitely no one. And finally, if simple appearance determines the essence of something, Sarsi must concede that the suns, moons and stars seen in still water and in mirrors are genuine suns, genuine moons and genuine stars.

After long experience with the human condition, it seems to me that when it comes to intellectual matters, the less people understand and know about something, the more resolutely they want to talk about it, while knowing and understanding many things makes people slower and less decisive in pronouncing judgement about something new.

Once, in a very solitary place, a man was born gifted by nature with great astuteness and extraordinary curiosity, and he raised several birds for pleasure as he very much enjoyed their songs. He observed with great amazement the beautiful skill with which they could form different songs at will, all very sweet, from the very same air that they breathed. It happened that one night near his house he heard a delicate sound, and being able to imagine only that it was some other small bird, he started out to capture it. But when he came to the road he found a young shepherd, who, blowing into some kind of hollowed wood, and moving his fingers over the wood – tightening and then opening certain holes that were there – drew from it diverse sounds similar to those of a bird, but in a very different manner. Astonished, and moved by his natural curiosity, he gave the shepherd a calf for the flute and returned home by himself. But knowing that if he had not happened to go out to the shepherd, he would never have learned that there were two ways to form sounds and sweet songs, he wanted to once again leave home, expecting thus to meet some other adventure. And it happened the following day that, passing near a small cottage, he heard a similar sound resound within, and to ascertain whether it was a flute or a bird, he went inside. He found a young boy who held a bow in his right hand, which sawed over some strings stretched over some kind of hollow wood, with the left hand supporting the instrument. He moved his fingers over it, so that without any breath he drew from it diverse and very sweet sounds. Now whoever shares his talents and his curiosity may judge his astonishment, seeing that he had arrived at two new ways of forming sounds and song, both unexpected. He thus began to believe that there might still be others in nature. And his wonder was great when, entering a certain temple, he looked inside the door to see who had made a sound and realized that it had emerged from the hooks and hinges when he opened the door. Another time, impelled by curiosity, he entered a tavern, believing that he would see someone lightly touching the strings of a violin with his bow, and saw instead someone rubbing his fingertip over the rim of a glass and creating a very sweet sound. He later observed that wasps, mosquitoes and flies do not form interrupted notes with their breathing, like the birds earlier, but make a constant sound through the very rapid beating of their wings.

His growing astonishment matched his diminishing belief that he knew how sounds were generated; nor would all his past experiences have been enough to make him understand or believe that crickets, though they do not fly, rub their wings together to produce their sweet and sonorous whistling. And when he believed that no other ways of forming sounds could be possible, after having observed, in addition to the things already mentioned, many organs, trumpets, fifes, stringed instruments and many others, including a certain little iron tongue which, suspended between the teeth, strangely uses the cavity of the mouth as a vehicle for resonance, and the breath for sound – when, I say, he thought he had seen it all, he found himself swept up more than ever in ignorance and wonder. Having captured a cicada in his hand, he could not lessen its very loud shriek by closing its mouth or stopping its wings, nor could he see it move its scales or any other part. Finally, lifting its chest, he saw some hard and thin sinews underneath. Believing that the noise derived from their shaking, he determined to quiet it by breaking them. But it was all in vain – he pushed his needle a little too deeply inside, and by piercing it he took away not only its voice but also its life, and so was not able to ascertain whether the song was produced that way. In this way his knowledge was reduced to diffidence, so that when asked how sounds are generated, he would generously reply that he knew some ways, but that he believed that there could be others, both unknown and unable to be conceived.

I could exhibit with many other examples the richness of nature in producing its effects in ways that we could not imagine without sense and experience showing them to us – and sometimes even these are not enough to make up for our incapacity. So if I do not know precisely how to determine the manner in which comets are produced, I ought to be excused, especially as I have never presumed to be able to do this, knowing that it may be in a way far removed from any of our imaginings. The difficulty of understanding how the song of the cicada is formed while it is singing in our hand should excuse me for not knowing how comets, at such great distances, are generated. I stop, therefore, at my first intention, which was to offer up the questions that seem to me to make our current opinions uncertain, and to propose some new considerations.

I can't help but marvel that Sarsi wants to persist in trying to prove through witnesses that which I can see at any time through experience. Witnesses are examined in past affairs that are doubtful and not permanent, and not in present, concrete affairs. So, it is necessary for the judge to seek to know by way of witnesses whether it is true that Peter wounded John last night, but not whether John was wounded, as he is able to see him directly and discover it for himself. But I say further that even in conclusions that we can only come to know via reason, I would have less respect for the testimony of many than for that of a few, since I am certain that the number of those who reason well in difficult matters is much less than those who reason badly. If reasoning about a difficult problem were like carrying a load, where many horses will carry more sacks of grain than one, I would agree that many reasoners would do more than one. But reasoning is like racing, and not like carrying, and a single Barbary horse will run more than a hundred Friesian draught horses. Therefore when Sarsi comes with such a multitude of authors, it does not seem to me that he strengthens his conclusions at all, but rather that he makes my case more noble, by showing that I have reasoned better than many well-renowned men.

While Sarsi says that he does not want to be someone who affronts the ancients by contradicting them and not believing their words, I say that I do not want to be someone very ignorant and ungrateful toward nature and God. As they have given me sense and reason, I have no desire to devote such great gifts to the fallacies of another man, and to blindly and foolishly believe what I hear. Nor do I wish to enslave the freedom of my intellect to someone who can err as easily as I can.

Galileo, Phases of Venus from
The Assayer, 1623

Between 1610 and 1611, Galileo observed Venus with his tele-
scope, paying particular attention to its phases (depicted here, in
his 1623 Assayer), which he reported in letters to a number of
correspondents, including to Kepler in anagram form (as we saw
in the preface to the Dioptrics). Though the general concept of
the phases of Venus was compatible with both Ptolemaic and
Copernican theory, the particular phases that Galileo observed
were incompatible with Ptolemaic theory and proved that Venus
orbited the sun. This did not necessarily prove Copernican the-
ory, however, as those phases were consistent with Tychonic
theory as well, in which Venus orbited the sun, which then
orbited a stationary earth. Galileo's observations of the phases
of Venus resolved another problem that had beset both Ptolem-
aic and Copernican theory. According to both models, Venus
should vary noticeably in brightness, yet it did not appear to do
so. This discrepancy between theory and observation was, in
fact, one of the primary reasons that Osiander had asserted that
no astronomical model should be accepted as physically true.
Galileo's observations revealed that Venus was in its crescent
phase when closer to the earth and was further away when full;
thus the amount of reflected light reaching the earth remained
approximately the same, accounting for the lack of observable
variation in brightness.

Kepler, Frontispiece, *Rudolphine Tables*, 1627

Kepler published the Rudolphine Tables, *based on Tycho Brahe's observational data, in 1627 after many years of computational work. Named in honour of Emperor Rudolph II, for whom both Tycho and Kepler had worked when they started on the tables, they were significantly more accurate than earlier attempts. As a frontispiece to the book, Kepler designed this image of the Temple of Astronomy. Urania, the muse of astronomy, is seated at the top of the dome, while the floor of the temple is a map of the heavens. The inside of the temple is a visual depiction of astronomical progress. It includes ten visible pillars, near which stand various astronomers and astronomical instruments. The pillars increase in sophistication from bare tree trunks to roughly hewn stone, to brick and finally to marble, paralleling the progress of astronomy from Babylonian times to the Greeks, and then to Copernicus and Kepler's own time. Hipparchus and Ptolemy stand next to the brick pillars, while Copernicus and Tycho Brahe stand next to the marble ones, with Tycho pointing at the ceiling, decorated with a depiction of his world system, and asking Copernicus 'What if it were thus?' Finally, Kepler depicted himself on the base of the temple, at the left, working on some calculations by candlelight, with a list of his books in front of him and the dome of the temple of astronomy before him.*

The Vatican Secretary to the Florentine Inquisitor, 24 May 1631*

Niccolò Riccardi was appointed Master of the Sacred Palace by Pope Urban VIII, which meant that he had authority over which books were given papal approval for printing. He had in fact been the one to grant permission to Galileo for the publication of The Assayer *in 1623. When Galileo asked Urban VIII to allow the printing of a new book on the Copernican system, the pope conditionally agreed, but Riccardi suggested some alterations to align with the pope's desires. Galileo had tentatively titled the book* On the Ebb and Flow of the Sea, *since he viewed the tides as strong evidence for a moving earth. The pope asked that the title be changed to reflect the more general aim of the book, which – along with the book itself – should clearly indicate that it was a consideration of both major world systems, rather than just Copernicanism, and that neither system was being endorsed as true but were, rather, discussed only hypothetically.*

Very Reverend and Most Honourable Father Inquisitor

Mr Galilei is thinking of publishing there a work of his, formerly entitled *On the Ebb and Flow of the Sea*, in which he discusses in a probable fashion the Copernican system and motion of the

* This translation is taken from *The Galileo Affair: A Documentary History*, translated by Maurice A. Finocchiaro (University of California Press, 1989).

earth, and he attempts to facilitate the understanding of that great natural mystery by means of this supposition, corroborating it in turn because of this usefulness. He came to Rome to show us the work, which I endorsed, with the understanding that certain adjustments would be made to it and shown back to us to receive the final approval for printing. As this cannot be done due to current restrictions on the roads and the risks for the manuscript, and since the author wants to complete the business there, Your Very Reverend Paternity can avail yourself of your authority and dispatch or not dispatch the book without depending in any way on my review. However, I want to remind you that Our Master thinks that the title and subject should not focus on the ebb and flow but absolutely on the mathematical examination of the Copernican position on the earth's motion, with the aim of proving that, if we remove divine revelation and sacred doctrine, the appearances could be saved with this supposition; one would thus be answering all the contrary indications which may be put forth by experience and by Peripatetic philosophy, so that one would never be admitting the absolute truth of this opinion, but only its hypothetical truth without the benefit of Scripture. It must also be shown that this work is written only to show that we do know all the arguments that can be advanced for this side, and that it was not for lack of knowledge that the decree was issued in Rome; this should be the gist of the book's beginning and ending, which I will send from here properly revised. With this provision the book will encounter no obstacle here in Rome, and Your Very Reverend Paternity will be able to please the author and serve the Most Serene Highness, who shows so much concern in this matter. I remind you that I am your servant and I beg you to honour me with your commands.

Your Most Devout Servant in the Lord,
Fra Niccolò Riccardi, Master of the Sacred Palace

Galileo, *Dialogue on the Great World Systems, Ptolemaic and Copernican,* Frontispiece and Title Page, 1632

Galileo's famous dialogue, approved with Riccardi's altera-tions (above), spanned four days and included three participants: Salviati, Galileo's spokesperson; Sagredo, a mostly neutral participant, though sympathetic to Salviati's arguments; and Simplicio, an Aristotelian. Further illustrating the different world systems, Galileo depicted three additional figures on the frontispiece, two ancient and one modern: Aristotle, Ptolemy and Copernicus, also in conversation. Behind them, however, Galileo hinted at whose side he would take: that of the mod-erns, rather than the ancients. He depicted fortifications, guns and a galley, all of which he referred to in the course of the Dialogue *to argue for a new approach to astronomy that relied on the physics of the everyday world, and that ultimately sup-ported the heliocentric system.*

DIALOGO
DI
GALILEO GALILEI LINCEO
MATEMATICO SOPRAORDINARIO

DELLO STVDIO DI PISA.

E Filosofo, e Matematico primario del

SERENISSIMO

GR. DVCA DI TOSCANA.

Doue ne i congreſſi di quattro giornate ſi diſcorre
ſopra i due

MASSIMI SISTEMI DEL MONDO
TOLEMAICO, E COPERNICANO;

*Proponendo indeterminatamente le ragioni Filoſofiche, e Naturali
tanto per l'vna, quanto per l'altra parte.*

CON PRI   VILEGI.

IN FIORENZA, Per Gio: Batiſta Landini MDCXXXII.

CON LICENZA DE' SVPERIORI.

Galileo, *Dialogue on the Great World Systems, Ptolemaic and Copernican*, 1632*

Galileo's Dialogue, *written in lucid and engaging Italian prose, focused on what Galileo called the 'two great world systems', the Ptolemaic and the Copernican, while ignoring the third, Tychonic system (the true serious alternative to Copernican theory by this time). The first day of the dialogue challenged the traditional dichotomy between the terrestrial and celestial realms by highlighting recently discovered evidence, including sunspots, comets, new stars and the surface of the moon. The second day focused on the daily rotation of the earth; Galileo pointed out that much as on a moving ship, there was simply no way to tell by experiment whether the earth was moving, since it imparted its motion to everything on it. The third day considered the annual motion of the sun, using evidence including the phases of Venus and the varying brightness of the planets. During the fourth day Galileo revealed what he wrongly judged to be a conclusive physical argument in favour of the earth's motion: the tides, which he explained using the earth's combined diurnal and annual motions. After doing all he could to argue for heliocentrism, Galileo ended his dialogue by having Simplicio (a name that sounded, likely not coincidentally, much like 'simpleton') echo the pope's admonition that God could do anything and therefore we cannot know which system is true.*

---

* This translation is taken from *Dialogue Concerning the Two Chief World Systems, Ptolemaic and Copernican*, by Galileo Galilei, translated by Stillman Drake, 2nd edn (University of California Press, 1967).

To the Discerning Reader

Several years ago there was published in Rome a salutary edict which, in order to obviate the dangerous tendencies of our present age, imposed a seasonable silence upon the Pythagorean opinion that the earth moves. There were those who impudently asserted that this decree had its origin not in judicious inquiry, but in passion none too well informed. Complaints were to be heard that advisers who were totally unskilled at astronomical observations ought not to clip the wings of reflective intellects by means of rash prohibitions.

Upon hearing such carping insolence, my zeal could not be contained. Being thoroughly informed about that prudent determination, I decided to appear openly in the theatre of the world as a witness of the sober truth. I was at that time in Rome; I was not only received by the most eminent prelates of that Court, but had their applause; indeed, this decree was not published without some previous notice of it having been given to me. Therefore I propose in the present work to show to foreign nations that as much is understood of this matter in Italy, and particularly in Rome, as transalpine diligence can ever have imagined. Collecting all the reflections that properly concern the Copernican system, I shall make it known that everything was brought before the attention of the Roman censorship, and that there proceed from this clime not only dogmas for the welfare of the soul, but ingenious discoveries for the delight of the mind as well.

To this end I have taken the Copernican side in the discourse, proceeding as with a pure mathematical hypothesis and striving by every artifice to represent it as superior to supposing the earth motionless – not, indeed, absolutely, but as against the arguments of some professed Peripatetics. These men indeed deserve not even that name, for they do not walk about; they are content to adore the shadows, philosophizing not with due circumspection but merely from having memorized a few ill-understood principles.

Three principal headings are treated. First, I shall try to show that all experiments practicable upon the earth are insufficient measures for proving its mobility, since they are indifferently adaptable to an earth in motion or at rest. I hope in so doing to reveal many observations unknown to the ancients. Secondly, the celestial phenomena will be examined, strengthening the Copernican hypothesis until it might seem that this must triumph absolutely. Here new reflections are adjoined which might be used in order to simplify astronomy, though not because of any necessity imposed by nature. In the third place, I shall propose an ingenious speculation. It happens that long ago I said that the unsolved problem of the ocean tides might receive some light from assuming the motion of the earth. This assertion of mine, passing by word of mouth, found loving fathers who adopted it as a child of their own ingenuity. Now, so that no stranger may ever appear who, arming himself with our weapons, shall charge us with want of attention to such an important matter, I have thought it good to reveal those probabilities which might render this plausible, given that the earth moves.

I hope that from these considerations the world will come to know that if other nations have navigated more, we have not theorized less. It is not from failing to take account of what others have thought that we have yielded to asserting that the earth is motionless, and holding the contrary to be a mere mathematical caprice, but (if for nothing else) for those reasons that are supplied by piety, religion, the knowledge of Divine Omnipotence and a consciousness of the limitations of the human mind.

I have thought it most appropriate to explain these concepts in the form of dialogues, which, not being restricted to the rigorous observance of mathematical laws, make room also for digressions which are sometimes no less interesting than the principal argument.

Many years ago I was often to be found in the marvellous city of Venice, in discussions with Signore Giovanni Francesco Sagredo, a man of noble extraction and trenchant wit. From Florence came Signore Filippo Salviati, the least of whose glories were the eminence of his blood and the magnificence of his fortune. His was a sublime intellect, which fed no more hungrily upon any pleasure than it did upon fine meditations. I often talked with these two of such matters in the presence of a certain Peripatetic philosopher whose greatest obstacle in apprehending the truth seemed to be the reputation he had acquired by his interpretations of Aristotle.

Now, since bitter death has deprived Venice and Florence of those two great luminaries in the very meridian of their years, I have resolved to make their fame live on in these pages, so far as my poor abilities will permit, by introducing them as interlocutors in the present argument. (Nor shall the good Peripatetic lack a place; because of his excessive affection toward the *Commentaries* of Simplicius, I have thought fit to leave him under the name of the author he so much revered, without mentioning his own.) May it please those two great souls, ever venerable to my heart, to accept this public monument of my undying love. And may the memory of their eloquence assist me in delivering to posterity the promised reflections.

It happened that several discussions had taken place casually at various times among these gentlemen, and had rather whetted than satisfied their thirst for learning. Hence very wisely they resolved to meet together on certain days during which, setting aside all other business, they might apply themselves more methodically to the contemplation of the wonders of God in the heavens and upon the earth. They met in the palace of the illustrious Sagredo; and, after the customary but brief exchange of compliments, Salviati commenced as follows.

First Day

SALVIATI: Yesterday we resolved to meet today and discuss as clearly and in as much detail as possible the character and the efficacy of those laws of nature which up to the present have been put forth by the partisans of the Aristotelian and Ptolemaic position on the one hand, and by the followers of the Copernican system on the other. Since Copernicus places the earth among the movable heavenly bodies, making it a globe like a planet, we may well begin our discussion by examining the Peripatetic steps in arguing the impossibility of that hypothesis; what they are, and how great is their force and effect. For this it is necessary to introduce into nature two substances which differ essentially. These are the celestial and the elemental, the former being invariant and eternal; the latter, temporary and destructible.

SIMPLICIO [*stating Aristotle's position*]: Bodies that are generable, corruptible, alterable, etc., are quite different from those that are ingenerable, incorruptible, inalterable, etc. The earth is generable, corruptible, alterable, etc., while celestial bodies are ingenerable, incorruptible, inalterable, etc. Therefore the earth is very different from the celestial bodies. [Further,] sensible experience shows that on earth there are continual generations, corruptions, alterations, etc., the like of which neither our senses nor the traditions or memories of our ancestors have ever detected in heaven; hence heaven is inalterable, etc., and the earth alterable, etc., and therefore different from the heavens.

SALVIATI: I declare that we do have in our age new events and observations such that if Aristotle were now alive, I have no doubt he would change his opinion. This is easily inferred from his own manner of philosophizing, for when he writes of considering the heavens inalterable, etc., because no new thing is seen to be generated there or any old one dissolved, he seems implicitly to let us understand that if he had seen any such event he would have reversed his opinion, and properly preferred the sensible experience to natural reason. Unless he had taken the senses into account, he would not have argued immutability from sensible mutations not being seen.

SIMPLICIO: Aristotle first laid the basis of his argument *a priori*, showing the necessity of the inalterability of heaven by means of natural, evident and clear principles. He afterwards supported the same *a posteriori*, by the senses and by the traditions of the ancients.

SALVIATI: What you refer to is the method he uses in writing his doctrine, but I do not believe it to be that with which he investigated it. Rather, I think it certain that he first obtained it by means of the senses, experiments and observations, to assure himself as much as possible of his conclusions. Afterwards he sought means to make them demonstrable. That is what is done for the most part in the demonstrative sciences.

Now getting back to the subject, I say that things which are being and have been discovered in the heavens in our own time are such that they can give entire satisfaction to all philosophers.

Excellent astronomers have [recently] observed many comets generated and dissipated in places above the lunar orbit, besides the two new stars of 1572 and 1604, which were indisputably beyond all the planets. And on the face of the sun itself, with the aid of the telescope, they have seen produced and dissolved dense and dark matter, appearing much like the clouds upon the earth; and many of these are so vast as to exceed not only the Mediterranean Sea, but all of Africa, with Asia thrown in.

SIMPLICIO: I have heard different opinions on this [latter] matter. Some say, 'They are stars which, like Venus and Mercury, go about the sun in their proper orbits, and in passing under it present themselves to us as dark; and because there are many of them, they frequently happen to collect together, and then again to separate.' Others believe them to be figments of the air; still others, illusions of the lenses; and still others, other things. And if this is not enough, there are more brilliant intellects who will find better answers.

SALVIATI: If what we are discussing were a point of law or of the humanities, in which neither true nor false exists, one might trust in subtlety of mind and readiness of tongue and in the greater experience of the writers, and expect him who excelled in those things to make his reasoning most plausible, and one might judge it to be the best. But in the natural sciences, whose conclusions are true and necessary and have nothing to do with human will, one must take care not to place oneself in the defence of error; for here a thousand Demostheneses and a thousand Aristotles would be left in the lurch by every mediocre wit who happened to hit upon the truth for himself. Therefore, Simplicio, give up this idea and this hope of yours that there may be men so much more learned, erudite and well read than the rest of us as to be able to make that which is false become true in defiance of nature.

Now I shall convey to you but two observed facts [about sunspots]. One is that many of these spots are seen to originate in the middle of the solar disc, and likewise many dissolve and vanish far from the edge of the sun, a necessary argument that they must be generated and dissolved. For without generation and corruption, they could appear there only by way of local motion, and they all ought to enter and leave by the very edge. The other observation, for those not in the rankest ignorance of perspective, is that from the changes of shape observed in the spots, and from their apparent changes in velocity, one must infer that the spots are in contact with the sun's body, and that, touching its surface, they are moved either with it or upon it, and in no sense revolve in circles distant from it. Their motion proves this by appearing to be very slow around the edge of the solar disc and quite fast toward its centre; the shapes of the spots prove the same by appearing very narrow around the sun's edge in comparison with how they look in the vicinity of the centre. For around the centre they are seen in their majesty and as they really are; but around the edge, because of the curvature of the spherical surface, they show themselves foreshortened. These diminutions of both motion and shape, for anyone who knows how to observe them and calculate diligently, correspond exactly to what ought to appear if the spots are contiguous to the sun, and hopelessly contradict their moving in distant circles, or even at small intervals from the solar body. This has been abundantly demonstrated by our mutual friend [i.e., Galileo] in his *Letters to [Marcus] Welser on the Solar Spots*.

SIMPLICIO: To tell the truth, I have not made such long and careful observations that I can qualify as an authority on the facts of this matter; but certainly, I wish to do so, and then to see whether I can once more succeed in reconciling what experience presents to us with what Aristotle teaches. For obviously two truths cannot contradict one another.

SALVIATI: It is better Aristotelian philosophy to say, 'Heaven is alterable because my senses tell me so,' than to say, 'Heaven is inalterable because Aristotle was so persuaded by reasoning.' Add to this that we possess a better basis for reasoning about celestial things than Aristotle did. He admitted such perceptions to be very difficult for him by reason of the distance from his senses, and conceded that one whose senses could better represent them would be able to philosophize about them with more certainty. Now we, thanks to the telescope, have brought the heavens thirty or forty times closer to us than they were to Aristotle, so that we can discern many things in them that he could not see; among other things these sunspots, which were absolutely invisible to him. Therefore we can treat of the heavens and the sun more confidently than Aristotle could.

SAGREDO: I can put myself in Simplicio's place and see that he is deeply moved by the overwhelming force of these conclusive arguments. But seeing on the other hand the great authority that Aristotle has gained universally; considering the number of famous interpreters who have toiled to explain his meanings; and observing that the other sciences, so useful and necessary to mankind, base a large part of their value and reputation upon Aristotle's credit; Simplicio is confused and perplexed, and I seem to hear him say, 'Who would there be to settle our controversies if Aristotle were to be deposed? What other author should we follow in the schools, the academies, the universities? What philosopher has written the whole of natural philosophy, so well arranged, without omitting a single conclusion? Ought we to desert that structure under which so many travellers have recuperated? Should we destroy that haven, that Prytaneum where so many scholars have taken refuge so comfortably; where, without exposing themselves to the inclemencies of the air, they can acquire a complete knowledge of the universe by merely turning over a few pages? Should that fort be levelled where one may abide in safety against all enemy assaults?'

I pity him no less than I should some fine gentleman who, having built a magnificent palace at great trouble and expense, employing hundreds and hundreds of artisans, and then beholding it threatened with ruin because of poor founda-tions, should attempt, in order to avoid the grief of seeing the walls destroyed, adorned as they are with so many lovely murals; or the columns fall, which sustain the superb galleries, or the gilded beams; or the doors spoiled, or the pediments and the marble cornices, brought in at so much cost – should attempt, I say, to prevent the collapse with chains, props, iron bars, buttresses and shores.

Second Day

SAGREDO: As I recall it, yesterday's discourse may be summa-rized as a preliminary examination of the two following opinions as to which is the more probable and reasonable. The first holds the substance of the heavenly bodies to be ingenerable, incorruptible, inalterable, invariant and, in a word, free from all mutations except those of situation, and accordingly to be a quintessence most different from our gen-erable, corruptible, alterable bodies. The other opinion, removing this disparity from the world's parts, considers the earth to enjoy the same perfection as other integral bodies of the universe; in short, to be a movable and a moving body no less than the moon, Jupiter, Venus, or any other planet. Later many detailed parallels were drawn between the earth and the moon. More comparisons were made with the moon than with other planets, perhaps from our having more and better sensible evidence about the former by reason of its lesser dis-tance. And having finally concluded this second opinion to have more likelihood than the other, it seems to me that our next step should be to examine whether the earth must be con-sidered immovable, as most people have believed up to the present, or mobile, as many ancient philosophers believed and as others of more recent times consider it; and, if movable, what its motion may be.

SALVIATI: Now I know and recognize the signposts along our road. But before starting in again and going ahead, I ought to tell you that I question this last thing you have said, about our having concluded in favour of the opinion that the earth is endowed with the same properties as the heavenly bodies. For I did not conclude this, just as I am not deciding upon any other controversial proposition. My intention was only to adduce those arguments and replies, as much on one side as on the other – those questions and solutions which others have thought of up to the present time (together with a few which have occurred to me after long thought) – and then to leave the decision to the judgement of others.

SAGREDO: I allowed myself to be carried away by my own sentiments and, believing that what I felt in my heart ought to be felt by others too, I made that conclusion universal which should have been kept particular. This really was an error on my part, especially as I do not know the views of Simplicio, here present.

SIMPLICIO: I confess that all last night I was meditating on yesterday's material, and truly I find it to contain many beautiful considerations which are novel and forceful. Still, I am much more impressed by the authority of so many great authors, and in particular . . . You shake your head, Sagredo, and smile as if I had uttered some absurdity.

SAGREDO: I merely smile, but believe me, I am hardly able to keep from laughing, because I am reminded of a situation that I witnessed not many years ago together with some friends of mine. One day I was at the home of a very famous doctor in Venice. It happened on this day that he was investigating the source and origin of the nerves. The anatomist showed that the great trunk of nerves, leaving the brain and passing through the nape, extended on down the spine and then branched out through the whole body, and that only a single strand as fine as a thread arrived at the heart. Turning to a gentleman whom he knew to be a Peripatetic philosopher, he asked this man whether he was at last satisfied and convinced that the nerves originated in the brain and not in the heart. The philosopher, after considering for awhile, answered: 'You have made me see this matter so plainly and palpably that if Aristotle's text were not contrary to it,

stating clearly that the nerves originate in the heart, I should
be forced to admit it to be true.'

SIMPLICIO: [But] Aristotle acquired his great authority only
because of the strength of his proofs and the profundity of his
arguments. One must have a grasp of the whole scheme, and
be able to combine this passage with that, collecting together
one text here and another very distant from it. There is no
doubt that whoever has this skill will be able to draw from
his books demonstrations of all that can be known; for every
single thing is in them.

SALVIATI: It is the followers of Aristotle who have crowned
him with authority, not he who has usurped or appropriated
it to himself. And since it is handier to conceal oneself under
the cloak of another than to show one's face in open court,
they dare not in their timidity get a single step away from
him, and rather than put any alterations into the heavens of
Aristotle, they want to deny out of hand those that they see
in nature's heaven.

SAGREDO: Such people remind me of that sculptor who, having transformed a huge block of marble into the image of a
Hercules or a thundering Jove, I forget which, and having
with consummate art made it so lifelike and fierce that it
moved everyone with terror who beheld it, he himself began
to be afraid, though all its vivacity and power were the work
of his own hands; and his terror was such that he no longer
dared affront it with his mallet and chisel. Oh, the inexpressible baseness of abject minds! To make themselves slaves
willingly; to accept decrees as inviolable; to place themselves
under obligation and to call themselves persuaded and convinced by arguments so 'powerful' and 'clearly conclusive'
that they themselves cannot tell the purpose for which they
were written, or what conclusion they serve to prove!

SALVIATI: But we had better get back to shore, lest we enter into a
boundless ocean and not get out of it all day. Let the beginning
of our reflections be the consideration that whatever motion
comes to be attributed to the earth must necessarily remain
imperceptible to us and as if non-existent, so long as we look
only at terrestrial objects; for as inhabitants of the earth, we

consequently participate in the same motion. But, on the other hand, it is indeed just as necessary that it display itself very generally in all other visible bodies and objects which, being separated from the earth, do not take part in this movement. So, the true method of investigating whether any motion can be attributed to the earth, and, if so, what it may be, is to observe and consider whether bodies separated from the earth exhibit some appearance of motion which belongs equally to all. Now there is one motion which is most general and supreme over all, and it is that by which the sun, moon and all other planets and fixed stars – in a word, the whole universe, the earth alone excepted – appear to be moved as a unit from east to west in the space of twenty-four hours. This, insofar as first appearances are concerned, may just as logically belong to the earth alone as the rest of the universe, since the same appearances would prevail as much in the one situation as in the other.

First, let us consider only the immense bulk of the starry sphere in contrast with the smallness of the terrestrial globe, which is contained in the former so many millions of times. Now if we think of the velocity of motion required to make a complete rotation in a single day and night, I cannot persuade myself that anyone could be found who would think it the more reasonable and credible thing that it was the celestial sphere which did the turning, and the terrestrial globe which remained fixed.

SAGREDO: If, through the whole variety of effects that could exist in nature as dependent upon these motions, all the same consequences followed indifferently to a hair's breadth from both positions, still my first general impression of them would be this: I should think that anyone who considered it more reasonable for the whole universe to move in order to let the earth remain fixed would be more irrational than one who should climb to the top of your cupola just to get a view of the city and its environs, and then demand that the whole countryside should revolve around him so that he would not have to take the trouble to turn his head.

SALVIATI: Despite much thinking about it, I have not been able to find any difference, so it seems to me that I have found that there can be no difference; hence I think it vain to seek one further. For consider: motion, in so far as it is and acts as motion, to that extent exists relatively to things that lack it; and among things which all share equally in any motion, it does not act, and is as if it did not exist. Thus the goods with which a ship is laden leaving Venice, pass by Corfu, by Crete, by Cyprus and go to Aleppo. Venice, Corfu, Crete, etc. stand still and do not move with the ship; but as to the sacks, boxes and bundles with which the boat is laden and with respect to the ship itself, the motion from Venice to Syria is as nothing, and in no way alters their relation among themselves. This is so because it is common to all of them and all share equally in it. If, from the cargo in the ship, a sack were shifted from a chest one single inch, this alone would be more of a movement for it than the two-thousand-mile journey made by all of them together.

It is obvious, then, that motion which is common to many moving things is idle and inconsequential to the relation of these movables among themselves, nothing being changed among them, and that it is operative only in the relation that they have with other bodies lacking that motion, among which their location is changed. Now, having divided the universe into two parts, one of which is necessarily movable and the other motionless, it is the same thing to make the earth alone move, and to move all of the rest of the universe, so far as concerns any result which may depend upon such movement. For the action of such a movement is only in the relation between the celestial bodies and the earth, which relation alone is changed. Now if precisely the same effect follows whether the earth is made to move and the rest of the universe stay still, or the earth alone remains fixed while the whole universe shares one motion, who is going to believe that nature (which by general agreement does not act by means of many things when it can do so by means of few) has chosen to make an immense number of extremely large bodies move with inconceivable velocities, to achieve what could have been done by a moderate movement of one single body around its own centre?

And let us redouble the difficulty with another very great one, which is this. If this great motion is attributed to the heavens, it has to be made in the opposite direction from the specific motion of all the planetary orbs, of which each one incontrovertibly has its own motion from west to east, this being very gentle and moderate, and must then be made to rush the other way; that is, from east to west, with this very rapid diurnal motion. Whereas by making the earth itself move, the contrariety of motions is removed, and the single motion from west to east accommodates all the observations and satisfies them all completely.

The improbability is shown for a third time in the relative disruption of the order which we surely see existing among those heavenly bodies whose circulation is not doubtful, but most certain. This order is such that the greater orbits complete their revolutions in longer times, and the lesser in shorter. And this very harmonious trend will not be altered a bit if the earth is made to move on itself in twenty-four hours. But if the earth is desired to remain motionless, it is necessary, after passing from the brief period of the moon to the other consecutively larger ones, and ultimately to that of Mars, in two years, and the greater one of Jupiter in twelve, and from this to the still larger one of Saturn, whose period is thirty years – it is necessary, I say, to pass on beyond to another incomparably larger sphere, and make this one finish an entire revolution in twenty-four hours. But by giving mobility to the earth, order becomes very well observed among the periods; from the very slow sphere of Saturn one passes on to the entirely immovable fixed stars, and manages to escape a fourth difficulty necessitated by supposing the stellar sphere to be movable. This difficulty is the immense disparity between the motions of the stars, some of which would be moving very rapidly in circles and others very slowly in little tiny circles, according as they are located further from or closer to the poles. Not only will the size of the circles and consequently the velocities of motions of these stars be very diverse from the orbits and motions of some others but (and this shall be the fifth difficulty) the same stars will keep changing their circles and their velocities, since those which two thousand years ago were on the celestial equator, and which consequently described great circles with their motion, are found in our time to be many degrees distant, and must be made slower in motion and reduced to moving in smaller circles.

For anyone who reasons soundly, the unlikelihood is increased – and this is the sixth difficulty – by the incomprehensibility of what is called the 'solidity' of that very vast sphere in whose depths are firmly fixed so many stars which, without changing place in the least among themselves, come to be carried around so harmoniously with such a disparity of motions. If, however, the heavens are fluid (as may much more reasonably be believed) so that each star roves around in it by itself, what law will regulate their motion so that, as seen from the earth, they appear as if made into a single sphere? For this to happen it seems to me that it is as much more effective and convenient to make them immovable than to have them roam around, as it is easier to count the myriad tiles set in a courtyard than to number the troop of children running around on them.

Finally, for the seventh objection, if we attribute the diurnal rotation to the highest heaven, then this has to be made of such strength and power as to carry with it the innumerable host of fixed stars, all of them vast bodies and much larger than the earth, as well as to carry along the planetary orbs despite the fact that the two move naturally in opposite ways. Besides this, one must grant that the element of fire and the greater part of the air are likewise hurried along, and that only the little body of the earth remains defiant and resistant to such power. This seems to me to be most difficult; I do not understand why the earth, a suspended body balanced on its centre and indifferent to motion or to rest, placed in and surrounded by an enclosing fluid, should not give in to such force and be carried around too. We encounter no such objections if we give the motion to the earth, a small and trifling body in comparison with the universe, and hence unable to do it any violence.

SAGREDO: When all things can proceed in most perfect harmony without introducing other huge and unknown spheres; without other movements or imparted speedings; with every sphere having only its simple motion, unmixed with contrary movements; and with everything taking place in the same direction, as must be the case if all depend upon a single principle, why reject the means of doing this, and give assent to such outlandish and laboured conditions?

SIMPLICIO: The point is to find a simple and ready means.

SAGREDO: This seems to me to be found, and quite elegantly. Make the earth the Primum Mobile; that is, make it revolve upon itself in twenty-four hours in the same way as all the other spheres. Then, without its imparting such a motion to any other planet or star, all of them will have their risings, settings and, in a word, all their other appearances.

SIMPLICIO: The crucial thing is being able to move the earth without causing a thousand inconveniences.

SALVIATI: All inconveniences will be removed as you propound them. Up to this point, only the first and most general reasons have been mentioned which render it not entirely improbable that the daily rotation belongs to the earth rather than the rest of the universe. Nor do I set these forth to you as inviolable laws, but merely as probable reasons. For I understand very well that one single experiment or conclusive proof to the contrary would suffice to overthrow both these and a great many other probable arguments.

SAGREDO: Simplicio, begin producing those difficulties that seem to you to contradict this new arrangement of the universe.

SIMPLICIO: It will be better for you to bring them up, for having given them greater study you will have them readier at hand, and in great number too.

SALVIATI: As the strongest reason of all is adduced that of heavy bodies, which, falling down from on high, go by a straight and vertical line to the surface of the earth. This is considered an irrefutable argument for the earth being motionless. For if it made the diurnal rotation, a tower from whose top a rock was let fall, being carried by the whirling of the earth, would travel many hundreds of yards to the east in the time the rock would consume in its fall, and the rock ought to strike the earth that distance away from the base of the tower. This effect they support with another experiment, which is to drop a lead ball from the top of the mast of a boat at rest, noting the place where it hits, which is close to the foot of the mast; but if the same ball is dropped from the same place when the boat is moving, it will strike at that distance from the foot of the mast which the boat will have run during the time of fall of the lead, and for no other reason than that the natural movement of the ball when set free is in a straight line toward the centre of the earth.

This argument is fortified with the experiment of a projectile sent a very great distance upward; this might be a ball shot from a cannon aimed perpendicular to the horizon. In its flight and return this consumes so much time that in our latitude the cannon and we would be carried together many miles eastward by the earth, so that the ball, falling, could never come back near the gun, but would fall as far to the west as the earth had run on ahead.

They add moreover the third and very effective experiment of shooting a cannon ball to the east, and then another one with equal charge at the same elevation to the west; the shot toward the west ought to range a great deal further out than the other one to the east. For when the ball goes toward the west, and the cannon, carried by the earth, goes east, the ball ought to strike the earth at a distance from the cannon equal to the sum of the two motions, one made by itself to the west, and the other by the gun, carried by the earth, toward the east. On the other hand, from the trip made by the ball shot toward the east it would be necessary to subtract that which was made by the cannon following it. Suppose, for example, that the journey made by the ball in itself was five miles and that the earth in that latitude travelled three miles during the flight of the ball; in the shot toward the west, the ball would fall to earth eight miles distant from the gun – that is, its own five toward the west and the gun's three to the east. But the shot toward the east would range no further than two miles, which is all that remains after subtracting from the five of the shot the three of the gun's motion toward the same place. Now experiment shows the shots to fall equally; therefore the cannon is motionless, and consequently the earth is, too. Not only this, but shots to the south or north likewise confirm the stability of the earth; for they would never hit the mark that one had aimed at, but would always slant toward the west because of the travel that would be made toward the east by the target, carried by the earth while the ball was in the air. And not merely shots along the meridians, but even those made to the east or west would not range truly; for the easterly shots would carry high and the westerly low whenever they were aimed point-blank. Since the shots in both directions take the path of a tangent – that is, a line parallel to the horizon – and the horizon is always falling away to the east and rising in the west if the diurnal motion belongs to the earth (which is why the eastern stars appear to rise and the western stars to set), it follows that the target to the east would drop away under the shot, wherefore the shot would range high, and the rising of the western target would make the shot to the west low. Hence in no direction would shooting ever be accurate; and since experience is contrary to this, it must be said that the earth is immovable.

SIMPLICIO: Oh, these are excellent arguments, to which it will be impossible to find a valid answer.

SALVIATI: Perhaps they are new to you?

SIMPLICIO: Yes, indeed, and now I see with how many elegant experiments nature graciously wishes to aid us in coming to the recognition of the truth. Oh, how well one truth accords with another, and how all co-operate to make themselves indomitable!

SAGREDO: What a shame there were no cannons in Aristotle's time! With them he would indeed have battered down ignorance, and spoken without the least hesitation concerning the universe.

SALVIATI: It suits me very well that these arguments are new to you, for now you will not remain of the same opinion as most Peripatetics, who believe that anyone who departs from Aristotle's doctrine must therefore have failed to understand his proofs. But you will certainly see further novelties; you will hear the followers of the new system producing observations, experiments and arguments against it more forcible than those adduced by Aristotle and Ptolemy and the other opponents of the same conclusions. Thus you will become assured that it is not through ignorance or inexperience that they have learned to adhere to such opinions.

SAGREDO: This is the time for me to tell you a few of the things that happened to me when I first began to hear these opinions spoken of. I was then a youth who had scarcely finished the course in philosophy, giving this up in order to apply myself to other activities. It happened that a certain foreigner from Rostock, whose name I believe was Christian Wursteisen, a supporter of the Copernican opinion, arrived in these parts and gave two or three lectures in an academy on this subject. He had a throng of hearers, more from the novelty of the subject than for any other reason, I think. I did not attend them, having formed a definite impression that this opinion could be nothing but solemn foolery. Later, asking about it from some who had gone, I heard them all making a joke of it except one, who told me that the matter was not entirely ridiculous. Since I considered this person an intelligent man and rather conservative, I was sorry that I had not gone; and from then on, as I happened from time to time to meet anyone who held the Copernican opinion, I asked him whether he had always believed in it. Among all the many whom I questioned, I found not a single one who did not tell me that he had long been of the contrary opinion, but had come over to this one, moved and persuaded by the force of its arguments. Examining them one by one then, to see how well they had mastered the arguments on the other side, I found them all to have these ready at hand, so that I could not truly say that they had forsaken that position out of ignorance or vanity or, so to speak, to show off their cleverness. On the other hand, so far as I questioned the Peripatetics and the Ptolemaics (for out of curiosity I asked many of them) how much they had studied Copernicus's book, I found very few who had so much as seen it, and none who I believed understood

it. Moreover I tried to find out from the followers of the Peripatetic doctrine whether any of them had ever held the other
opinion, and likewise found none.

From this, considering that everyone who followed the
opinion of Copernicus had at first held the opposite, and was
very well informed concerning the arguments of Aristotle and
Ptolemy, and that, on the other hand, none of the followers
of Ptolemy and Aristotle had been formerly of the Copernican opinion and had left that to come round to Aristotle's
view – considering these things, I say, I commenced to believe
that one who forsakes an opinion, which he imbibed with his
milk and which is supported by multitudes, to take up another
that has few followers and is rejected by all the schools, and
that truly seems to be a gigantic paradox, must of necessity be
moved, not to say compelled, by the most effective arguments. This made me very curious to get to the bottom of this
matter. And I consider it great fortune to have met you two,
from whom without any trouble I can hear everything that
has been said – perhaps all that can be said – on this subject,
certain that I ought by virtue of your reasonings to be lifted
out of doubt and put into a position of certainty.

But let us hear the rest of the arguments favourable to his
opinion so that we may proceed with their testing, refining
them in the crucible and weighing them in the assayer's
balance.

SALVIATI: Before going further I must tell you, Sagredo, that I
act the part of Copernicus in our arguments and wear his
mask. As to the internal effects upon me of the arguments
which I produce in his favour, I want you to be guided not by
what I say when we are in the heat of acting out our play, but
after I have put off the costume, for perhaps then you shall
find me different from what you saw of me on the stage.

In replying, those who make the earth movable answer
that the cannon and the ball which are on the earth share its
motion, or rather that all of them together have the same
motion naturally. Therefore the ball does not start from rest
at all, but to its motion about the centre joins one of projection upward which neither removes nor impedes the former.

In such a way, following the general eastward motion of the earth, it keeps itself continually over the same gun during both its rise and its return. You will see the same thing happen by making the experiment on a ship with a ball thrown perpendicularly upward from a catapult. It will return to the same place whether the ship is moving or standing still.

SAGREDO: For my part I am fully satisfied, and I understand perfectly that anyone who will impress upon his mind this general communication to all terrestrial things of the diurnal motion (which suits them all naturally, just as in the ancient idea it was considered that rest with respect to the centre suited them) will discern without any trouble the fallacy and the equivocation that make the arguments appear conclusive.

There remains for me only that doubt which I hinted at before, about the flight of birds. Since these have the lively faculty of moving at will in a great many ways, and of keeping themselves for a long time in the air, separated from the earth and wandering about with the most irregular turnings, I am not entirely able to see how among such a great mixture of movements they can avoid becoming confused and losing the original common motion. Once having been deprived of it, how could they make up for this or compensate for it by flying, and keep up with all the towers and trees which run with such a precipitous course toward the east? I say 'precipitous', because for the great circle of the globe it is little less than a thousand miles an hour, while I believe that the swallow in flight makes no more than fifty.

SALVIATI: The fact is that the birds' own motion – I mean that of flight – has nothing to do with the universal motion, from which it receives neither aid nor hindrance. What keeps that motion unaltered in the birds is the air itself through which they wander. This, following naturally the whirling of the earth, takes along the birds and everything else that is suspended in it, just as it carries the clouds. So the birds do not have to worry about following the earth, and so far as that is concerned they could remain for ever asleep.

This seems to me the place to show you a way to test [these ideas] very easily. Shut yourself up with some friend in the main cabin below decks on some large ship, and have with you there some flies, butterflies and other small flying animals. Have a large bowl of water with some fish in it; hang up a bottle that empties drop by drop into a wide vessel beneath it. With the ship standing still, observe carefully how the little animals fly with equal speed to all sides of the cabin. The fish swim indifferently in all directions; the drops fall into the vessel beneath; and, in throwing something to your friend, you need throw it no more strongly in one direction than another, the distances being equal; jumping with your feet together, you pass equal spaces in every direction. When you have observed all these things carefully (though there is no doubt that when the ship is standing still everything must happen in this way), have the ship proceed with any speed you like, so long as the motion is uniform and not fluctuating this way and that. You will discover not the least change in all the effects named, nor could you tell from any of them whether the ship was moving or standing still. In jumping, you will pass on the floor the same spaces as before, nor will you make larger jumps toward the stern than toward the prow even though the ship is moving quite rapidly, despite the fact that during the time that you are in the air the floor under you will be going in a direction opposite to your jump. In throwing something to your companion, you will need no more force to get it to him whether he is in the direction of the bow or the stern, with yourself situated opposite. The droplets will fall as before into the vessel beneath without dropping toward the stern, although while the drops are in the air the ship runs many spans. The fish in their water will swim toward the front of their bowl with no more effort than toward the back, and will go with equal ease to bait placed anywhere around the edges of the bowl. Finally the butterflies and flies will continue their flights indifferently toward every side, nor will it ever happen that they are concentrated toward the stern, as if tired out from keeping up with the course of the ship, from which they will have been separated during long intervals by keeping themselves in the air. And if smoke is made by burning some incense, it will be seen going up in the form of a little cloud, remaining still and moving no more toward one side than the other. The cause of all these correspondences of effects is the fact that the ship's motion is common to all the things contained in it, and to the air also. That is why I said you should be below decks; for if this took place above in the open air, which would not follow the course of the ship, more or less noticeable differences would be seen in some of the effects noted. No doubt the smoke would fall as much behind as the air itself. The flies likewise, and the butterflies, held back by the air, would be unable to follow the ship's motion if they were separated from it by a perceptible distance. But keeping themselves near it, they would follow it without effort or hindrance; for the ship, being an unbroken structure, carries with it a part of the nearby air.

SAGREDO: Although it did not occur to me to put these observations to the test when I was voyaging, I am sure that they would take place in the way you describe. In confirmation of this I remember having often found myself in my cabin wondering whether the ship was moving or standing still; and sometimes at a whim I have supposed it going one way when its motion was the opposite. Still, I am satisfied so far, and convinced of the worthlessness of all experiments brought forth to prove the negative rather than the affirmative side as to the rotation of the earth.

Third Day

SALVIATI: We shall next consider the annual movement generally attributed to the sun, but then, first by Aristarchus of Samos and later by Copernicus, removed from the sun and transferred to the earth.

Let us assume out of respect for Aristotle that the universe (of the magnitude of which we have no sensible information beyond the fixed stars), like anything that is spherical in shape and moves circularly, has necessarily a centre for its shape and for its motion. Being certain, moreover, that within the stellar sphere there are many orbs one inside another, with their stars which also move circularly, our question is this: which is it more reasonable to believe and to say; that these included orbs move around the same centre as the universe does, or around some other one which is removed from that? Now you, Simplicio, say what you think about this matter.

SIMPLICIO: If we could stop with this one assumption and were sure of not running into something else that would disturb us, I should think it would be much more reasonable to say that the container and the things it contained all moved around one common centre rather than different ones.

SALVIATI: Now if it is true that the centre of the universe is that point around which all the orbs and world bodies (that is, the planets) move, it is quite certain that not the earth, but the sun, is to be found at the centre of the universe. Hence, as for this first general conception, the central place is the sun's, and the earth is to be found as far away from the centre as it is from the sun.

SIMPLICIO: How do you deduce that it is not the earth, but the sun, which is at the centre of the revolutions of the planets?

SALVIATI: This is deduced from most obvious and therefore most powerfully convincing observations. The most palpable of these, which excludes the earth from the centre and places the sun there, is that we find all the planets closer to the earth at one time and further from it at another. The differences are so great that Venus, for example, is six times as distant from us at its furthest as at its closest, and Mars soars nearly eight times as high in the one state as in the other. You may thus see whether Aristotle was not some trifle deceived in believing that they were always equally distant from us.

SIMPLICIO: But what are the signs that they move around the sun?

SALVIATI: This is reasoned out from finding the three outer planets – Mars, Jupiter and Saturn – always quite close to the earth when they are in opposition to the sun, and very distant when they are in conjunction with it. This approach and recession is of such moment that Mars when close looks sixty times as large as when it is most distant. Next, it is certain that Venus and Mercury must revolve around the sun, because of their never moving far away from it, and because of their being seen now beyond it and now on this side of it, as Venus's changes of shape conclusively prove. As to the moon, it is true that this can never separate from the earth in any way, for reasons that will be set forth more specifically as we proceed.

SAGREDO: I have hopes of hearing still more remarkable things arising from this annual motion of the earth than were those which depended upon its diurnal rotation.

SALVIATI: You will not be disappointed, for as to the action of the diurnal motion upon celestial bodies, it was not and could not be anything different from what would appear if the universe were to rush speedily in the opposite direction. But this annual motion, mixing with the individual motions of all the planets, produces a great many oddities which in the past have baffled all the greatest men in the world.

Now returning to these first general conceptions, I repeat that the centre of the celestial rotation for the five planets, Saturn, Jupiter, Mars, Venus and Mercury, is the sun; this will hold for the earth too, if we are successful in placing that in the heavens. Then as to the moon, it has a circular motion around the earth, from which as I have already said it cannot be separated; but this does not keep it from going around the sun along with the earth in its annual movement.

SIMPLICIO: I am not yet convinced of this arrangement at all. Perhaps I should understand it better from the drawing of a diagram, which might make it easier to discuss.

SALVIATI: That shall be done. But for your greater satisfaction and your astonishment, too, I want you to draw it yourself. You will see that however firmly you may believe yourself not to understand it, you do so perfectly, and just by answering my questions you will describe it exactly. So take a sheet of paper and the compasses; let this page be the enormous expanse of the universe, in which you have to distribute and arrange its parts as reason shall direct you. And first, since you are sure without my telling you that the earth is located in this universe, mark some point at your pleasure where you intend this to be located, and designate it by means of some letter.

SIMPLICIO: Let this be the place of the terrestrial globe, marked A.

SALVIATI: Very well. I know in the second place that you are aware that this earth is not inside the body of the sun, nor even contiguous to it, but is distant from it by a certain space. Therefore assign to the sun some other place of your choosing, as far from the earth as you like, and designate that as O.

SIMPLICIO: Here I have done it; let this be the sun's position, marked O.

SALVIATI: These two established, I want you to think about placing Venus in such a way that its position and movement can conform to what sensible experience shows us about it. Hence you must call to mind, either from past discussions or from your own observations, what you know happens with this star. Then assign it whatever place seems suitable for it to you.

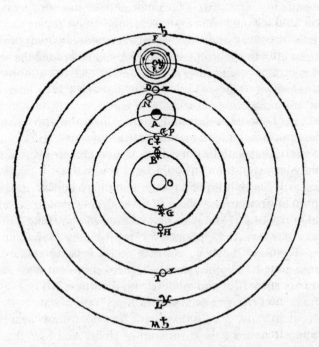

SIMPLICIO: I shall assume that those appearances are correct which you have related and which I have read also in the booklet of theses; that is, that this star never recedes from the sun beyond a certain definite interval of forty degrees or so; hence it not only never reaches opposition to the sun, but not even quadrature, nor so much as a sextile aspect. Moreover, I shall assume that it displays itself to us about forty times as large at one time than at another; greater when, being retrograde, it is approaching evening conjunction with the sun, and very small when it is moving forward toward morning conjunction, and furthermore that when it appears very large, it reveals itself in a horned shape, and when it looks very small it appears perfectly round. These appearances being correct, I say, I do not see how to escape affirming that this star revolves in a circle around the sun, in such a way that this circle cannot possibly be said to embrace and contain within itself the earth, nor to be beneath the sun (that is, between the sun and the earth), nor yet beyond the sun. Such a circle cannot embrace the earth because then Venus would sometimes be in opposition to the sun; it cannot be beneath the sun, for then Venus would appear sickle-shaped at both conjunctions; and it cannot be beyond the sun, since then it would always look round and never horned. Therefore for its lodging I shall draw the circle CH around the sun, without having this include the earth.

SALVIATI: Venus provided for, it is fitting to consider Mercury, which, as you know, keeping itself always around the sun, recedes therefrom much less than Venus. Therefore consider what place you should assign to it.

SIMPLICIO: There is no doubt that, imitating Venus as it does, the most appropriate place for it will be a smaller circle, within this one of Venus and also described about the sun. A reason for this, and especially for its proximity to the sun, is the vividness of Mercury's splendour surpassing that of Venus and all the other planets. Hence on this basis we may draw its circle here and mark it with the letters BG.

SALVIATI: Next, where shall we put Mars?

SIMPLICIO: Mars, since it does come into opposition with the sun, must embrace the earth with its circle. And I see that it must also embrace the sun; for, coming into conjunction with the sun, if it did not pass beyond it but fell short of it, it would appear horned as Venus and the moon do. But it always looks round; therefore its circle must include the sun as well as the earth. And since I remember your having said that when it is in opposition to the sun it looks sixty times as large as when in conjunction, it seems to me that this phenomenon will be well provided for by a circle around the sun embracing the earth, which I draw here and mark DI. When Mars is at the point D, it is very near the earth and in opposition to the sun, but when it is at the point I, it is in conjunction with the sun and very distant from the earth. And since the same appearances are observed with regard to Jupiter and Saturn (although with less variation in Jupiter than in Mars, and with still less in Saturn than in Jupiter), it seems clear to me that we can also accommodate these two planets very neatly with two circles, still around the sun. This first one, for Jupiter, I mark EL; the other, higher, for Saturn, is called FM.

SALVIATI: So far you have comported yourself uncommonly well. And since, as you see, the approach and recession of the three outer planets is measured by double the distance between the earth and the sun, this makes a greater variation in Mars than in Jupiter because the circle DI of Mars is smaller than the circle EL of Jupiter. Similarly, EL here is smaller than the circle FM of Saturn, so the variation is still less in Saturn than in Jupiter, and this corresponds exactly to the appearances. It now remains for you to think about a place for the moon.

SIMPLICIO: Following the same method (which seems to me very convincing), since we see the moon come into conjunction and opposition with the sun, it must be admitted that its circle embraces the earth. But it must not embrace the sun also, or else when it was in conjunction it would not look horned but always round and full of light. Besides, it would never cause an eclipse of the sun for us, as it frequently does, by getting in between us and the sun. Thus one must assign to it a circle around the earth, which shall be this one, NP, in such a way that when at P it appears to us here on the earth

A as in conjunction with the sun, which sometimes it will eclipse in this position. Placed at N, it is seen in opposition to the sun, and in that position it may fall under the earth's shadow and be eclipsed.

SALVIATI: Now what shall we do, Simplicio, with the fixed stars? Do we want to sprinkle them through the immense abyss of the universe, at various distances from any predetermined point, or place them on a spherical surface extending around a centre of their own so that each of them will be the same distance from that centre?

SIMPLICIO: I had rather take a middle course, and assign to them an orb described around a definite centre and included between two spherical surfaces – a very distant concave one and another, closer and convex, between which are placed at various altitudes the innumerable host of stars. This might be called the universal sphere, containing within it the spheres of the planets which we have already designated.

SALVIATI: Well, Simplicio, what we have been doing all this while is arranging the world bodies according to the Copernican distribution, and this has now been done by your own hand. Moreover, you have assigned their proper movements to them all except the sun, the earth and the stellar sphere. To Mercury and Venus you have attributed a circular motion around the sun without embracing the earth. Around the same sun you have caused the three outer planets, Mars, Jupiter and Saturn, to move, embracing the earth within their circles. Next, the moon cannot move in any way except around the earth and without embracing the sun. And in all these movements you likewise agree with Copernicus himself. It now remains to apportion three things among the sun, the earth and the stellar sphere: the state of rest, which appears to belong to the earth; the annual motion through the zodiac, which appears to belong to the sun; and the diurnal movement, which appears to belong to the stellar sphere, with all the rest of the universe sharing in it except the earth. And since it is true that all the planetary orbs (I mean Mercury, Venus, Mars, Jupiter and Saturn) move around the sun as a centre, it seems most reasonable for the state of rest to belong to the sun rather than to the earth – just as it does for the centre of any movable sphere to remain fixed, rather than some other point of it remote from the centre.

Next as to the earth, which is placed in the midst of moving objects – I mean between Venus and Mars, one of which makes its revolution in nine months and the other in two years – a motion requiring one year may be attributed to it much more elegantly than a state of rest, leaving the latter for the sun. And such being the case, it necessarily follows that the diurnal motion, too, belongs to the earth. For if the sun stood still, and the earth did not revolve upon itself but merely had the annual movement around the sun, our year would consist of no more than one day and one night; that is, six months of day and six months of night, as was remarked once previously. See, then, how neatly the precipitous motion of each twenty-four hours is taken away from the universe, and how the fixed stars (which are so many suns) agree with our sun in enjoying perpetual rest. See also what great simplicity is to be found in this rough sketch, yielding the reasons for so many weighty phenomena in the heavenly bodies.

SAGREDO: I see this very well indeed. But just as you deduce from this simplicity a large probability of truth in this system, others may, on the contrary, make the opposite deduction from it. If this very ancient arrangement of the Pythagoreans is so well accommodated to the appearances, they may ask (and not unreasonably) why it has found so few followers in the course of centuries; why it has been refuted by Aristotle himself, and why even Copernicus is not having any better luck with it in these latter days.

SALVIATI: Sagredo, if you had suffered even a few times, as I have so often, from hearing the sort of follies that are designed to make the common people contumacious and unwilling to listen to this innovation (let alone assent to it), then I think your astonishment at finding so few men holding this opinion would dwindle a good deal. It seems to me that we can have little regard for imbeciles who take it as a conclusive proof in confirmation of the earth's motionlessness, holding them firmly in this belief, when they observe that they cannot dine today at Constantinople and sup in Japan, or for those who are positive that the earth is too heavy to climb up over the sun and then fall headlong back down again. There is no need to bother about such men as these, whose name is legion, or to take notice of their fooleries. Neither need we try to convert men who define by generalizing and cannot make room for distinctions, just in order to have such fellows for our company in very subtle and delicate doctrines. Besides, with all the proofs in the world what would you expect to accomplish in the minds of people who are too stupid to recognize their own limitations?

No, Sagredo, my surprise is very different from yours. You wonder that there are so few followers of the Pythagorean opinion, whereas I am astonished that there have been any up to this day who have embraced and followed it. Nor can I ever sufficiently admire the outstanding acumen of those who have taken hold of this opinion and accepted it as true; they have through sheer force of intellect done such violence to their own senses as to prefer what reason told them over that which sensible experience plainly showed them to the contrary. For the arguments against the whirling of the earth that we have already examined are very plausible, as we have seen; and the fact that the Ptolemiacs and Aristotelians and all their disciples took them to be conclusive is indeed a strong argument of their effectiveness. But the experiences which overtly contradict the annual movement are indeed so much greater in their apparent force that, I repeat, there is no limit to my astonishment when I reflect that Aristarchus and Copernicus were able to make reason so conquer sense that, in defiance of the latter, the former became mistress of their belief.

SAGREDO: Then we are about to encounter still further strong attacks against this annual movement?

SALVIATI: We are, and such obvious and sensible ones that were it not for the existence of a superior and better sense than natural and common sense to join forces with reason, I much question whether I, too, should not have been much more recalcitrant toward the Copernican system than I have been since a clearer light than usual has illuminated me.

A while ago I sketched for you an outline of the Coperni-can system, against the truth of which the planet Mars launches a ferocious attack. For if it were true that the dis-tances of Mars from the earth varied as much from minimum to maximum as twice the distance from the earth to the sun, then when it is closest to us its disc would have to look sixty times as large as when it is most distant. Yet no such difference is to be seen. Rather, when it is in opposition to the sun and close to us, it shows itself as only four or five times as large as when, at conjunction, it becomes hidden behind the rays of the sun.

Another and greater difficulty is made for us by Venus, which, if it circulates around the sun as Copernicus says, would be now beyond it and now on this side of it, receding from and approaching toward us by as much as the diameter of the circle it describes. Then when it is beneath the sun and very close to us, its disc ought to appear to us a little less than forty times as large as when it is beyond the sun and near conjunction. Yet the difference is almost imperceptible.

Add to these another difficulty; for if the body of Venus is intrinsically dark, and, like the moon, it shines only by illu-mination from the sun, which seems reasonable, then it ought to appear horned when it is beneath the sun, as the moon does when it is likewise near the sun – a phenomenon which does not make itself evident in Venus. For that reason, Coper-nicus declared that Venus was either luminous in itself or that its substance was such that it could drink in the solar light and transmit this through its entire thickness in order that it might look resplendent to us. In this manner Copernicus par-doned Venus its unchanging shape, but he said nothing about its small variation in size; much less of the requirements of Mars. I believe this was because he was unable to rescue to his own satisfaction an appearance so contradictory to his view; yet being persuaded by so many other reasons, he main-tained that view and held it to be true.

Besides these things, to have all the planets move around together with the earth, the sun being the centre of their rota-tions, then the moon alone disturbing this order and having its own motion around the earth (going around the sun in a year together with the earth and the whole elemental sphere) seems in some way to upset the whole order and to render it improbable and false.

These are the difficulties which make me wonder at Aristarchus and Copernicus. They could not have helped noticing them, without having been able to resolve them; nevertheless they were confident of that which reason told them must be so in the light of many other remarkable observations. Thus they confidently affirmed that the structure of the universe could have no other form than that which they had described. Then there are other very serious but beautiful problems which are not easy for ordinary minds to resolve, but which were seen through and explained by Copernicus; these we shall put off until we have answered the objections of people who show themselves hostile to this position.

Coming now to the explanations and replies to the three grave objections mentioned, I say that the first two are not only not contrary to the Copernican system, but that they absolutely favour it, and greatly. For both Mars and Venus do show themselves variable in the assigned proportions, and Venus does appear horned when beneath the sun, and changes her shape in exactly the same way as the moon.

SAGREDO: But if this was concealed from Copernicus, how is it revealed to you?

SALVIATI: These things can be comprehended only through the sense of sight, which nature has not granted so perfect to men that they can succeed in discerning such distinctions. Rather, the very instrument of seeing introduces a hindrance of its own. But in our time it has pleased God to concede to human ingenuity an invention so wonderful as to have the power of increasing vision four, six, ten, twenty, thirty and forty times, and an infinite number of objects which were invisible, either because of distance or extreme minuteness, have become visible by means of the telescope.

SAGREDO: But Venus and Mars are not objects that are invisible because of any distance or small size. We perceive these by simple natural vision. Why, then, do we not discern the differences in their sizes and shapes?

SALVIATI: In this the impediment of our eyes plays a large part, as I have just hinted to you. On account of that, bright distant objects are not represented to us as simple and plain, but are festooned with adventitious and alien rays which are so long and dense that the bare bodies are shown as expanded ten, twenty, a hundred, or a thousand times as much as would appear to us if the little radiant crown which is not theirs were removed.

SAGREDO: Now I recall having read something of the sort, but I don't remember whether it was in the *Letters on Sunspots* or in *The Assayer* by our friend. It would be a good thing, in order to refresh my memory as well as to inform Simplicio, who perhaps has not read those works, to explain to us in more detail how the matter stands. For I should think that a knowledge of this would be most essential to an understanding of what is now under discussion.

SIMPLICIO: Everything that Salviati is presently setting forth is truly new to me. Frankly, I had no interest in reading those books, nor up till now have I put any faith in the newly introduced optical device. Instead, following in the footsteps of other Peripatetic philosophers of my group, I have considered as fallacies and deceptions of the lenses those things which other people have admired as stupendous achievements. If I have been in error, I shall be glad to be lifted out of it; and, charmed by the other new things I have heard from you, I shall listen most attentively to the rest.

SALVIATI: The confidence which men of that stamp have in their own acumen is as unreasonable as the small regard they have for the judgements of others. It is a remarkable thing that they should think themselves better able to judge such an instrument without ever having tested it, than those who have made thousands and thousands of experiments with it and make them every day. But let us forget about such headstrong people, who cannot even be censured without doing them more honour than they deserve.

It will not be difficult to understand how it might be that Mars, when in opposition to the sun and therefore seven or more times as close to the earth as when it is near conjunction, looks to us scarcely four or five times as large in the former state as in the latter. Nothing but irradiation is the cause of this. For if we deprive it of the adventitious rays, we shall find it enlarged in exactly the proper ratio. And to remove its head of hair from it, the telescope is the unique and supreme means. Enlarging its disc nine hundred or a thousand times, it causes this to be seen bare and bounded like that of the moon, and in the two positions varying in exactly the proper proportion.

Next in Venus, which at its evening conjunction when it is beneath the sun ought to look almost forty times as large as in its morning conjunction, and is seen as not even doubled, it happens in addition to the effects of irradiation that it is sickle-shaped, and its horns, besides being very thin, receive the sun's light obliquely and therefore very weakly. So that because it is small and feeble, it makes its irradiations less ample and lively than when it shows itself to us with its entire hemisphere lighted. But the telescope plainly shows us its horns to be as bounded and distinct as those of the moon, and they are seen to belong to a very large circle, in a ratio almost forty times as great as the same disc when it is beyond the sun, toward the end of its morning appearances.

SAGREDO: Oh Nicolaus Copernicus, what a pleasure it would have been for you to see this part of your system confirmed by so clear an experiment!

SALVIATI: Yes, but how much less would his sublime intellect be celebrated among the learned! For as I said before, we may see that with reason as his guide he resolutely continued to affirm what sensible experience seemed to contradict. I cannot get over my amazement that he was constantly willing to persist in saying that Venus might go around the sun and be more than six times as far from us at one time as at another, and still look always equal, when it should have appeared forty times larger.

It remains for us to remove what would seem to be a great objection to the motion of the earth. This is that though all the planets turn about the sun, the earth alone is not solitary like the others, but goes together in the company of the moon and the whole elemental sphere around the sun in one year, while at the same time the moon moves around the earth every month. Here one must once more exclaim over and exalt the admirable perspicacity of Copernicus, and simultaneously regret his misfortune at not being alive in our day. For, now, Jupiter removes this apparent anomaly of the earth and moon moving conjointly. We see Jupiter, like another earth, going around the sun in twelve years accompanied not by one but by four moons, together with everything that may be contained within the orbits of its four satellites.

SAGREDO: And what is the reason for your calling the four Jovian planets 'moons'?

SALVIATI: That is what they would appear to be to anyone who saw them from Jupiter. For they are dark in themselves, and receive their light from the sun; this is obvious from their being eclipsed when they enter into the cone of Jupiter's shadow. And since only that hemisphere of theirs is illuminated which faces the sun, they always look entirely illuminated to us who are outside their orbits and closer to the sun; but to anyone on Jupiter they would look completely lighted only when they were at the highest points of their circles. In the lowest part – that is, when between Jupiter and the sun – they would appear horned from Jupiter. In a word, they would make for Jovians the same changes of shape which the moon makes for us Terrestrials.

Now you see how admirably these three notes harmonize with the Copernican system, when at first they seemed so discordant with it. From this, Simplicio will be much better able to see with what great probability one may conclude that not the earth, but the sun, is the centre of rotation of the planets. And since this amounts to placing the earth among the world bodies which indubitably move about the sun (above Mercury and Venus but beneath Saturn, Jupiter and Mars), why will it not likewise be probable, or perhaps even necessary, to admit that it also goes around?

SIMPLICIO: These events are so large and so conspicuous that it is impossible for Ptolemy and his followers not to have had knowledge of them. And having had, they must also have found a way to give reasons sufficient to account for such sensible appearances; congruous and probable reasons, since they have been accepted for so long by so many people.

SALVIATI: You argue well, but you must know that the principal activity of pure astronomers is to give reasons just for the appearances of celestial bodies, and to fit to these and to the motions of the stars such a structure and arrangement of circles that the resulting calculated motions correspond with those same appearances. They are not much worried about admitting anomalies that might in fact be troublesome in other respects. Copernicus himself writes, in his first studies, of having rectified astronomical science upon the old Ptolemaic assumptions, and corrected the motions of the planets in such a way that the computations corresponded much better with the appearances, and vice versa. But this was still taking them separately, planet by planet. He goes on to say that when he wanted to put together the whole fabric from all individual constructions, there resulted a monstrous chimera composed of mutually disproportionate members, incompatible as a whole. Thus, however well the astronomer might be satisfied merely as a calculator, there was no satisfaction and peace for the astronomer as a scientist. And since he very well understood that although the celestial appearances might be saved by means of assumptions essentially false in nature it would be very much better if he could derive them from true suppositions, he set himself to inquiring diligently whether any one among the famous men of antiquity had attributed to the universe a different structure from that of Ptolemy's which is commonly accepted. Finding that some of the Pythagoreans had, in particular, attributed the diurnal rotation to the earth, and others the annual revolution as well, he began to examine under these two new suppositions the appearances and peculiarities of the planetary motions, all of which he had readily at hand. And seeing that the whole then corresponded to its parts with wonderful simplicity, he embraced this new arrangement, and in it he found peace of mind.

Now I should like to know whether Tycho or any of his disciples has ever tried to investigate in any way whether any phenomenon is perceived in the stellar sphere by which one might boldly affirm or deny the annual motion of the earth.

SAGREDO: I should answer 'no' for them, they having had no need to do so, since Copernicus himself says that there is no such variation there; and they, arguing ad hominem, grant this to him. Then on this assumption they show the improbability which follows from it; namely, it would be required to make the sphere so immense that in order for a fixed star to look as large as it does, it would actually have to be so immense in bulk as to exceed the earth's orbit – a thing which is, as they say, entirely unbelievable.

SALVIATI: Now it might be that there is a variation, but that it is not looked for; or that because of its smallness, or through lack of accurate instruments, it was not known by Copernicus. Hence it would be a good thing to investigate with the greatest possible precision whether one could really observe such a variation as ought to be perceived in the fixed stars, assuming an annual motion of the earth. This is a thing which I firmly believe has not been done by anyone up to the present. Not only that, but perhaps, as I said, few people have well understood what it is that should be looked for.

SAGREDO: I have listened to your discourse with great pleasure, and with profit too; now, to make sure that I have understood everything, I shall state briefly the heart of your conclusions. It seems to me that you have explained to us two sorts of differing appearances as being those which because of the annual motion of the earth we might observe in the fixed stars. One is their variation in apparent size as we, carried by the earth, approach them or recede from them; the other (which likewise depends upon this same approach and retreat) is their appearing to us to be now more elevated and now less so on the same meridian. Besides this you tell us (and I thoroughly understand) that these two alterations do not occur equally in all stars, but to a greater extent in some, to a lesser in others, and not at all in still others. The approach and retreat by which the same star ought to appear larger at one time and smaller at another is imperceptible and practically non-existent for stars which are close to the pole of the ecliptic, but it is great for the stars placed in the ecliptic itself, being intermediate for those in between. The reverse is true of the other alteration; that is, the elevation or lowering is nil for stars along the ecliptic and large for those encircling the pole of the ecliptic, being intermediate for those in the middle. Furthermore, both these alterations are more perceptible in the closest stars, less sensible in those more distant, and would ultimately vanish for those extremely remote.

Fourth Day

SALVIATI: Up to this point the indications of [the earth's] mobility have been taken from celestial phenomena, seeing that nothing which takes place on the earth has been powerful enough to establish the one position any more than the other. This we have already examined at length by showing that all terrestrial events from which it is ordinarily held that the earth stands still and the sun and the fixed stars are moving would necessarily appear just the same to us if the earth moved and the others stood still. Among all sublunary things it is only in the element of water (as something which is very vast and is not joined and linked with the terrestrial globe as are all its solid parts, but is rather, because of its fluidity, free and separate and a law unto itself) that we may recognize some trace or indication of the earth's behaviour in regard to motion and rest. After having many times examined for myself the effects and events, partly seen and partly heard from other people, which are observed in the movements of the water; after, moreover, having read and listened to the great follies which many people have put forth as causes for these events, I have arrived at two conclusions which were not lightly to be drawn and granted. Certain necessary assumptions having been made: these are that if the terrestrial globe were immovable, the ebb and flow of the oceans could not occur naturally; and that when we confer upon the globe the movements just assigned to it, the seas are necessarily subjected to an ebb and flow agreeing in all respects with what is to be observed in them.

Two sorts of movement may be conferred upon a vessel so that the water contained in it acquires the property of running first toward one end and then toward the other, and rises and sinks there. The first would occur when one end is lowered and then the other, for under those conditions the water, running toward the depressed part, rises and sinks alternately at either end. But since this rising and sinking is nothing but a retreat from and an approach toward the centre of the earth, this sort of movement cannot be attributed to concavities in the earth itself as containing vessels of the waters. For such containers could not have parts able to approach toward or retreat from the centre of the terrestrial globe by any motion whatever that might be assigned to the latter.

The other sort of motion would occur when the vessel was moved without being tilted, advancing not uniformly but with a changing velocity, being sometimes accelerated and sometimes retarded. From this variation it would follow that the water (being contained within the vessel but not firmly adhering to it as do its solid parts) would because of its fluidity be almost separate and free, and not compelled to follow all the changes of its container. Thus the vessel being retarded, the water would retain a part of the impetus already received, so that it would run toward the forward end, where it would necessarily rise. On the other hand, when the vessel was speeded up, the water would retain a part of its slowness and would fall somewhat behind while becoming accustomed to the new impetus, remaining toward the back end, where it would rise somewhat.

We have already said that there are two motions attributed to the terrestrial globe; the first is annual, made by its centre along the circumference of its orbit about the ecliptic in the order of the signs of the zodiac (that is, from west to east), and the other is made by the globe itself revolving around its own centre in twenty-four hours (likewise from west to east) around an axis which is somewhat tilted, and not parallel to that of its annual revolution. From the composition of these two motions, each of them in itself uniform, I say that there results an uneven motion in the parts of the earth, now accelerated and now retarded by the additions and subtractions of the diurnal rotation upon the annual revolution.

Now if it is true (as is indeed proved by experience) that the acceleration and retardation of motion of a vessel makes the contained water run back and forth along its length, and rise and fall at its extremities, then who will make any trouble about granting that such an effect may – or rather, must – take place in the ocean waters? For their basins are subjected to just such alterations; especially those which extend from west to east, in which direction the movement of these basins is made.

Now this is the most fundamental and effective cause of the tides, without which they would not take place.

SAGREDO: It seems to me that you have done a great deal by opening the first portal to such lofty speculations. In your first general proposition, which seems to me to admit of no refutation, you have explained very persuasively why it would be impossible for the observed movements to take place in the ordinary course of nature if the basins containing the waters of the seas were standing still, and that on the other hand such alterations of the seas would necessarily follow if one assumed the movements attributed by Copernicus to the terrestrial globe for quite other reasons. If you had given us no more, this alone seems to me to excel by such a large margin the trivialities which others have put forth that just to think of those once more makes me ill. And I am much astonished that among men of sublime intellect, of whom there have been plenty, none have been struck by the incompatibility between the reciprocating motion of the contained waters and the immobility of the containing vessels, a contradiction which now seems so obvious to me.

SALVIATI: Among all the great men who have philosophized about this remarkable effect, I am more astonished at Kepler than at any other. Despite his open and acute mind, and though he has at his fingertips the motions attributed to the earth, he has nevertheless lent his ear and his assent to the moon's dominion over the waters, to occult properties and to such puerilities.

SAGREDO: In the conversations of these four days we have, then, strong evidences in favour of the Copernican system, among which three have been shown to be very convincing – those taken from the stoppings and retrograde motions of the planets, and their approaches toward and recessions from the earth; second, from the revolution of the sun upon itself, and from what is to be observed in the sunspots; and third, from the ebbing and flowing of the ocean tides.

SALVIATI: To these there may perhaps be added a fourth, from the fixed stars, since by extremely accurate observations of these there may be discovered those minimal changes that Copernicus took to be imperceptible.

Now, since it is time to put an end to our discourses, it remains for me to beg you that if later, in going over the things that I have brought out, you should meet with any difficulty or any question not completely resolved, you will excuse my deficiency because of the novelty of the concept and the limitations of my abilities; then because of the magnitude of the subject; and finally because I do not claim and have not claimed from others that assent which I myself do not give to this invention, which may very easily turn out to be a most foolish hallucination and a majestic paradox.

To you, Sagredo, though during my arguments you have shown yourself satisfied with some of my ideas and have approved them highly, I say that I take this to have arisen partly from their novelty rather than from their certainty, and even more from your courteous wish to afford me by your assent that pleasure which one naturally feels at the approbation and praise of what is one's own. And as you have obligated me to you by your urbanity, so Simplicio has pleased me by his ingenuity. Indeed, I have become very fond of him for his constancy in sustaining so forcibly and so undauntedly the doctrines of his master. And I thank you, Sagredo, for your most courteous motivation, just as I ask pardon of Simplicio if

I have offended him sometimes with my too-heated and opinionated speech. Be sure that in this I have not been moved by any ulterior purpose, but only by that of giving you every opportunity to introduce lofty thoughts, that I might be the better informed.

SIMPLICIO: You need not make any excuses; they are superfluous, and especially so to me, who, being accustomed to public debates, have heard disputants countless times not merely grow angry and get excited at each other, but even break out into insulting speech and sometimes come very close to blows.

As to the discourses we have held, and especially this last one concerning the reasons for the ebbing and flowing of the ocean, I am really not entirely convinced; but from such feeble ideas of the matter as I have formed, I admit that your thoughts seem to me more ingenious than many others I have heard. I do not therefore consider them true and conclusive; indeed, keeping always before my mind's eye a most solid doctrine that I once heard from a most eminent and learned person, and before which one must fall silent, I know that if asked whether God in His infinite power and wisdom could have conferred upon the watery element its observed reciprocating motion using some other means than moving its containing vessels, both of you would reply that He could have, and that He would have known how to do this in many ways which are unthinkable to our minds. From this I forthwith conclude that, this being so, it would be excessive boldness for anyone to limit and restrict the Divine power and wisdom to some particular fancy of his own.

SALVIATI: An admirable and angelic doctrine, and well in accord with another one, also Divine, which, while it grants to us the right to argue about the constitution of the universe (perhaps in order that the working of the human mind shall not be curtailed or made lazy) adds that we cannot discover the work of His hands. Let us, then, exercise these activities permitted to us and ordained by God, that we may recognize and thereby so much the more admire His greatness, however much less fit we may find ourselves to penetrate the profound depths of His infinite wisdom.

SAGREDO: And let this be the final conclusion of our four days' arguments, after which if Salviati should desire to take some interval of rest, our continuing curiosity must grant that much to him. But this is on condition that when it is more convenient for him, he will return and satisfy our desires – mine in particular – regarding the problems set aside and noted down by me to submit to him at one or two further sessions, in accordance with our agreement. Above all, I shall be waiting impatiently to hear the elements of our Academician's new science of natural and constrained local motions. Meanwhile, according to our custom, let us go and enjoy an hour of refreshment in the gondola that awaits us.

Campanella, Letter to Galileo, 1632

Campanella wrote this letter to Galileo immediately after reading his Dialogue on the Great World Systems, *and it allows us to see the ways a supporter of Galileo's understood the book just before the affair of the trial began. Campanella clearly perceived the work as a persuasive defence of the Copernican system, and praised Galileo both for the style of his writing and for the force of his arguments. The changes Campanella saw in the world around him – philosophical, astronomical and political – signalled to him, he wrote, that he was living through the beginning of a new age.*

Rome, 5 August 1632

Most Illustrious and Excellent Lord,

I received Your Lordship's *Dialogue* from Lord Magalotti in July, as Your Lordship had predicted to me on 17 May, and I did not write to you immediately because it seemed better to read them first. Everyone plays his part admirably; Simplicio offers amusement for this philosophy comedy, and also shows the folly of his sect, the instability and obstinacy of their discourse, and how far it extends. You certainly need not envy Plato's writing! Salviati is a great Socrates who helps others give birth instead of giving birth himself; and Sagredo is a free mind who, without being tainted by the schools, judges them all with great sagacity. I liked everything in it, and I see how much more force the argument of Copernicus has. Regarding the movement of the sea, I am not at all in agreement with you for now. I would also like Your Lordship to bring to light soon your little work on motion, because I sense from what you say here that it will be very useful for philosophizing.

I defend this book against everyone with respect to the decree against the motion of the earth. These new approaches to ancient truths, new worlds, new stars, new systems, new nations, etc. are the beginning of a new age.

Fra Tommaso Campanella
Your true friend and servant

Galileo, Trial Documents, 1632–1633*

Pope Urban VIII, elected in 1623, was an admirer of Galileo, and Galileo had dedicated his 1623 Assayer *to him. In conversations with Galileo in 1624, he seems to have indicated that a hypothetical discussion of Copernicanism was acceptable, as the Vatican secretary also suggested in his review of Galileo's earlier draft of his* Dialogue *in 1631. Yet immediately after Galileo's* Dialogue *was published, complaints began to emerge. Upon review of Galileo's Inquisition file, the Special Injunction of February 1616 was found, in which Galileo was said to have been warned by Cardinal Bellarmine not to discuss Copernicanism in any way whatsoever – seemingly even hypothetically. Bellarmine was no longer alive, Galileo's signature was not on the document, and Galileo's own recollections of their conversations differed, as did other documents, to suggest that hypothetical discussion was allowed. The trial of Galileo revolved in part around this question – just what had been banned earlier, in 1616 – and around Galileo's own attitude, and whether he had genuinely treated Copernicanism hypothetically, or only pretended to do so while arguing in favour of it as true. While Galileo was formally threatened with torture over the course of his trial, he was never actually tortured, contrary to later myths. He was eventually found guilty and sentenced to house arrest for the remainder of his life, and his* Dialogue *was banned. Though some versions of the story end with Galileo muttering to the inquisitors about*

---

* Translations taken from *The Galileo Affair: A Documentary History*, translated by Maurice A. Finocchiaro (University of California Press, 1989).

the earth 'and yet it moves' after his forced abjuration, this,
too, is likely a myth.

Special Commission Report on the *Dialogue*,
September 1632

We think that Galileo may have overstepped his instructions by
asserting absolutely the earth's motion and the sun's immobility
and thus deviating from hypothesis; that he may have wrongly
attributed the existing ebb and flow of the sea to the non-existent
immobility of the sun and motion of the earth, which are the
main things; and that he may have been deceitfully silent about
an injunction given him by the Holy Office in the year 1616,
whose tenor is: 'that he abandon completely the above-mentioned
opinion that the sun is the centre of the world and the earth
moves, nor henceforth hold, teach, or defend it in any way whatever, orally or in writing; otherwise the Holy Office would start
proceedings against him. He acquiesced in this injunction and
promised to obey.' One must now consider how to proceed,
both against the person and concerning the printed book.

In point of fact:

1. Galileo did come to Rome in the year 1630, and he brought
    and showed his original manuscript to be reviewed for
    printing. Though he had been ordered to discuss the Coper-
    nican system only as a pure mathematical hypothesis, one
    found immediately that the book was not like this, but that
    it spoke absolutely, presenting the reasons for and against
    though without deciding. Thus the Master of the Sacred
    Palace determined that the book be reviewed and be
    changed to the hypothetical mode: it should have a preface
    to which the body would conform, which would describe
    this manner of proceeding, and which would prescribe it to
    the whole dispute to follow, including the part against the
    Ptolemaic system, carried on merely ad hominem and to
    show that in reproving the Copernican system the Holy
    Congregation had heard all the arguments.

2. To follow this through, the book was given for review with these orders to Father Raffaello Visconti, an associate of the Master of the Sacred Palace, since he was a professor of mathematics. He reviewed it and emended it in many places, informing the Master about others disputed with the author, which the Master took out without further discussion. Having approved it for the rest, Father Visconti was ready to give it his endorsement to be placed at the beginning of the book as usual, if the book were to be printed in Rome as it was then presumed. We have written to the Inquisitor to send this endorsement to us and are expecting to receive it momentarily, and we have also sent for the original so that we can see the corrections made.

3. The Master of the Sacred Palace wanted to review the book himself, but since the author complained about the unusual practice of a second revision and about the delay, to facilitate the process it was decided that before sending it to press the Master would see it page by page. In the meantime, to enable the author to negotiate with printers, he was given the imprimatur for Rome, the book's beginning was compiled, and printing was expected to begin soon.

4. The author then went back to Florence, and after a certain period he petitioned to print it in that city. The Master of the Sacred Palace absolutely denied the request, answering by saying that the original should be brought back to him to make the last revision agreed upon and that without this he for his part would have never given permission to print it. The reply was that the original could not be sent because of the dangers of loss and the plague. Nevertheless, after the intervention of His Highness there, it was decided that the Master of the Sacred Palace would remove himself from the case and refer it to the Inquisitor of Florence: the Master would describe to him what was required for the correction of the book and would leave him the decision to print it or not, so that he would be using his authority without any responsibility on the part of the Master's office. Accordingly, he wrote to the Inquisitor the letter whose copy is appended here, labelled A, dated 24 May 1631, received

and acknowledged by the Inquisitor with the letter labelled B, where he says he entrusted the book to Father Stefani, Consultant to the Holy Office there. Then a brief composition of the book's preface was sent to the Inquisitor, so that the author would incorporate it with the whole, would embellish it in his own way, and would make the ending of the *Dialogue* conform with it. A copy of the sketch that was sent is enclosed labelled C, and a copy of the accompanying letter is enclosed labelled D.

5. After this the Master of the Sacred Palace was no longer involved in the matter, except when, the book having been printed and published without his knowledge, he received the first few copies and held them in customs, seeing that the instructions had not been followed. Then, upon orders from Our Master, he had them all seized where it was not too late and diligence made it possible to do so.

6. Moreover, there are in the book the following things to consider, as specific items of indictment:

   i. That he used the imprimatur for Rome without permission and without sharing the fact of the book's being published with those who are said to have granted it.

   ii. That he had the preface printed with a different type and rendered it useless by its separation from the body of the work; and that he put the 'medicine of the end' in the mouth of a fool and in a place where it can only be found with difficulty, and then he had it approved coldly by the other speaker by merely mentioning but not elaborating the positive things he seems to utter against his will.

   iii. That many times in the work there is a lack of and deviation from hypothesis, either by asserting absolutely the earth's motion and the sun's immobility, or by characterizing the supporting arguments as demonstrative and necessary, or by treating the negative side as impossible.

   iv. He treats the issue as undecided and as if one should await rather than presuppose the resolution.

v. The mistreatment of contrary authors and those most used by the Holy Church.

vi. That he wrongly asserts and declares a certain equality between the human and the divine intellect in the understanding of geometrical matters.

vii. That he gives as an argument for the truth the fact that Ptolemaics occasionally become Copernicans, but the reverse never happens.

viii. That he wrongly attributed the existing ebb and flow of the sea to the non-existent immobility of the sun and motion of the earth.

All these things could be emended if the book were judged to have some utility which would warrant such a favour.

7. In 1616 the author had from the Holy Office the injunction that 'he abandon completely the above-mentioned opinion that the sun is the centre of the world and the earth moves, nor henceforth hold, teach, or defend it in any way whatever, orally or in writing; otherwise the Holy Office would start proceedings against him'. He acquiesced in this injunction and promised to obey.

Cardinal Barberini to the Florentine Nuncio, 25 September 1632

Because some suspicious things were discovered in Galileo's works, out of regard for the Most Serene Grand Duke His Holiness appointed a special commission to examine them and see whether one could avoid bringing them to the attention of the Sacred Congregation of the Holy Office; the commission met five times, considered everything carefully, and decided that one could not avoid bringing the business to the attention of the Congregation. To show his goodwill toward His Highness, His Holiness sent this information to His Highness's Lord Ambassador, who in the name of His Highness has implored His Holiness not to forward the case to the Congregation. To the person who brought him the news, the Ambassador said that the decision made little sense in view of the fact that the book had been seen and approved by the Master of the Sacred Palace; but it was pointed out to him that if the book actually contains errors, these should not for that reason be allowed to circulate. When told all this, His Excellency was bound to secrecy by the Holy Office, but he was given permission to convey the information to the Most Serene Grand Duke, who was bound to secrecy in the same way.

Thus the book was brought before the Congregation of the Holy Office, and everything was considered with the utmost seriousness. It was decided to order the Father Inquisitor of your city to call Galileo and in the name of His Holiness give him an injunction to appear before the Father Commissary of the Holy Office for the whole month of October; the Inquisitor is supposed to make Galileo promise to obey this injunction in the presence of witnesses, so that, if he refuses to accept and obey it, they could be examined . . .

Excerpts from Galileo's First Deposition, 12 April 1633

Galileo, son of the late Vincenzio Galilei, Florentine, seventy years old, who, having taken a formal oath to tell the truth, was asked by the Fathers the following:

Q: *Whether he knows or can guess the reason why he was ordered to come to Rome.*

A: I imagine that the reason why I have been ordered to present myself to the Holy Office in Rome is to account for my recently printed book. I imagine this because of the injunction to the printer and to myself, a few days before I was ordered to come to Rome, not to issue any more of these books, and similarly because the printer was ordered by the Father Inquisitor to send the original manuscript of my book to the Holy Office in Rome.

Q: That he explain the character of the book on account of which he thinks he was ordered to come to Rome.

A: It is a book written in dialogue form, and it treats of the constitution of the world, that is, of the two chief systems, and the arrangement of the heavens and the elements.

Q: Whether he was in Rome other times, especially in the year 1616, and for what occasion.

A: I was in Rome in the year 1616; then I was here in the second year of His Holiness Urban VIII's pontificate; and lastly I was here three years ago, the occasion being that I wanted to have my book printed. The occasion for my being in Rome in the year 1616 was that, having heard objections to Nicolaus Copernicus's opinion on the earth's motion, the sun's stability, and the arrangement of the heavenly spheres, in order to be sure of holding only holy and Catholic opinions, I came to hear what was proper to hold in regard to this topic.

Q: Whether he came of his own accord or was summoned, what the reason was why he was summoned, and with which person or persons he discussed the above-mentioned topics.

A: In 1616 I came to Rome of my own accord, without being summoned, for the reason I mentioned. In Rome I discussed this matter with some cardinals who oversaw the Holy Office at that time, especially with Cardinals Bellarmine, Aracoeli, San Eusebio, Bonsi, and d'Ascoli.

Q: What specifically he discussed with the above-mentioned cardinals.

A: The occasion for discussing with the said cardinals was that they wanted to be informed about Copernicus's doctrine, his book being very difficult to understand for those who are not professional mathematicians and astronomers. In particular they wanted to understand the arrangement of the heavenly spheres according to Copernicus's hypothesis, how he places the sun at the centre of the planets' orbits, how around the sun he places next the orbit of Mercury, around the latter that of

Venus, then the moon around the earth, and around this Mars, Jupiter and Saturn; and in regard to motion, he makes the sun stationary at the centre and the earth turn on itself and around the sun, that is, on itself with the diurnal motion and around the sun with the annual motion.

Q: *Since, as he says, he came to Rome to be able to have the resolution and the truth regarding the above, what then was decided about this matter.*

A: Regarding the controversy which centred on the above-mentioned opinion of the sun's stability and earth's motion, it was decided by the Holy Congregation of the Index that this opinion, taken absolutely, is repugnant to Holy Scripture and is to be admitted only suppositionally, in the way that Copernicus takes it.

Q: *Whether he was then notified of the said decision, and by whom.*

A: I was indeed notified of the said decision of the Congregation of the Index, and I was notified by Lord Cardinal Bellarmine.

Q: *What the Most Eminent Bellarmine told him about the said decision, whether he said anything else about the matter, and if so what.*

A: Lord Cardinal Bellarmine told me that Copernicus's opinion could be held suppositionally, as Copernicus himself had held it. His Eminence knew that I held it suppositionally, namely in the way that Copernicus held it, as you can see from an answer by the same Lord Cardinal to a letter of Father Master Paolo Antonio Foscarini, Provincial of the Carmelites; I have a copy of this, and in it one finds these words: 'I say that it seems to me that Your Paternity and Mr Galileo are proceeding prudently by limiting yourselves to speaking suppositionally and not absolutely.' This letter by the said Lord Cardinal is dated 12 April 1615. Moreover, he told me that otherwise, namely taken absolutely, the opinion could be neither held nor defended.

Q: What was decided and then made known to him precisely in the month of February 1616.

A: In the month of February 1616, Lord Cardinal Bellarmine told me that since Copernicus's opinion, taken absolutely, was contrary to Holy Scripture, it could be neither held nor defended, but it could be taken and used suppositionally. In conformity with this I keep a certificate by Lord Cardinal Bellarmine himself, dated 26 May 1616, in which he says that Copernicus's opinion cannot be held or defended, being against Holy Scripture. I present a copy of this certificate, and here it is.

And he showed a sheet of paper with twelve lines of writing on one side only, beginning 'We, Robert Cardinal Bellarmine, have' and ending 'on this 26th day of May 1616', signed 'The same mentioned above, Robert Cardinal Bellarmine.' This evidence was accepted and marked with the letter B. Then he added: I have the original of this certificate with me in Rome, and it is written all in the hand of the above-mentioned Lord Cardinal Bellarmine.

And having been told that the said injunction, given to him then in the presence of witnesses, states that he cannot in any way whatever hold, defend, or teach the said opinion, he was asked whether he remembers how and by whom he was so ordered.

A: I do not recall that this injunction was given me any other way than orally by Lord Cardinal Bellarmine. I do remember that the injunction was that I could not hold or defend, and maybe even that I could not teach. I do not recall, further, that there was the phrase in any way whatever, but maybe there was; in fact, I did not think about it or keep it in mind, having received a few months thereafter Lord Cardinal Bellarmine's certificate dated 26 May which I have presented and in which is explained the order given to me not to hold or defend the said opinion. Regarding the other two phrases in the said injunction now mentioned, namely not to teach and in any way whatever, I did not retain them in my memory, I think because they are not contained in the said certificate, which I relied upon and kept as a reminder.

Q: Whether, after the issuing of the said injunction, he obtained any permission to write the book identified by himself, which he later sent to the printer.

A: After the above-mentioned injunction I did not seek permission to write the above-mentioned book which I have identified, because I do not think that by writing this book I was contradicting at all the injunction given me not to hold, defend, or teach the said opinion, but rather that I was refuting it.

Q: Whether, when he asked the above-mentioned Master of the Sacred Palace for permission to print the above-mentioned book, he revealed to the same Most Reverend Father Master the injunction previously given to him concerning the directive of the Holy Congregation, mentioned above.

A: When I asked him for permission to print the book, I did not say anything to the Father Master of the Sacred Palace about the above-mentioned injunction because I did not judge it necessary to tell it to him, having no scruples since with the said book I had neither held nor defended the opinion of the earth's motion and sun's stability; on the contrary, in the said book I show the contrary of Copernicus's opinion and show that Copernicus's reasons are invalid and inconclusive.

With this the deposition ended, and he was assigned a certain room in the dormitory of the officials, located in the Palace of the Holy Office, in lieu of prison, with the injunction not to leave it without special permission, under penalty to be decided by the Holy Congregation; and he was ordered to sign below and was sworn to silence.

Inchofer's Report on the *Dialogue*, 17 April 1633

I am of the opinion that Galileo not only teaches and defends the immobility or rest of the sun or centre of the universe, around which both the planets and the earth revolve with their own motions, but also that he is vehemently suspected of firmly adhering to this opinion, and indeed that he holds it.

On the question of Galileo's mental attitude, it is certain on the basis of the reasons given under each heading that he teaches, defends and holds the opinion of the earth's motion and the sun's immobility and central position; in addition, however, all these things are shown very powerfully by that somewhat lengthy treatise of the same Galileo which, before the publication of the present *Dialogue*, he presented to the Grand Duke of Florence in support of his cause and in which he not only proved Copernicus's opinion but established it by solving scriptural difficulties as well as he could. Moreover, in discussing the passages from Scripture, especially those about the sun's motion, he did his best to show that Scripture speaks with a meaning adapted to popular opinion and that in reality the sun does not move. Then he ridiculed those who are strongly committed to the common scriptural interpretation of the sun's motion as if they were small-minded, unable to penetrate the depth of the issue, half-witted and almost idiotic. I have read this composition, and, if I am not deceived, here in Rome it passed through the hands of quite a few. These things are said in confirmation of previous claims.

I am of the opinion not only that Galileo teaches and defends the view of Pythagoras and Copernicus but also, if we consider the manner of proceeding, of reasoning, and then of expression, that he is vehemently suspected of firmly adhering to it, and indeed that he holds it.

Reasons why it appears that Galileo teaches, defends and holds the opinion of the earth's motion:

1. It is beyond question that Galileo teaches the earth's motion in writing. Indeed, his whole book speaks for itself. Nor can one teach in any other way those of future generations and those who are absent than through writing or by tradition.
2. A function of a teacher is to transmit those concepts of a discipline which he regards easier and more effective for acquiring willing and well-disposed disciples, especially in the case of an allegedly new discipline, which entices curious minds with wonder. How expert and adroit Galileo shows

himself to be in this sort of activity is something that the whole book makes clear to the careful reader.

3. Furthermore, one who teaches those who oppose his doctrine tries, as much as he can, to answer their objections and to expose their difficulties and even falsehoods. In this whole work Galileo argues for nothing more than to establish the doctrine of the earth's turning and to thoroughly defeat the contrary doctrine.

4. Next, something extraordinary happens, namely that Galileo attributes to the earth's turning all kinds of effects visible in nature, whose true causes are not unknown and have been found by others, as if it were the only genuine and proper cause; and so he teaches ad nauseam that the phenomena of sunspots, tides and terrestrial magnetism are of this sort. This, most assuredly, is a sign not only of wishing to teach but of actually teaching, even explaining many topics neither Copernicus nor his followers had thought about, as the author himself wishes to claim.

5. Galileo additionally deplores the fact that this opinion is understood by so few and that so many are committed to the long-established opinion; thus, he tries to have Simplicio unlearn it and, in his person, draw to his own opinion, if he can, all Peripatetics. Without a doubt he acts with the solicitude of a diligent teacher who desires to have and to make disciples. As St Augustine says in his commentary on Psalm 118 (section 17), to teach is nothing more than to impart knowledge, which is so connected to the teaching that one cannot exist without the other; if so, then it is clear that Galileo really and truly teaches this opinion, and indeed that under the name of 'the Academician' he acts as a teacher of those to whom he introduces the speakers in his Dialogue. Nor is the process of teaching or learning ever easier than when doctrines are expounded by means of a dialogue, as is well known from countless examples of great men.

Enough about the first point concerning the teaching of the doctrine in writing. Actually, that the same thing is not new for Galileo appears evident from the little book published a long time before, in which he is praised and defended for the same doctrine.

As regards the second point, that is, whether he defends it, though it can be easily deduced from what has been said, nevertheless undoubtedly the answer appears affirmative in this way:

1. Because if someone is said to defend an opinion he merely supports without refutation or destruction of the contrary view, how much more can this be said of one who defends it so as to wish the contrary completely destroyed? Hence, in law, to defend is sometimes equated with to attack.
2. Because Copernicus, satisfied with a simple system, merely explained celestial phenomena in terms of this hypothesis by what, so he thought, was an easier method; whereas Galileo, having searched for many additional arguments, confirms Copernicus's discoveries and introduces new ones, and this is a twofold defence.
3. Because this time Galileo's principal aim was to attack Father Christoph Scheiner, who more recently than anyone else had written against the Copernicans. But this is precisely to defend and to wish to strengthen the opinion of the earth's motion, lest perhaps it be ruined when attacked by others.
4. Because there is no better means of defence, indeed the strongest, than the one used by Galileo, namely, to present contrary arguments and in the process to refute and weaken them so that they seem without force, without reason and, accordingly, without any ability and judgement on the part of one's opponents.
5. Because if he had undertaken this discussion only with the purpose of engaging in a disputation and exercising the mind, he would not have waged such arrogant war against Ptolemaics and Aristotelians, nor would he have ridiculed Aristotle and his followers so insolently; rather he should have modestly proposed arguments for investigating and establishing the truth, and not impugning it, which granted he did not understand. Enough about the second point concerning the defence published in writing. From

these things one can also infer a conclusion concerning the defence carried on orally.

As regards the third point, that is, whether Galileo holds the opinion of the earth's physical motion so that he might be proved guilty of regarding it as true, an affirmative answer may be seen in two ways: first, by means of necessary consequences; second, by Galileo's own words, which are absolute and categorical, or certainly equivalent. Moreover, I submit that the mind of a speaker is tied to his words, and there is no value in any nice-sounding declarations he may be wont to make lest he be viewed as sinning against the Decree. Indeed my judgement will be based on facts indicating the contrary. But let us come to the evidence.

1. Because the reason which he pretends moved him to writing is empty and frivolous and insufficient to induce a wise man to undertake such a work; that is, that foreigners had murmured against the Decree and denounced the consultants to the Holy Congregation to be ignorant of astronomy. On this matter I did not see any publications by a single foreigner in which there is any mention of the Decree, or of the consultants, of whom even the designation is unknown to them. It is certain that among Catholics no one would have dared. And then, if this reason moved Galileo, why did he not undertake to defend the Decree and the Holy Congregation together with its consultants? Indeed, this would have been arguably preferable for him in order to respond to the cause for which he was writing. But this is so far from Galileo's preference that he tries to strengthen the Copernican opinion with new arguments of which foreigners would never think in this connection; and he writes in Italian, certainly not to extend the hand to foreigners or other learned men, but rather to entice to that view common people in whom errors very easily take root.

2. Anyone who discusses something for the sake of the argument, and not because he really is convinced of it, proceeds problematically either by laying down neither part as more

certain than the other or, having rejected the one, by adhering to the other which he regards as more certain. Galileo everywhere makes pronouncements by means of theorems and solid demonstrations (in his view), as if he wanted the opinion of the earth's immobility to be completely discarded.

3. Galileo promises to proceed in the manner of a mathematical hypothesis, but a mathematical hypothesis is not established by physical and necessary conclusions. For example, a mathematician assumes the existence of an infinite line, and from this he concludes that a triangle made of infinite lines is of infinite capacity; nevertheless, he neither proves nor believes that an infinite line exists, taking infinity in a strict sense. So Galileo should have posited the earth's motion as something he intended to analyse deductively, not as something to be proved true by destroying the opposite view, as indeed he does in the entire work.

4. Theologians inquire whether God exists, not because a Christian theologian doubts that he exists, but in order to show that, even independently of faith, God's existence can be proved with many arguments intelligible to us (as they say) and by refuting arguments to the contrary. If Galileo wanted to proceed hypothetically, he should have produced the arguments that seem to indicate the earth's motion; then, after criticizing them, he should have either supposed or proved the contrary, certainly not confuted it. And, indeed, I say this for the case where one is not proceeding purely mathematically, but rather, as Galileo does, with physical disputations intermingled; otherwise a mere supposition suffices for the mathematician, without assuming and requiring any demonstration of it.

5. Philosophers also inquire whether the world could have existed from eternity; yet no Christian says that it has existed from eternity, but merely that, assuming it existed from eternity, such and such consequences would follow necessarily or probably. So, in order to restrict himself to a pure mathematical hypothesis, Galileo did not have to prove absolutely that the earth moves, but only to conceive

its motion in the imagination without assuming it physically, and thereby explain celestial phenomena and derive
the numerical details of the various motions.

6. If Galileo did not firmly adhere to the view of the earth's
motion as if he believed it true, he would not have fought
so vigorously in its favour; nor would he have regarded so
despicably those who believe the opposite, as if he thought
them not to be numbered among human beings. What
Catholic ever conducted such a bitter dispute against heretics, even regarding a truth of faith, as Galileo does against
those who maintain the earth's immobility, especially when
not attacked by anybody? Surely, if this is not to defend an
opinion to which someone adheres firmly, then, disregarding matters of faith, I do not know how you would discern
any sign that a person is of this or that opinion, except that
he defends it with every effort.

7. If Galileo had attacked someone in particular who may not
have argued very strongly for the earth's immobility or may
not have completely convinced the Copernicans, many things
could be interpreted in a favourable light on his behalf. But
since he declares war on everybody and regards as dwarfs all
who are not Pythagorean or Copernican, it is clear enough
what he has in mind; this is especially so since he praises
greatly and prefers to others William Gilbert, a perverse heretic and a quarrelsome and sophistical defender of this
opinion. And so these are all the particular reasons which for
me render Galileo vehemently suspected of being of the opinion that the earth physically moves. Without doubt nowhere
in this whole work does it emerge that he feels otherwise.
Though he sometimes says that he does not want to make a
decision at all, he does this because, after the wounds inflicted
upon his targets, he wants to ensure that he not be viewed as
having wounded them deliberately.

Commissary General to Cardinal Barberini, 28 April 1633

Most Eminent, Most Reverend, and Most Honourable Lord and Patron:

Yesterday, in accordance with the orders of His Holiness, I reported on Galileo's case to the Most Eminent Lords of the Holy Congregation by briefly relating its current state. Their Lordships approved what has been done so far, and then they considered various difficulties in regard to the manner of continuing the case and leading it to a conclusion; for in his deposition Galileo denied what can be clearly seen in the book he wrote, so that if he were to continue in his negative stance it would become necessary to use greater rigour in the administration of justice and less regard for all the ramifications of this business.

Finally I proposed a plan, namely that the Holy Congregation grant me the authority to deal extrajudicially with Galileo, in order to make him understand his error and, once having recognized it, to bring him to confess it. The proposal seemed at first too bold, and there did not seem to be much hope of accomplishing this goal as long as one followed the road of trying to convince him with reasons; however, after I mentioned the basis on which I proposed this, they gave me the authority.

In order not to lose time, yesterday afternoon I had a discussion with Galileo, and, after exchanging innumerable arguments and answers, by the grace of the Lord I accomplished my purpose: I made him grasp his error, so that he clearly recognized that he had erred and gone too far in his book; he expressed everything with heartfelt words, as if he were relieved by the knowledge of his error; and he was ready for a judicial confession. However, he asked me for a little time to think about the way to render his confession honest, for in regard to the substance he will hopefully proceed as mentioned above.

I have not communicated this to anyone else, but I felt obliged to inform Your Eminence immediately, for I hope His Holiness and Your Eminence will be satisfied that in this manner the case is brought to such a point that it may be settled without difficulty. The Tribunal will maintain its reputation; the culprit can be treated with benignity; and, whatever the final outcome, he will know the favour done to him, with all the consequent satisfaction one wants in this. I am thinking of examining him today to obtain the said confession; after obtaining it, as I hope, the only thing left for me will be to question him about his intention and allow him to present a defence. With this done, he could be granted imprisonment in his own house, as Your Eminence mentioned.

Galileo's Second Deposition, 30 April 1633

Called personally to the hall of the Congregations, in the presence and with the assistance of those mentioned above and of myself, the above-mentioned Galileo Galilei, who has since then petitioned to be heard, having sworn an oath to tell the truth, was asked by the Fathers the following:

Q: *That he state whatever he wished to say.*

A: For several days I have been thinking continuously and directly about the interrogations I underwent on the 16th of this month, and in particular about the question whether sixteen years ago I had been prohibited, by order of the Holy Office, from holding, defending and teaching in any way whatever the opinion, then condemned, of the earth's motion and sun's stability. It dawned on me to reread my printed *Dialogue*, which over the last three years I had not even looked at. I wanted to check very carefully whether, against my purest intention, through my oversight, there might have fallen from my pen not only something enabling readers or superiors to infer a defect of disobedience on my part, but also other details through which one might think of me as a transgressor of the orders of the Holy Church. Being at liberty, through the generous approval of superiors, to send one of my servants for errands, I managed to get a copy of my book, and I started to read it with the greatest concentration and to examine it in the most detailed manner. Not having seen it for so long, I found it almost a new book by another author. Now, I freely confess that it appeared to me in several places to be written in such a way that a reader, not aware of my intention, would have had reason to form the opinion that the arguments for the false side, which I intended to confute, were so stated as to be capable of convincing because of their strength, rather than being easy to answer. In particular, two arguments, one based on sunspots and the other on the tides, are presented favourably to the reader as being strong and powerful, more than would seem proper for someone who deemed them to be inconclusive and wanted to confute them, as indeed I inwardly and truly did and do hold them to be inconclusive and refutable. As an excuse for myself, within myself, for having fallen into an error so foreign to my intention, I was not completely satisfied with saying that when one presents arguments for the opposite side with the intention of confuting them, they must be explained in the fairest way and not be made out of straw to the disadvantage of the opponent, especially when one is writing in dialogue form. Being dissatisfied with this excuse, as I said, I resorted to that of the natural gratification everyone feels for his own subtleties and for showing himself to be cleverer than the average man, by finding ingenious and apparent considerations of probability even in favour of false propositions. Nevertheless – even though, to use Cicero's words, 'I am more desirous of glory than is suitable' – if I had to write out the same arguments now, there is no doubt I would weaken them in such a way that they could not appear to exhibit a force which they really and essentially lack. My error then was, and I confess it, one of vain ambition, pure ignorance and inadvertence. This is as much as I need to say on this occasion, and it occurred to me as I reread my book. With this, having obtained his signature, and having sworn him to silence, the Fathers formally concluded the hearing. I, Galileo Galilei, have testified as above.

And returning after a little, he said: And for greater confirmation that I neither did hold nor do hold as true the condemned opinion of the earth's motion and sun's stability, if, as I desire,

I am granted the possibility and the time to prove it more clearly, I am ready to do so. The occasion for it is readily available since in the book already published the speakers agree that after a certain time they should meet again to discuss various physical problems other than the subject already dealt with. Hence, with this pretext to add one or two other Days, I promise to reconsider the arguments already presented in favour of the said false and condemned opinion and to confute them in the most effective way that the blessed God will enable me. So I beg this Holy Tribunal to co-operate with me in this good resolution, by granting me the permission to put it into practice. And again he signed. I, Galileo Galilei, affirm the above.

Galileo's Defence, Third Deposition, 10 May 1633

In an earlier interrogation, I was asked whether I had informed the Most Reverend Father Master of the Sacred Palace about the private injunction issued to me sixteen years ago by order of the Holy Office – 'not to hold, defend, or teach in any way whatever' the opinion of the earth's motion and sun's stability – and I answered No. Since I was not asked the reason why I did not inform him, I did not have the opportunity to say anything else. Now it seems to me necessary to mention it, in order to prove the absolute purity of my mind, always averse to using simulation and deceit in any of my actions. I say, then, that at that time some of my enemies were spreading the rumour that I had been called by the Lord Cardinal Bellarmine in order to abandon some opinions and doctrines of mine, that I had had to abjure, that I had also received punishments for them, etc., and so I was forced to resort to His Eminence and to beg him to give me a certificate explaining why I had been called. I received this certificate, written by his own hand, and it is what I attach to the present statement. In it one clearly sees that I was only told not to hold or defend Copernicus's doctrine of the earth's motion and sun's stability; but one cannot see any trace that, besides this general pronouncement applicable to all, I was given any other special order. Having the reminder of this authentic certificate, handwritten by the one who issued the order himself, I did not try later to recall or give any other thought to the words used to give me orally the said injunction, to the effect that one cannot defend or hold, etc.; thus, the two phrases besides 'holding' and 'defending' which I hear are contained in the injunction given to me and recorded, that is, 'teaching' and 'in any way whatever', struck me as very new and unheard. I do not think I should be mistrusted about the fact that in the course of fourteen or sixteen years I lost any memory of them, especially since I had no need to give the matter any thought, having such a valid reminder in writing. Now, when those two phrases are removed and we retain only the other two mentioned in the attached certificate, there is no reason to doubt that the order contained in it is the same as the injunction issued by the decree of the Holy Congregation of the Index. From this I feel very reasonably excused for not notifying the Father Master of the Sacred Palace of the injunction given to me in private, the latter being the same as the one of the Congregation of the Index. Given that my book was not subject to more stringent censures than those required by the decree of the Index, I followed the surest and most effective way to protect it and purge it of any trace of blemish. It seems to me that this is very obvious, since I handed it over to the supreme Inquisitor at a time when many books on the same subjects were being prohibited solely on account of the above-mentioned decree. From the things I am saying, I think I can firmly hope that the idea of my having knowingly and willingly disobeyed the orders given me will be given no credence by the Most Eminent and Most Prudent Lord judges. Thus, those flaws that can be seen scattered in my book were not introduced through the cunning of an insincere intention, but rather through the vain ambition and satisfaction of appearing clever above and beyond the average among popular writers; this was an inadvertent result of my writing, as I confessed in another deposition of mine. I am ready to make amends and compensate for this flaw by every possible means, whenever I may be either ordered or allowed by Their Most Eminent Lordships. Finally, I am left with asking you to consider the pitiable state of ill health to which I am reduced, due to ten months of constant mental distress, and the discomforts of a long and tiresome journey in the most awful season and at the age of seventy; I feel I have lost the greater part of the years which my previous state of health promised me. I am encouraged to do this by the faith I have in the clemency and kindness of heart of the Most Eminent Lordships, my judges; and I hope that if their sense of justice perceives anything lacking among so many ailments as adequate punishment for my crimes, they will, I beg them, condone it out of regard for my declining old age, which I humbly also ask them to consider. Equally, I want them to consider my honour and reputation against the slanders of those who hate me, and I hope that when the latter insist on disparaging my reputation, the Most Eminent Lordships will take it as evidence why it became necessary for me to obtain from the Most Eminent Lord Cardinal Bellarmine the certificate attached herewith.

Galileo's Fourth Deposition, 21 June 1633

Called personally to the hall of Congregations in the palace of the Holy Office in Rome, fully in the presence of the Reverend Father Commissary-General of the Holy Office, assisted by the Reverend Father Prosecutor, etc., Galileo Galilei, Florentine, mentioned previously, having sworn an oath to tell the truth, was asked by the Fathers the following:

Q: *Whether he had anything to say.*

A: I have nothing to say.

Q: *Whether he holds or has held, and for how long, that the sun is the centre of the world and the earth is not the centre of the world but moves also with diurnal motion.*

A: A long time ago, that is, before the decision of the Holy Congregation of the Index, and before I was issued that injunction, I was undecided and regarded the two opinions, those of Ptolemy and Copernicus, as disputable, because either the one or the other could be true in nature. But after the above-mentioned decision, assured by the prudence of the authorities, all my uncertainty stopped, and I held, as I still hold, as very true and undoubted Ptolemy's opinion, namely the stability of the earth and the motion of the sun.

Having been told that he is presumed to have held the said opinion after that time, from the manner and procedure in which the said opinion is discussed and defended in the book he published after that time, indeed from the very fact that he wrote and published the said book, therefore he was asked to freely tell the truth whether he holds or has held that opinion.

A: In regard to my writing of the *Dialogue* already published, I did not do so because I held Copernicus's opinion to be true. Instead, deeming only to be doing a beneficial service, I explained the physical and astronomical reasons that can be advanced for one side and for the other; I tried to show that none of these, neither those in favour of this opinion or that, had the strength of a conclusive proof and that therefore to proceed with certainty one had to resort to the determination of more subtle doctrines, as one can see in many places in the *Dialogue*. So for my part I conclude that I do not hold and, after the determination of the authorities, I have not held the condemned opinion.

Having been told that from the book itself and the reasons advanced for the affirmative side, namely that the earth moves and the sun is motionless, he is presumed, as it was stated, that he holds Copernicus's opinion, or at least that he held it at the time, therefore he was told that unless he decided to proffer the truth, one would have recourse to the remedies of the law and to appropriate steps against him.

A: I do not hold this opinion of Copernicus, and I have not held it after being ordered by injunction to abandon it. For the rest, here I am in your hands; do as you please.

And he was told to tell the truth, otherwise one would have recourse to torture.

A: I am here to obey, but I have not held this opinion after the determination was made, as I said.

And since nothing else could be done for the execution of the decision, after he signed he was sent to his place.

Sentence, 22 June 1633

Whereas you, Galileo, son of the late Vincenzio Galilei, Florentine, aged seventy years, were denounced to this Holy Office in 1615 for holding as true the false doctrine taught by some that the sun is the centre of the world and motionless and the earth moves even with diurnal motion; for having disciples to whom you taught the same doctrine; for being in correspondence with some German mathematicians about it; for having published some letters entitled *On Sunspots*, in which you explained the same doctrine as true; for interpreting Holy Scripture according to your own meaning in response to objections based on Scripture which were sometimes made to you; and whereas later we received a copy of an essay in the form of a letter, which was said to have been written by you to a former disciple of yours and which in accordance with Copernicus's position contains various propositions against the authority and true meaning of Holy Scripture; and whereas this Holy Tribunal wanted to remedy the disorder and the harm which derived from it and which was growing to the detriment of the Holy Faith, by order of His Holiness and the Most Eminent and Most Reverend Lord Cardinals of this Supreme and Universal Inquisition, the Assessor Theologians assessed the two propositions of the sun's stability and the earth's motion as follows: that the sun is the centre of the world and motionless is a proposition which is philosophically absurd and false, and formally heretical, for being explicitly contrary to Holy Scripture; That the earth is neither the centre of the world nor motionless but moves even with diurnal motion is philosophically equally absurd and false, and theologically at least erroneous in the Faith. Whereas however we wanted to treat you with benignity at that time, it was decided at the Holy Congregation held in the presence of His Holiness on 25 February 1616 that the Most Eminent Lord Cardinal Bellarmine would order you to abandon this false opinion completely; that if you refused to do this, the Commissary of the Holy Office would give you an injunction to abandon this doctrine, not to teach it to others, not to defend it, and not to treat of it; and that if you did not acquiesce in this injunction, you should be imprisoned.

To execute this decision, the following day at the palace of and in the presence of the above-mentioned Most Eminent Lord Cardinal Bellarmine, after being informed and warned in a friendly way by the same Lord Cardinal, you were given an injunction by the then Father Commissary of the Holy Office in the presence of a notary and witnesses to the effect that you must completely abandon the said false opinion, and that in the future you could neither hold, nor defend, nor teach it in any way whatever, either orally or in writing; having promised to obey, you were dismissed. Furthermore, in order to completely eliminate such a pernicious doctrine, and not let it creep any further to the great detriment of Catholic truth, the Holy Congregation of the Index issued a decree which prohibited books treating of such a doctrine and declared it false and wholly contrary to the divine and Holy Scripture.

And whereas a book has appeared here lately, printed in Florence last year, whose inscription showed that you were the author, the title being *Dialogue by Galileo Galilei on the Great World Systems, Ptolemaic and Copernican*; and whereas the Holy Congregation was informed that with the printing of this book the false opinion of the earth's motion and sun's stability was being disseminated and taking hold more and more every day, the said book was diligently examined and found to violate explicitly the above-mentioned injunction given to you; for in the same book you have defended the said opinion already condemned and so declared to your face, although in the said book you try by means of various subterfuges to give the impression of leaving it undecided and labelled as probable; this is still a very serious error since there is no way an opinion declared and defined contrary to divine Scripture may be probable.

Therefore, by our order you were summoned to this Holy Office, where, examined under oath, you acknowledged the book as written and published by you. You confessed that about ten or twelve years ago, after having been given the injunction mentioned above, you began writing the said book, and that then you asked for permission to print it without explaining to those who gave you such permission that you were under the injunction of not holding, defending, or teaching such a doctrine in any way whatever. Likewise, you confessed that in several places the exposition of the said book is expressed in such a way that a reader could get the idea that the arguments given for the false side were effective enough to be capable of convincing, rather than being easy to refute. Your excuses for having committed an error, as you said so foreign from your intention, were that you had written in dialogue form, and everyone feels a natural satisfaction for one's own subtleties and showing oneself sharper than the average man by finding ingenious and apparently probable arguments even in favour of false propositions.

Having been given suitable terms to present your defence, you produced a certificate in the handwriting of the Most Eminent Lord Cardinal Bellarmine, which you said you obtained to defend yourself from the calumnies of your enemies, who were claiming that you had abjured and had been punished by the Holy Office. This certificate says that you had neither abjured nor been punished, but only that you had been notified of the declaration made by His Holiness and published by the Holy Congregation of the Index, whose content is that the doctrine of the earth's motion and sun's stability is contrary to Holy Scripture and so can be neither defended nor held. Because this certificate does not contain the two phrases of the injunction, namely 'to teach' and 'in any way whatever', one is supposed to believe that in the course of fourteen or sixteen years you had lost any recollection of them, and that for this same reason you had been silent about the injunction when you applied for the licence to publish the book. Furthermore, one is supposed to believe that you point out all of this not to excuse the error, but in order to have it attributed to conceited ambition rather than to malice. However, the said certificate you produced in your defence aggravates your case further since, while it says that the said opinion is contrary to Holy Scripture, yet you dared to treat of it, defend it and show it as probable; nor are you helped by the licence you artfully and cunningly extorted since you did not mention the injunction you were under.

Because we did not think you had said the whole truth about your intention, we deemed it necessary to proceed against you by a rigorous examination. Here you answered in a Catholic manner, though without prejudice to the above-mentioned matters confessed by you and deduced against you about your intention.

Therefore, having seen and seriously considered the merits of your case, together with the above-mentioned confessions and excuses and with any other reasonable matter worth seeing and considering, we have come to the final sentence against you given below.

We say, pronounce, sentence and declare that you, the above-mentioned Galileo, because of the things deduced in the trial and confessed by you as above, have rendered yourself according to this Holy Office vehemently suspected of heresy, namely of having hèld and believed a doctrine which is false and contrary to the divine and Holy Scripture: that the sun is the centre of the world and does not move from east to west, and the earth moves and is not the centre of the world, and that one may hold and defend as probable an opinion after it has been declared and defined contrary to Holy Scripture. Consequently, you have incurred all the censures and penalties imposed and promulgated by the sacred canons and all particular and general laws against such delinquents. We are willing to absolve you from them provided that first, with a sincere heart and unfeigned faith, in front of us you abjure, curse and detest the above-mentioned errors and heresies, and every other error and heresy contrary to the Catholic and Apostolic Church, in the manner and form we will prescribe to you. Furthermore, so that this serious and pernicious error and transgression of yours does not remain completely unpunished, and so that you will be more cautious in the future and an example for others to abstain from similar crimes, we order that the book *Dialogue by Galileo Galilei* be prohibited by public edict. We condemn you to formal imprisonment in this Holy Office at our pleasure. As a salutary penance we impose on you to recite the seven penitential Psalms once a week for the next three years. And we reserve the authority to moderate, change, or condone wholly or in part the above-mentioned penalties and penances. This we say, pronounce, sentence, declare, order and reserve by this or any other better manner or form that we reasonably can or shall think of.

Galileo's Abjuration, 22 June 1633

I, Galileo, son of the late Vincenzio Galilei of Florence, seventy years of age, arraigned personally for judgement, kneeling before you Most Eminent and Most Reverend Cardinals Inquisitors-General against heretical depravity in all of Christendom, having before my eyes and touching with my hands the Holy Gospels, swear that I have always believed, I believe now, and with God's help I will believe in the future all that the Holy Catholic and Apostolic Church holds, preaches and teaches. However, whereas, after having been judicially instructed with injunction by the Holy Office to abandon completely the false opinion that the sun is the centre of the world and does not move and the earth is not the centre of the world and moves, and not to hold, defend, or teach this false doctrine in any way whatever, orally or in writing; and after having been notified that this doctrine is contrary to Holy Scripture; I wrote and published a book in which I treat of this already condemned doctrine and adduce very effective reasons in its favour, without refuting them in any way; therefore, I have been judged vehemently suspected of heresy, namely of having held and believed that the sun is the centre of the world and motionless and the earth is not the centre and moves. Therefore, desiring to remove from the minds of Your Eminences and every faithful Christian this vehement suspicion, rightly conceived against me, with a sincere heart and unfeigned faith I abjure, curse and detest the above-mentioned errors and heresies, and in general each and every other error, heresy and sect contrary to the Holy Church; and I swear that in the future I will never again say or assert, orally or in writing, anything which might cause a similar suspicion about me; on the contrary, if I should come to know any heretic or anyone suspected of heresy, I will denounce him to this Holy Office, or to the Inquisitor or Ordinary of the place where I happen to be. Furthermore, I swear and promise to comply with and observe completely all the penances which have been or will be imposed upon me by this Holy Office; and should I fail to keep any of these promises and oaths, which God forbid, I submit myself to all the penalties and punishments imposed and promulgated by the sacred canons and other particular and general laws against similar delinquents. So help me God and these Holy Gospels of His, which I touch with my hands. I, the above-mentioned Galileo Galilei, have abjured, sworn, promised and obliged myself as above; and in witness of the truth I have signed with my own hand the present document of abjuration and have recited it word for word in Rome, at the convent of the Minerva, this twenty-second day of June 1633. I, Galileo Galilei, have abjured as above, by my own hand.

René Descartes, *The World, or a Treatise on Light*, c.1629–1633

René Descartes (1596–1650) was a French philosopher and mathematician who embraced the so-called 'mechanical philosophy', an increasingly popular approach to nature in the seventeenth century. Though Aristotle had spoken of different elements that had distinct forms and behaviours, mechanists typically believed that all observable phenomena in nature could be explained via an inert, undifferentiated matter in motion. Descartes differed from some proponents of this approach by arguing against the possibility of a vacuum, or void, in nature. He combined these views in his The World, or a Treatise on Light, *which set out to develop some basic laws of nature through which not only the behaviour of corpuscles – bits of matter – but also all of celestial physics, even light itself, could be explained. In the book, he embraced heliocentrism, and argued that the cosmos consisted of fluid-like vortices of matter with a sun or star at each centre, around which planets revolved. The book was not published until 1664, after Descartes's death, for reasons Descartes discussed in his* Letters to Mersenne.

Chapter 6: Description of a New World, and the Qualities of the Matter of Which It Is Composed

Allow your thoughts to leave this world for a little while and come see another, completely new one, which I will bring into being alongside it in imaginary spaces. The philosophers tell us that these spaces are infinite and they must be believed, since it is they themselves who made them. But in order that this infinity not hinder and embarrass us, let us not try to go all the way until the end; let us enter only so much that we can lose sight of all the creatures that God created five or six thousand years ago. After we have stopped there, in some place we've determined, let us suppose that God creates anew so much matter all around us that, however far we may extend our imagination, it no longer perceives any place that is empty.

Now since we take the liberty of pretending as it suits us, let us attribute to it, if you please, a nature in which there is nothing at all that anyone cannot know as perfectly as possible. To that effect, let us expressly suppose that it does not have the form of earth, or fire, or air; or any more particular one like wood, stone, or metal; or the qualities of being hot and cold, dry or moist, light or heavy; or of having some taste, smell, sound, colour, light, or the like; for we might say that in the nature of all of those there is something that is obviously not known to everyone.

Let us also not think, on the other hand, that it is the prime matter of the philosophers, which has been stripped of all its forms and qualities so well that nothing remains that can be clearly understood. Rather, let us conceive of it as a real, perfectly solid body that equally fills the full length, breadth and depth of this great space in the middle of which we have stopped our thought. Thus each of its parts always occupies a part of this space and is so proportioned to its size that it could not fill a greater one nor shrink into a lesser one, nor suffer another to find a place there while it remains.

Further, let us add that this matter can be divided into all the parts and according to all the shapes that we can imagine, and that each of its parts is capable of receiving in itself all the movements that we can also conceive. And let us suppose still further that God actually divides it into many such parts, some larger and some smaller, some of one shape and some of another, however it pleases us to concoct them. This does not mean that He separates them one from the other such that there is some void in between them; rather, let us think that the whole distinction

He creates consists in the diversity of motions that He gives them. He makes it so that from the moment they are first created some begin to move in one direction and others in another; some more quickly, and others more slowly (or even, if you like, not at all); and so that they continue their movement afterwards according to the ordinary laws of nature. For God has so marvellously established these laws that – even if we suppose that He creates nothing more than what I have said, and even if He does not impose on it any order or proportion, but instead creates the most confused and muddled chaos that the poets could describe – they are sufficient to make the parts of this chaos unravel by themselves and arrange themselves in such good order that they will have the form of a most perfect world, in which one can see not only light, but also all the other things, both general and particular, that appear in this true world.

Chapter 7: On the Laws of Nature of This New World

But I do not want to delay any longer in telling you how Nature alone can disentangle the confused chaos of which I have spoken, and what laws God has imposed on it.

Know first, then, that by Nature I do not here mean some Goddess or other sort of imaginary power, but that I use this word to signify Matter itself, understood with all the qualities that I have attributed to it, and under the condition that God continues to preserve it in the same way that He created it. For from this alone – that He thus continues to preserve it – it follows of necessity that there will be many changes in its parts which cannot, it seems to me, be properly attributed to the action of God, because that does not change. I therefore attribute them to Nature, and the rules according to which these changes occur I call the Laws of Nature.

I will specify here two or three of the principal rules according to which God causes the Nature of this new world to act and which will be enough, I believe, to understand all the others.

The first is that each particular part of matter always continues to be in the same state unless some meeting with others compels it to change that state. That is to say: if it has some size, it will never become smaller, unless others divide it; if it is round or square, it will never change that shape unless others force it to do so; if it is stopped in some place, it will never leave it unless others drive it out; and if it has once begun to move, it will always continue with equal force until others stop or slow it down.

I suppose as a second rule that, when a body pushes another, it cannot give it any motion unless at the same time it loses as much of its own; nor can it deprive it of motion unless its own increases as much. This rule, joined with the preceding one, agrees very well with all the experiments in which we see that a body begins or ceases to move because it is pushed or stopped by some other.

I will add for the third rule: that when a body moves, even though its movement is most often along a curved line, and even though it must in some way be circular, nevertheless each of its particular parts tends to continue its own movement in a straight line. And so their action, i.e. their inclination to move, is different from their motion.

For example, if we turn a wheel on its axle, although all its parts go in a circle – because, being joined one to the other, they cannot go otherwise – nevertheless their inclination is to go straight, as clearly appears if by chance one becomes detached from the others. For as soon as it is free, its movement ceases to be circular and continues in a straight line.

Similarly, when a stone is swung around in a slingshot, not only does it go straight as soon as it departs, but also, during all the time it is in the sling, it presses against the middle of the sling and makes the cord taut, clearly showing thereby that it always is inclined to go in a straight line, and that it goes in circles only by constraint.

I will content myself with informing you that besides the three laws that I have explained here, I do not want to suppose any others except those that follow inseparably from those eternal truths on which mathematicians are accustomed to basing their most certain and evident demonstrations. And so that there is no exception, I will add to our suppositions, if you please, that God will never cause any miracle, and will in no way disturb the ordinary course of nature.

However, following this, I do not promise to offer exact demonstrations of all the things I will say. It will be enough if I open a way for you to find them yourselves, when you take the trouble to look for them. Most minds are repelled when things are made too easy for them. And to set a table here that suits you, I must use shadow as well as bright colours. So I will content myself with continuing the description that I have begun, as though having no other intention than to tell you a fable.

Chapter 8: On the Formation of the Sun and the Stars of the New World

First, because there is no void at all in this new world, it was impossible for all the parts of matter to move in a straight line. Rather, being more or less equal and as able to easily divert each other, they must all have come together in some circular motions. And still, because we suppose that God first moved them differently, we must not think that they all come together to revolve around a single centre, but around many different ones, which we can imagine as variously situated with respect to each other.

Imagine, for example, that the points S, E, ε and A are the centres of which I speak and that all matter contained in the space F G G F is a heaven that revolves around the sun, marked S, while all that of the space H G G H is another that revolves around the star marked ε, and the same for the others. Thus, there are as many different heavens as there are stars, and as their number is indefinite, so too is that of the heavens. Thus, too, the firmament is nothing other than the surface, without thickness, separating all these heavens from one another.

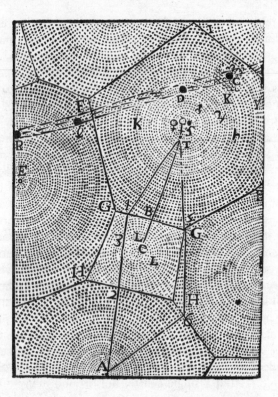

Chapter Ten: On the Planets in General, and, in Particular, on the Earth and Moon

There are likewise several things to note concerning the planets. The first is that even though they all tend toward the centre of the heavens that contain them, this does not mean that they can ever reach the interior of those centres. For, as I have already said earlier, the sun and the other fixed stars occupy them.

Rather, just as we experience that boats following the course of a river never move as quickly as the water that carries them, nor do the largest of them move so quickly as the smallest, so too the planets follow the course of the matter of the sky without resistance, and move as it does, though not quite as quickly.

Likewise the inequality of their motion must have some relation to that found between the size of their mass and the smallness of the parts of the sky that surround them. The reason for this is that, generally speaking, the larger a body is the easier it is for it to communicate a part of its motion to other bodies, and the more difficult it is for others to communicate any part of theirs to it. For even though several small bodies joined together against a larger one can have just as much force as it, nevertheless they can never make it move as quickly in all the directions that they move because, while they agree in some of their motions which they communicate to it, they necessarily differ in others at the same time, which they cannot communicate to it.

Now two things follow from this that seem to me worthy of consideration. The first is that the matter of the heaven must be able not only to turn the planets around the sun, but also around their own centres (except when there is some particular cause that prevents them from doing so), and thus that it must compose small heavens around them, which move in the same direction as the larger one. And the second is that if two planets meet, unequal in size but disposed to take their course in the sky at the same distance from the sun, the smaller of these, having a faster motion than the larger, will have to join the small sky around the larger one, and turn continually with it.

I do not add here how it is possible to find a greater number of planets joined together that take their course around one another, like those which the new astronomers have observed around Jupiter and Saturn. For I did not undertake to say everything.

Descartes, Letters to Mersenne, 1633–1634

When in 1633 Descartes learned of the condemnation of Galileo, he was both dismayed and worried, as his own treatise, The World, *clearly endorsed the heliocentric views that seemed to be at the heart of Galileo's conflict with the Church. He felt that the earth's motion was so central a part of the work that he could not 'detach it without rendering the rest entirely defective', and so opted to withdraw the work from publication and suppress it entirely, though he published the optical material in it later in two works of 1637:* Dioptrics *and* Meteors.

End of November, 1633

I had proposed to send you my *World* as a gift, and no more than fifteen days ago I was still fully resolved to send you at least part of it, if the whole could not be transcribed in time. But I will tell you that I inquired recently whether Galileo's *System of the World* was available in Leiden and Amsterdam, because I thought I had learned that it had been printed in Italy last year. I was informed that it was true that it had been printed, but that all the copies were burned in Rome at the same time, and that Galileo had been condemned to some fine. This surprised me so greatly that I have almost decided to burn my papers, or at least not to let anyone see them. For I can't imagine that he, who is Italian and, as I understand it, quite well liked by the Pope, could have been made into a criminal for anything except that he undoubtedly wanted to establish the movement of the earth, which I know had been formerly censured by some Cardinals. But I thought I had heard that it was still being taught publicly, even in Rome. And I confess that if it is false then all the foundations of my philosophy are also false, for it clearly is demonstrated from them. And it is so linked with all the parts of my treatise that I cannot detach it without rendering the rest entirely defective. But for all the world I would not want to put out a discourse in which would be found a single world that the Church condemned, so I preferred to suppress it rather than to allow it to appear in a compromised way. And perhaps you will be very glad to be free from the trouble of reading evil things. There are so many opinions in philosophy that are possible and can be upheld in disputes that if mine are not more certain, and cannot be approved without controversy, I never want to publish them. Meanwhile, I beg you to tell me what you know of the Galileo affair.

February 1634

I want to completely suppress the treatise that I wrote, and sacrifice almost all of my work of four years, to render complete obedience to the Church and its position on the movement of the earth. But I still have not seen that the Pope or the Council has ratified this prohibition, which was made only by the Congregation of Cardinals established for the censorship of books. I would be very happy to learn what is now held in France, and whether their authority was sufficient to make this an article of faith. I was told that the Jesuits helped in the condemnation of Galileo, and all of Father Scheiner's book [*Rosa Ursina*] shows that they are not his friends. But the observations in that book furnish so many proofs that deprive the sun of the motions typically attributed to it that I cannot believe that Father Scheiner himself, in his soul, does not believe the opinion of Copernicus. This astonishes me so much that I dare not write my feelings about it. As for me, I seek only rest and peace of mind, which are goods that cannot be possessed by those with animosity or ambition. Still, I do not remain idle. But now I think only of instructing myself, and consider myself little capable of instructing others, especially those who, having built their reputation on false opinions, would be afraid of losing it should the truth be discovered.

April 1634

I learned from your letter that the last one I wrote to you had been lost, although I thought I had sent it quite securely. I told you in it all the reasons that prevented me from sending you my treatise, which I do not doubt you would find so legitimate that, rather than blaming me for my resolve to prevent anyone from seeing it, you would, on the contrary, be the first one to exhort me to do just that if I was not already so resolved. You no doubt know that Galileo was recently taken up by the Inquisitors of Faith and that his opinion concerning the motion of the earth was condemned as heretical. Now I will tell you that all the things that I explained in my treatise, among them this opinion on the motion of the earth, depended so much on each other that it is clear that if one is false, then all the reasoning I used has no force. Although I thought they were supported by very certain and evident demonstrations, I would not, for all the world, want to support them against the authority of the Church. I know very well that one might say that not everything that the Inquisitors of Rome decide is immediately an article of faith, and that it first must be passed by a Council. But I am not so in love with my own thoughts as to want to make use of such details in order to maintain them. I only desire to live in peace, and to continue the life that I have begun by choosing the motto 'He who lives well, hides well'.

I am astonished that a man of the Church would dare to write about the earth's motion, however he excuses himself. For I have seen the condemnation of Galileo, printed in Liège on 20 September 1633, which had these words: 'although he pretended it was proposed by him hypothetically'. So they seem to imply that we cannot use this hypothesis even hypothetically in astronomy. Thus I do not dare to tell anyone of my thoughts on this subject. But seeing that this censure has not yet been authorized by the Pope or a Council, but only by a particular congregation of Cardinal Inquisitors, I do not entirely lose hope that it will end up like the Antipodes, which were formerly condemned almost in the same way. And so with time my *World* may yet appear.

Galileo, *Discourses and Demonstrations on Two New Sciences*, 1638

While under house arrest after his trial, Galileo was not allowed to publish on cosmological issues. Instead, he turned back to his earlier work on terrestrial physics, and in 1638 published a book on matter and motion, called Discourses and Demonstrations on Two New Sciences. *Though the book's subject matter did not explicitly violate the terms of his sentence, he had it published in Leiden, Holland, where the reach of the Inquisition was weaker. Much like his earlier* Dialogue on the Great World Systems, *the* Two New Sciences *was published as a dialogue held over the course of four days between Salviati, Sagredo and Simplicio, the latter of whom represents the Aristotelian point of view. The characters discuss a book written by 'our Author', who represents Galileo himself, and from which Salviati reads aloud; the text of the 'book' from which he reads is written in Latin, while the discussion of the three interlocuters is in Italian. The third day of the dialogue focuses on uniform and naturally accelerated motion, and the fourth day focuses on projectile motion. In the former, Galileo describes how the distance travelled in naturally accelerated motion is proportional to the square of the time; in the latter, he describes how projectile motion is parabolic. Both ideas would be important in Newton's later* Principia.

Third Day

SALVIATI [*reading the text by 'our author'*]: 'Theorem II, Proposition II: If something movable falls from rest in a uniformly accelerated motion, the distances described by it during any period are related to the times in a duplicate ratio; that is, they are in the ratio of the squares of those same times.

'Let the flow of time from any first instant A be represented by the line AB, in which two times, AD and AE are taken. Let the line HI represent the space in which the moveable object descends from point H, the beginning of its motion, uniformly accelerated. Let the space HL represent that passed through in the first time, AD, and HM be the space passed through in time AE. I say that the space MH to the space HL is in the duplicate ratio of the time EA to the time AD, or that the spaces MH and HL have the same ratio as the squares of EA and AD.'

SIMPLICIO: I still remain doubtful whether this is the acceleration that nature uses in the motion of falling heavy bodies, and therefore, for my sake and that of others like me, it seems to me that it would have been opportune here to bring in some experiments – there are said to have been many – which agree with the conclusions demonstrated.

SALVIATI: As a true person of science, you ask a very reasonable question, and so it is customary and appropriate to do in the sciences that apply mathematical demonstrations to natural conclusions, as can be seen in perspective, astronomy, mechanics, music and others, which, with sensible experiments, confirm the principles that are the foundations of their whole structure. And therefore I do not want it to seem superfluous if we take time to discuss this first and paramount foundation, on which the whole immense machine of infinite conclusions rest, of which only a small part are posed in this book by the author, who has made a great effort to open a door that has hitherto been closed to speculative minds. As far as experiments, therefore: the author did not neglect to make them, and to ensure that the acceleration of naturally falling bodies follows the aforementioned proportion; I found myself testing it in the following way many times, in his company:

In a beam of wood about 12 cubits long, half a cubit wide and three fingers deep, a small channel slightly wider than a finger was hollowed out on the small side. It was made very straight, and in order for it to be perfectly smooth and polished, a sheepskin parchment was glued into it, and in was dropped a hard, polished round bronze ball. The wooden beam was situated by raising one of its ends a cubit or two above the other, and the ball was allowed to descend along the channel, while we noted, in a manner I will describe below, the time that it took up as it fell. We repeated this same act many times to make sure of the quantity of time, and there was never a difference of even a tenth-part of the beat of a pulse. Having completed the operation and established it as precise, we made the same ball go down only one quarter of the channel, and we measured the time of its descent, and always found it exactly half of what it was previously. We did the same experiment with other divisions, and examined the time for whole length compared with the time for half, or other divisions, and in experiments repeated a hundred times, we always found that the spaces crossed were as the squares of the times, and this was for all inclinations of the plane, that is, the channel, into which the ball was dropped. We observed too that the times of descent for the different inclinations maintain exactly that ratio between them which, we will find later, was assigned and demonstrated by the Author.

As for the measurement of time, a large bucket of water was attached above a thin tube, which was welded below it; through it poured a thin stream of water, which was collected in a small glass during the time that the ball descended into the channel, or into part of it. We collected the particles of water each time and weighed them with a very exact balance, which gave us the differences and ratios of their weights. Those gave us the difference and proportions of their times, and with such accuracy that, as I have said, these operations, repeated over and over many times, did not differ by any notable amount.

SIMPLICIO: I would have received great satisfaction from being present at such experiments; but being sure of your diligence in performing them, and your faithfulness in relating them, I accede and admit them as very certain and true.

SALVIATI: We can therefore resume our reading, and continue forward.

Fourth Day

SALVIATI: Here is the text of our author:

'On the motion of projectiles: We have observed above the properties of uniform motion and of naturally accelerated motion over planes of any inclination. Now I will endeavour to put forward and establish with firm demonstrations some of the principal properties worthy of study that belong to a movable object with a composite, twofold motion, namely uniform and naturally accelerated. This is the sort of motion we call projectile, whose origin I constitute thusly:

'Imagine a moveable object cast upon a horizontal plane without any friction. Now it is evident from that which was said at greater length already, that it will have a uniform and perpetual motion along the plane, if the plane is extended infinitely. But if it is bounded and raised, then the moveable object, which I imagine to be heavy, will reach the edge of the plane and pass over it, and will add to its previous uniform and perpetual motion a downward propensity by its own weight. From this a certain motion will emerge composed of a horizontal uniform motion and a naturally accelerated vertical motion: this I call projection. We will show some of its properties, the first of which is:

'Theorem I, Proposition I: A projectile which is directed by a motion composed of a uniform horizontal motion and a naturally accelerated vertical motion describes a semi-parabolic path as it moves.'

SAGREDO: Please, Salviati, for my sake and also, I believe, for Simplicio, pause here a little; for I have not gone so far into geometry that I have studied Apollonius, except insofar as I know that he deals with these parabolas and other conic sections, without knowledge of which I do not think we can understand the demonstrations of other propositions that depend on them. And because already in this first beautiful proposition the author suggests that the line described by the projectile must be parabolic, and since I imagine that we are only dealing with these lines, it is absolutely necessary to have a complete understanding, if not of all their properties, then at least with those necessary for the present science.

SALVIATI: Sagredo humbles himself excessively, making knowledge now seem new that not long ago he admitted as well known – that is, in the discussion on resistances when we needed the knowledge of a certain proposition from Apollonius, which did not give him any difficulty.

SAGREDO: It may be either that I knew it by chance or that I assumed it that once, as much as I needed to for that treatise. But here, where I imagine that I have to understand all the demonstrations about these lines, we must not, as they say, gulp it all down, wasting time and effort.

SIMPLICIO: And also, though I believe that Sagredo is well equipped for all his needs compared to me, in my case even the very first terms are new. For although our philosophers have dealt with the matter of the motion of projectiles, it does not seem to me that they defined the lines described by them, except very generally to say that they are always curved, unless the projection is vertically upwards. But if the little geometry that I have learned from Euclid since we had our other discussions is not enough to make me able to understand the necessary knowledge for the following demonstrations, then it will be necessary for me to rely on simply believing the propositions, but not understanding them.

SALVIATI: On the contrary, I want you to understand them thanks to the Author of the work himself, who, when he allowed me to see this work of his, demonstrated to me two principal properties of parabolas without any necessary foreknowledge, since I did not have the books of Apollonius ready at that time. These are the only ones we need for the present discussion, and they are also proved well by Apollonius, but after many other steps, such that it would take a long time to arrive at them. I want us to shorten the journey considerably, deriving the first property immediately from the pure and simple generation of the parabola, and from this immediately demonstrating the second.

[*Here follows a discussion of parabolas and their properties.*]

SIMPLICIO: You proceed too quickly in your demonstrations, and it seems to me you always assume that all the propositions of Euclid are as familiar and ready to me as his first axioms, which they are not.

SALVIATI: Indeed, all sophisticated mathematicians assume that the reader has at least the *Elements* of Euclid at the ready.

John Wilkins, *A Discovery of a World in the Moon*, 1638

John Wilkins (1614–1672) was an English theologian, bishop and natural philosopher, and a founding member of the Royal Society of London for Improving Natural Knowledge. Written for a broad general audience when he was only twenty-four, Wilkins's Discovery of a World in the Moon *embraced the Copernican and Galilean argument that the earth and the moon were similar, that the earth itself was a planet, and that the moon, as Kepler had suggested, might be inhabited. Both the 1638* Discovery *and the 1640* Discourse Concerning a New Planet, *Wilkins's expansion of the 1638 book, were highly popular. Bernard de Fontenelle translated the* Discovery *into French in 1656, and it served as an inspiration for his own 1686* Conversations on the Plurality of Worlds.

The First Proposition, by way of Preface. That the strangeness of this opinion is no sufficient reason why it should be rejected, because other certain truths have been formerly esteemed ridiculous, and great absurdities entertained by common consent.

I may justly expect to be derided by most, and to be believed by few or none; especially since this opinion seems to carry in it so much strangeness, so much contradiction to the general consent of others. But however, I am resolved that this shall not be any discouragement, since I know that it is not the common opinion of others that can either add or detract from the truth.

For,

1. Other truths have been formerly esteemed altogether as ridiculous as this can be.
2. Gross absurdities have been entertained by general opinion.

So that from that which I have said may be gathered thus much.

1. That a new truth may seem absurd and impossible not only to the vulgar, but to those also who are otherwise wise men and excellent scholars; and hence it will follow that every new thing which seems to oppose common principles is not presently to be rejected, but rather to be pried into with a diligent inquiry, since there are many things which are yet hid from us, and reserved for future discovery.
2. That it is not the commonness of an opinion that can privilege it for a truth; the wrong way is sometime a well-beaten path, whereas the right way (especially to hidden truths) may be less trodden and more obscure.

True indeed, the strangeness of this opinion will detract much from its credit; but yet we should know that nothing is in itself strange, since every natural effect has an equal dependence upon its cause, and with the like necessity doth follow from it, so that 'tis our ignorance which makes things appear so, and hence it comes to pass that many more evident truths seem incredible to such who know not the causes of things: you may as soon persuade some Country peasant that the Moon is made of green Cheese (as we say) as that 'tis bigger than his Cart-wheel since both seem equally to contradict his sight, and he has not reason enough to lead him further than his senses.

I must needs confess, though I had often thought with myself that it was possible there might be a world in the Moon, yet it seemed such an uncouth opinion that I never durst discover it, for fear of being counted singular and ridiculous, but afterwards having read *Plutarch, Galileus, Keplar,* with some others, and finding many of mine own thoughts confirmed by such strong authority, I then concluded that it was not only possible there might be, but probable that there was another habitable world in that Planet. In the prosecuting of this assertion I shall first endeavour to clear the way from such doubts as may hinder the speed or ease of further progress; and because the suppositions implied in this opinion may seem to contradict the principles of reason or faith, it will be requisite that I first remove this scruple shewing the conformity of them to both these, and proving those truths that may make way for the rest, which I shall labour to perform in the second, third, fourth and fifth Chapters, and then proceed to confirm such propositions, which do more directly belong to the main point in hand.

Proposition 2. That a plurality of worlds does not contradict any principle of reason or faith.

'Tis not the endeavour of *Moses* or the Prophets to discover any Mathematical or Philosophical subtleties, but rather to accommodate themselves to vulgar capacities, and ordinary speech, as nurses are wont to use their infants. True indeed, *Moses* is there to handle the history of the Creation, but 'tis observed that he does not anywhere meddle with such matters as were very hard to be apprehended, for being to inform the common people as well as others, he does it after a vulgar way, as it is commonly noted, declaring the original chiefly of those things which were obvious to the sense, and being silent of other things which then could not well be apprehended.

Proposition 3. That the heavens do not consist of any such pure matter which can privilege them from the like change and corruption, as these inferior bodies are liable unto.

Though *Aristotle*'s consequence were sufficient, when he proved that the heavens were not corruptible, because there have not any changes been observed in it, yet this by the same reason must be as prevalent, that the Heavens are corruptible, because there have been so many alterations observed there; but of these together with a further confirmation of this proposition, I shall have occasion to speak afterwards; In the mean space, I will refer the Reader to that work of *Scheiner* a late Jesuit which he titles his *Rosa Ursina*, where he may see this point concerning the corruptibility of the Heavens largely handled and sufficiently confirmed.

There are some other things, on which I might here take an occasion to enlarge myself, but because they are directly handled by many others, and do not immediately belong to the chief matter in hand, I shall therefore refer the Reader to their authors, and omit any large proof of them myself, as desiring all possible brevity.

1. The first is this: That there are no solid Orbs. If there be a habitable world in the Moon (which I now affirm) it must follow, that her Orb is not solid as *Aristotle* supposed; and if not her, why any of the other. I rather think that they are all of a fluid (perhaps aereous) substance.

2. There is no element of fire, which must be held with this opinion here delivered; for if we suppose a world in the Moon, then it will follow, that the sphere of fire, either is not there where 'tis usually placed in the concavity of his Orb, or else that there is no such thing at all, which is most probable since there are not any such solid Orbs, that by their swift motion might heat and enkindle the adjoining air, which is imagined to be the reason of that element. Concerning this see *Cardan, Iohannes Pena* that learned Frenchman, the noble *Tycho*, with diverse others who have purposely handled this proposition.

3. I might add a third, *viz.* that there is no Music of the spheres, for if they be not solid, how can their motion cause any such sound as is conceived? However the world would have no great loss in being deprived of this Music, unless at sometimes we had the privilege to hear it.

Proposition 4. That the Moon is a solid, compacted, opacous body.

I shall not need to stand long in the proof of this proposition, since it is a truth already agreed on by the general consent of the most and the best Philosophers.

1. It is solid in opposition to fluid, as is the air, for how otherwise could it beat back the light which it receives from the Sun?

2. It is compact, and not a spongy and porous substance.

3. It is opacous, not transparent or diaphanous like Crystal or glass, as *Empedocles* thought, who held the Moon to be a globe of pure congealed air, like hail enclosed in a sphere of fire, for then, Why does she not always appear in the full? Since the light is dispersed through all her body? How can the interposition of her body so darken the Sun, or cause such great eclipses as have turned day into night?

Proposition 5. That the Moon has not any light of her own.

'Tis agreed upon by all sides, that this Planet receives most of her light from the Sun, but the chief controversy is, whether or not she has any of her own? The greater multitude affirm this. But I shall first show, that this light cannot be her own.

   If you ask now what the reason may be of that light which we discern in the darker part of the new Moon, I answer, 'tis reflected from our earth which returns as great a brightness to that Planet, as it receives from it.

Proposition 6. That there is a world in the Moon, has been the direct opinion of many ancient, with some modern Mathematicians, and may probably be deduced from the tenets of others.

The opinion which I have here delivered was more directly proved by *Maeslin, Keplar* and *Galilaeus,* each of them late writers, and famous men for their singular skill in Astronomy. As for those works of *Maeslin* and *Keplar* wherein they do more expressly treat of this opinion, I have not yet had the happiness to see them. However their opinions appear plain enough from their own writings, and the testimony of others concerning them.

   'Tis true indeed, in many things they do but trifle, but for the main scope of those discourses, 'tis as manifest they seriously meant it, as any indifferent Reader may easily discern; otherwise sure *Campanella* would never have writ an apologie for him. And besides 'tis very likely if it had been but a jest, *Galilaeus* would never have suffered so much for it as afterwards he did. But as for the knowledge which he pretends, you may guess what it was by his confidence (I say not presumption) in other assertions, and his boldness in them may well derogate from his credit in this.

In my following discourse I shall most insist on the observation of *Galilaeus*, the inventor of that famous perspective, whereby we may discern the Heavens hard by us, whereby those things which others have formerly guessed at are manifested to the eye, and plainly discovered beyond exception or doubt, of which admirable invention, these latter ages of the world may justly boast, and for this expect to be celebrated by posterity. 'Tis related of *Eudoxus*, that he wished himself burnt with *Phaeton*, so he might stand over the Sun to contemplate its nature; had he lived in these days, he might have enjoyed his wish at an easier rate, and scaling the heavens by this glass, might plainly have discerned what he so much desired. *Keplar*, considering those strange discoveries which this perspective had made, could not choose but cry out in a rapture of admiration, 'O telescope, much-knowing and more precious than any sceptre: is not one who holds you in his right hand ordained as king and lord of the works of God?'

Now if you would know what might be done by this glass, in the sight of such things as were nearer at hand, the same Author will tell you, when he says, that by it those things which could scarce at all be discerned by the eye at the distance of a mile and a half, might plainly and distinctly be perceived for sixteen *Italian* miles, and that as they were really in themselves, without any transposition or falsifying at all. So that what the ancient Poets were fain to put in a fable, our more happy age hath found out in a truth, and we may discern as far with these eyes which *Galilaeus* hath bestowed upon us, as *Lynceus* could with those which the Poets attributed unto him.

I have now cited such Authors both ancient and modern, who have directly maintained the same opinion. I told you likewise in the proposition that it might probably be deduced from the tenets of others: such were *Aristarchus*, *Philolaus* and *Copernicus*, with many other later writers who assented to their hypothesis, so *Ioach. Rheticus, David Origanus Lansbergius, Guil. Gilbert* and (if I may believe *Campanella*), very many others, both English and French, all who affirmed our Earth to be one of the Planets, and the Sun to be the Centre of all, about which the heavenly bodies did move, and how horrid soever this may seem at the first, yet is it likely enough to be true, nor is there any maxim or observation in Opticks (says *Pena*) that can disprove it.

Now if our earth were one of the Planets (as it is according to them) then why may not another of the Planets be an earth?

Thus have I shewed you the truth of this proposition: Before I proceed further, 'tis requisite that I inform the Reader, what method I shall follow in the proving of this chief assertion, that there is a World in the Moon.

Proposition 7. That those spots and brighter parts which by our sight may be distinguished in the Moon, do show the difference betwixt the Sea and Land in that other World.

Proposition 8. The spots represent the Sea, and the brighter parts the Land.

Proposition 9. That there are high Mountains, deep valleys and spacious plains in the body of the Moon.

Proposition 10. That there is an Atmosphaera, or an orb of gross vaporous air, immediately encompassing the body of the Moon.

Proposition 11. That as their world is our Moon, so our world is their Moon.

Proposition 12. That 'tis probable there may be such Meteors belonging to that world in the Moon, as there are with us.

Proposition 13. That 'tis probable there may be inhabitants in this other World, but of what kind they are is uncertain.

Proposition 14. That 'tis possible for some of our posterity to find out a conveyance to this other world, and if there be inhabitants there, to have commerce with them.

'Twas a great While, ere the Planets were Distinguished from the fixed Stars, and some time after that, ere the Morning and Evening Star were Found to be the same. And in greater space (I doubt not) but this also, and other Excellent Mysteries will be Discovered. Time, who has always been the Father of new Truths, and has Revealed unto us many things, which our Ancestors were Ignorant of, will also Manifest to our Posterity, that which we now desire but cannot know. Time will come, when the Endeavours of after Ages shall bring such things to Light as now lie hid in Obscurity. Arts are not yet come to their Solstice. But the Industry of Future Times, Assisted with the Labours of their Forefathers, may reach that Height which we could not Attain to. As we now wonder at the Blindness of our Ancestors, who were not able to Discern such things as seem Plain and Obvious unto us, so will our Posterity Admire our Ignorance in Perspicuous matters.

Francis Godwin, *The Man in the Moon, or A Discourse of a Voyage Thither by Domingo Gonsales*, 1638

Francis Godwin (1562–1633) was an English bishop whose publications were mainly historical and theological, aside from this one: a short work about a fantastical voyage to the moon in a flying machine pulled by 'gansas' (wild swans), published posthumously and pseudonymously. Godwin's book explicitly supported the earth's motion as described by Copernicus, whom he mentioned by name, as well as the theories of William Gilbert and Johannes Kepler, who are not named explicitly. A highly popular book – the rest of the century saw another twenty-five editions of the work, and it was translated from English into several languages – this was one way that non-scholarly readers might have come into contact with some of the latest astronomical ideas, including heliocentric theory.

Before I come to relate our extraordinary Voyage of *Domingo Gonsales* to the World in the Moon, I will make a Halt at St *Hellens*, or *Hellena*, which is now possessed by the Honourable *East India* Company. There is no Island in the World so far distant from the Continent or main Land as this.

It is in this Island that the Scene of that notable Fancy, called, *The Man in the Moon, or a Discourse of a Voyage Thither*, by *Domingo Gonsales* is laid, written by a learned Bishop, says the ingenious Bishop Wilkins, who calls it a pleasant and well contrived Fancy, in his own Book, entitled, *A Discourse of the New World, tending to prove that it is possible there may be another habitable World in the Moon*; wherein among other curious Arguments he affirms, that this has been the direct Opinion of diverse ancient, and some modern Mathematicians, and may probably be deduced from the Tenets of others, neither does it contradict any Principle of Reason nor Faith; and that as their World is our Moon, so our World is theirs.

Now this small Tract having so worthy a Person to vouch for it, and many of our English Historians having published for Truth, what is almost as improbable as this, as Sir *John Mandavil* in his Travels and others, and this having what they are utterly destitute of, that is, Invention mixed with Judgement; and was judged worthy to be Licensed fifty years ago, and not since reprinted, whereby it would be utterly lost. I have thought fit to republish the Substance thereof, wherein the Author says he does not design to discourse his Readers into a Belief of each particular Circumstance, but expects that his new Discovery of a *new World*, may find little better Entertainment than *Columbus* had in his first Discovery of *America*, though yet that poor Espial betrayed so much Knowledge as has since increased to vast Improvements, and the then *Unknown* is now found to be of as large Extent as all the other *known World*; that there should be Antipodes was once thought as great a Paradox, as now that the Moon should be habitable. But the Knowledge of it may be reserved for this our discovering Age, wherein our Virtuosi can by their Telescopes gaze the Sun into Spots, and descry Mountains in the Moon. But this and much more must be left to the Critics, as well as the following Relation of our little Eye-witness and great Discoverer, which you shall have in his own *Spanish* Style, and delivered with that Grandeur and Thirst of Glory, which is generally imputed to that Nation.

It is known to all the Countries of Andaluzia, that I *Domingo Gonsales* was born of a noble Family in the renowned City of *Seville*. I lived like a Gentleman many Years very happily: At length a Quarrel arising between me and *Pedro Delgades*, a Gentleman and Kinsman of mine; it grew so high, that when no Mediation of Friends could prevail, we two went alone with our Swords into the Field, where it was my Chance to kill him, though a stout proper Man; but what I wanted in Strength I

Supplied in Courage, and my Agility countervailed for his Stat-ure. This being acted in Carmona, I fled to *Lisbon* and embarked in a stout Carrick bound for the *East Indies*.

In the *Indies* I thrived exceedingly, laying out my Stock in Dia-monds, Emeralds and Pearls, which I bought at such easy Rates, that my Stock safely arriving in *Spain* (as I understood it did) must needs yield ten for one. But having doubled *Cape Bona Esperanza* in my Way home, I fell dangerously sick, expediting nothing but Death, which had undoubtedly happened, but that we just then recovered the blessed Isle of St *Hellens*, the only Paradise I believe on Earth, for Healthfulness of Air, and Fruit-fulness of Soil, producing all Necessaries for the Life of Man. It is about 16 Leagues in Compass, and has no firm Land or Con-tinent within 300 Leagues, nay not so much as an Island within 100 Leagues of it, so that it may seem a Miracle of Nature, that out of so vast and tempestuous an Ocean, such a small Rock or Piece of Ground should arise and discover itself. On the South is a good Harbour, and near it diverse small Houses built by the *Portuguese* to accommodate Strangers, with a pretty Chapel handsomely beautified with a Tower, and Bell therein. Near it is a Stream of excellent fresh Water, divers handsome Walks, planted on both Sides with Orange, Lemon, Pomegranate, Almond Trees and the like, which bear Fruit all the Year, as do also divers others. There are Store of Garden herbs, with Wheat, Peas, Barley and most Kinds of Pulse; but it chiefly abounds with Cattle and Fowl, as Goats, Swine, Sheep, Partridges, wild Hens, Pheasants, Pigeons and wild Fowl beyond Credit; but especially about *February* and *March* are to be seen huge Flocks of a kind of wild Swans (whereof I shall have Occasion to speak more hereafter) who like our Cuckoos and Nightingales, go away at a certain Season, and are no more seen that Year.

On this happy Island did they set me ashore with a Negro to attend me, where I recovered my Health, and continued a whole Year, solacing myself for want of human Society with Birds and brute Beasts; Diego my Black moor was forced to live in a Cave at the West End of the Isle, for had we dwelt together, Victuals would not have been so plenty with us.

Upon the Shore, about the Mouth of our River, I found Store

of a kind of wild Swans feeding upon Prey, both of Fish and Birds, and which is more strange; having one Claw like an Eagle, and the other like a Swan. These Birds breeding here in infinite Numbers, I took thirty or forty of them young, and bred them up by Hand for Recreation; yet not without some Thoughts of that Experiment which I after put in Practice. These being strong and able to continue a great Flight, I taught them first to come at Call afar off, not using any Noise, but only showing them a white Cloth; and here I found it true what Plutarch affirms, That Creatures which eat Flesh are more docible than others. 'Tis wonderful to think what Tricks I taught them ere they were a Quarter old, among others I used them by Degrees to fly with Burdens, wherein I found them able beyond Belief, and a white Sheet being displayed to them by Diego, upon the Side of a Hill, they would carry from me to him Bread, Flesh, or whatever I pleased, and upon the like Calf come to me again. Having proceeded thus far, I consulted how to join a Number of them together, so as to carry a heavier Weight, which if I could compass, I might enable a Man to be carried safely in the Air from one Place to another. I puzzled my Wits extremely with this Thought, and upon Trial found, that if many were put to the bearing of one great Burthen, by reason it was impossible all of them, should rise together just at one Instant, the first that rise finding himself stayed by a Weight heavier than he could stir, would soon give over, and so the second, third and all the rest. I contrived at last a Way whereby each might rise with only his own Proportion of Weight, I fastened about each Gansa a little Pulley of Cork, and putting a String of a just Length through it, I fastened one End to a Block of almost eight Pounds' Weight, and tied a two-Pound Weight to the other End of the String, and then causing the Signal to be erected, they all arose together, being four in Number, and carried away my Block to the Place appointed. This hitting so luckily, I added two or three Birds more, and made Trial of their carrying a Lamb, whose Happiness I much envied, that he should be the first living Creature to partake of such an excellent Device.

At length after divers Trials, I was surprised with a great Longing to cause myself to be carried in the same Manner. Diego my Moor was likewise possessed with the same Desire, and had

I not loved him well, and wanted his Service, I should have resented his ambitious Thought; for I count it greater Honour to have been the first Flying Man, than to be another Neptune who first adventured to sail on the Sea. Yet seeming not to understand his Intention, I only told him, that all my Gansas were not strong enough to carry him, being a Man though of no great Bulk, yet twice heavier than myself. Having prepared all Necessaries, I one Time placed myself and all my Utensils on the Top of a Rock at the River's Mouth, and putting myself upon my Engine at full Sea, I caused Diego to advance the Signal, whereupon my Birds, twenty-five in Number, rose all at once, and carried me over lustily to the Rock on the other Side, being about a Quarter of a League. I chose this Time and Place, because if any Thing had fallen out contrary to Expectation, the worst that could happen was only falling into the Water; and being able to swim well, I hoped to receive little Hurt in my Fall. When I was once safe over, O how did my Heart even swell with Joy and Admiration at my own Invention; how often, did I wish myself in the Midst of Spain, that I might fill the World with the Fame of my Glory and Renown?

It was now the Season that these Birds take their Flight away, as our Cuckoos and Swallows do in *Spain* towards *Autumn*, and as I afterwards found, being mindful of their usual Voyage, just when I began to settle myself to take them in, they with one Consent rose up, and having no other higher Place to make toward, to my unspeakable Fear and Amazement, struck bolt upright, and never left towering upward, still higher and higher, for the Space, as I guessed, of an Hour, after which I thought they laboured less than before, till at length, ah wonderful! they remained immoveable, as steadily as if they had sat upon so many Perches; the Lines slacked, neither I, nor the Engine moved at all, but continued still, as having no Manner of Weight. I found then by Experience, what no Philosopher ever dreamt of, namely, that those Things we call heavy do not fall towards the Centre of the Earth as their natural Place, but are drawn by a secret Property of the Globe of the Earth, or rather something within it, as the Lodestone draws Iron which is within the Compass of its attractive Beams. For though my *Gansas* would continue unmoved, without being sustained by any Thing but the Air, as easily and quietly as a Fish in the Water, yet if they forced themselves never so little, it is impossible to imagine with what Swiftness they were carried, either Upward, Downward, or Sideways; I must ingenuously confess my Horror and Amazement in this Place was such, that had I not been armed with a true *Spanish* Resolution, I should certainly have died for Fear.

The next Thing that disturbed me was the Swiftness of the Motion, which was so extraordinary, that it almost stopped my Breath, if I should liken it to an Arrow out of a Bow, or a Stone thrown down from the Top of an high Tower, it would come vastly short of it; another Thing was exceeding troublesome to me, that is the Illusions of Devils and wicked Spirits, who the first Day of my Arrival came about me in great Numbers in the Likeness of Men and Women, wondering at me like so many Birds about an Owl, and speaking several Languages which I understood not, till at last I met with some that spoke good *Spanish*, some *Dutch*, and others *Italian*, all which I understood; and here I had only a Touch of the Sun's Absence once for a short Time, having him ever after in my Sight. Now though my *Gansas* were entangled in my Lines, yet they easily seized upon diverse Kinds of Flies and Birds, especially Swallows and Cuckoos, whereof there were Multitudes, even like Motes in the Sun, though I never saw them eat any Thing at all. I was much obliged to those, whether Men or Devils I know not, who among divers Discourses told me, 'If I would follow their Directions, I should not only be carried safe Home, but be assured to command at all Times all the Pleasures of that Place.' To which Motion, not daring to give a flat Denial, I desired Time to consider, and withal indebted them (though I felt no Hunger at all, which may seem strange) to help me to some Victuals, least I should starve in my Journey, so they readily brought me very good Flesh and Fish of several Sorts, and well dressed, but that it was extreme fresh without any Relish of Salt. Wines likewise I tasted of diverse Kinds as good as any in *Spain*, and Beer no better in all *Antwerp*. They advised me, that while I had Opportunity I should make my Provisions, telling me, that till the next *Thursday* they could help me to no more, at which Time they would find Means to carry me back, and set me safe in *Spain*, in any Place I would desire, provided I would become one of their Fraternity, and enter into such Covenants as they had made to their Captain and Master, whom they would not name: I answered civilly, 'I saw little Reason to rejoice in such an Offer, desiring them to be mindful of me as Occasion served'; so for that Time I was rid of them; having first, furnished my Pockets with as much Victuals as I could thrust in, among which I would be sure to find a Place for a small Bottle of good Canary.

I shall now declare the Quality of the Place wherein I was: The Clouds I perceived to be all under between me and the Earth. The Stars, because it was always Day, I saw at all Times alike, not shining bright, as we see in the Night upon Earth, but of a whitish Colour, like the Moon with us in the Day-Time, those that were seen, which were not many, shewed far greater than with us, yea, as I guessed no less than ten Times bigger: As for the Moon, being then within two Days of the Change, she appeared of an huge and dreadful Greatness. It is not to be forgot, that no Stars appeared but on that Part of the Hemisphere next the Moon, and the nearer to her, the larger they appeared again; whether I lay quiet and rested, or were carried in the Air, I perceived myself to be always directly between the Moon and the Earth, whereby 'tis plain, that my *Gansas* took their Way directly toward the Moon, and that when we rested, as we did at first for many Hours, either we were insensibly carried round about the Globe of the Earth, though I perceived no such Motion, or else that, according to the Opinion of *Copernicus*, the Earth is carried about, and turns round perpetually from West to East, leaving to the Planets only that Motion which the Astronomers call natural, and is not upon the Poles of the Equinoctial, commonly called the Poles of the World, but upon those of the Zodiac; the Air in that Place I found without any Wind, and exceeding temperate, neither Hot nor Cold, where neither the Sunbeams had any Object to reflect upon, nor the Earth and Water appear to affect the Air with their natural Quality of Coldness; as for the Philosophers attributing Heat and Moisture to the Air, I always esteemed it a Fancy: Lastly, I remember that after my Departure from the Earth, I never felt either Hunger or Thirst, whether the Purity of the Air, freed from the Vapours of the Earth and Water, might yield Nature sufficient Nourishment, or what else might be the Cause I cannot determine, but so I found it, though I was perfectly in Health both of Body and Mind, even above my usual Vigour.

Some Hours after the Departure of that Devilish Company, my *Gansas* began to bestir themselves, still directing their Course toward the Globe or Body of the Moon, making their Way with such incredible Swiftness, that I conceive they advanced little less than fifty Leagues in an Hour, in which Passage I observed three Things very remarkable, one that the further we went the less the Globe of the Earth appeared to us, and that of the Moon still larger: Again the Earth, which I had ever in mine Eye, seemed to made itself with a kind of Brightness like another Moon, and as we discern certain Spots or Clouds as it were in the Moon, so did I then see the like in the Earth; but whereas the Form of these Spots in the Moon are always the same, these on the Earth seemed by Degrees to change every Hour; the Reason whereof seems to be, that whereas the Earth according to his natural Motion (for such a Motion I am now satisfied she has according to the Opinion of *Copernicus*) turns round upon her own Axis every four-and-twenty Hours from West to East) I should at first see in the Middle of the Body of this new Star the Earth, a Spot like a Pear, with a Morsel bit out on one Side, in some Hours I should observe this Spot move away toward the East: This no doubt was the mainland of *Africa*; then might I perceive a great shining Brightness in that Place which continued about the same Time, and was questionless the vast *Atlantick* Ocean: After this succeeded a Spot almost Oval, just as we see *America* described in our Maps, then another immense Clearness, representing *Mare del zar* or the *South Sea*; lastly, a number of Spots like the Countries and Islands in the *East Indies*, so that it seemed to me no other than an huge mathematical Globe turned round leisurely before me, wherein successively all the Countries of our earthly World were within twenty-four Hours represented to my View, and this was all the Means I now had to number the Days, and reckon the Time.

I could now wish that Philosophers and Mathematicians would confess their own Blindness, who have hitherto made the World believe that the Earth has no Motion, and to confirm it, are forced to attribute to every one of the celestial Bodies two Motions directly contrary to each other, one from the East to the West, to be performed in twenty-four Hours with an impetuous rapid Motion; the other from West to East in several Proportions: O incredible Supposition! that those huge Bodies of the fixed Stars in the highest Orb, whereof they confess diverse, are above a hundred Times bigger than the whole Earth, should like so many Nails in a Cartwheel be whirled about in so short a Time; whereas it is many thousand Years, no less (say they) than thirty thousand, before that Orb finishes his Course from West to East, which they call his natural Motion; now whereas they allow their natural Course from West to East to every one of them, therein they do well; the Moon performs it in seven-and-twenty Days, the *Sun*, *Venus* and *Mercury* in a Year or thereabout, *Mars* in three Years, *Jupiter* in twelve, and *Saturn* in thirty. But to attribute to these celestial Bodies contrary Motions at once, is an absurd Conceit, and much more to imagine, that the same Orb wherein the fixed Stars are, whose natural Course takes up so many thousands of Years, should be turned about every twenty-four Hours. I will not go so far as *Copernicus*, who makes the Sun the Centre of the Earth and immoveable, neither will I be positive in any Thing, only this I say, allow the Earth its Motion, which these Eyes of mine can testify to be true, and all those Absurdities are removed, every one having only his own single and proper Motion.

But where am I? I promised a History, and am unawares turned Disputer. One Accident more befell me worth mentioning, that during my Stay I saw a kind of a reddish Cloud coming toward me, and continually approaching nearer, which at last I perceived was nothing but a huge Swarm of Locusts. He that reads the Discourses of learned Men concerning them, as *John Leo* of *Africa*, and others who relate that they are seen in the Air several Days before they fall on the Earth, and adds thereto this Experience of mine, will easily conclude, that they can come from no other Place than the Globe of the Moon. But now give

me leave to go on quietly in my Journey for eleven or twelve
Days, during all which Time I was carried directly toward the
Globe or Body of the Moon, with such a violent Whirling as is
inexpressible, for I cannot imagine a Bullet out of a Cannon
could make Way through the vaporous and muddy Air near the
Earth with half that Celerity; which is the more strange, since
my *Gansas* moved their Wings but now and then, and sometimes for a quarter of an Hour not at all, only holding them
stretched out, as we see Kites, and Eagles sometimes do for a
short Space; during which Pauses, I suppose they took their
Naps, and Times of Sleeping, for other Times I could perceive
they never had any; for myself I was so fastened to mine Engine,
that I durst slumber enough to serve my Turn, which I took with
as great Ease, as if I had lain on the Down-bed in *Spain*.

After eleven Days' Passage in this violent Flight, I perceived
we began to approach to another Earth (if I may so call it),
being the Globe or very Body of that Star which we call the
Moon. The first Difference I found between this and our Earth
was, that it appeared in its natural Colours, as soon as ever I
was free from the Attraction of the Earth; whereas with us, a
Thing a League or two from us, puts on that deadly Colour of
Blue. I then perceived also that this World was the greatest Part
covered with a huge mighty Sea, those Parts only being dry
Land which are to us somewhat darker than the rest of her
Body, I mean, what the Country People call, *The Man in the
Moon*, and that Part which shines so bright is another Ocean
besprinkled with Islands, which for their Smallness we cannot
discern so far off; so that the Splendour which appears to us in
the Night, is nothing but the Reflection of the Sunbeams
returned to us out of the Water as from a Looking-glass. How
much this disagrees with what our Philosophers teach in the
Schools is evident: But alas, how many of their Errors have
Time and Experience in this our Age, and among other vain
Conjectures, who have not hitherto believed the upper Region
of the Air to be very hot; as being next, forsooth, to the natural
Place of the Element of Fire; mere Vanities, Fancies and Dreams:
For after I was once free from the attractive Beams of that tyr-
annous Lodestone the Earth, I found the Air altogether serene,

without Winds, Rain, Mists, or Clouds, neither hot nor cold, but constantly pleasant, calm and comfortable, till my Arrival in that *New World of the Moon*; as for that Region of Fire, our Philosophers talk of, I heard no News of it, mine Eyes have sufficiently informed me there is no such Thing.

The Earth had now by turning about showed me all her Parts twelve Times, when I finished my Course; for when my Reckoning it seemed to be (as indeed it was) *Tuesday, 11 September*, at which Time the Moon being two Days old was in the twentieth Degree of *Libra*) my *Gansas* seemed by one Consent to stay their Course, and rested for certain Hours, after which they took their Flight, and in less than an Hour set me on the Top of an high Hill in that *Other World*, where many wonderful Things were presented to my Sight. For I observed first, that though the Globe of the Earth appeared much greater there than the Moon does to us even three Times bigger, yet all Things there were ten, twenty, yea thirty Times larger than ours; their Trees were thrice as high, and above five Times broader and thicker; so were their Herbs, Birds and Beasts, though I cannot well compare them to ours, because I found not any kind of Beast or Bird there which any way resembled ours, except Swallows, Nightingales, Cuckoos, Woodcocks, Bats and some kind of Wild Fowl: And likewise such Birds as my *Gansas*, all which, as I now perceived, spend their Time in their Absence from us, in that World, neither do they differ in any Thing from ours, but are the very same kind.

Wilkins, Frontispiece, *A Discourse Concerning A New World & Another Planet*, 1640

This elaborate image is the frontispiece to the 1640 updating of Wilkins's 1638 Discovery of a World in the Moon. *Its elements are drawn from a number of previously published works by both Wilkins and others, including the cosmological diagram at the top of the page, drawn from the title page of the 1638* Discovery. *It depicts a heliocentric system in which, as in Digges's 1576 image, the outer, bounded sphere of the fixed stars has been replaced with an unbounded region of infinitely extended stars. Each sphere beneath the fixed stars is shown with a figure from myth – Ceres and Proserpina, in the case of the earth and moon, suggesting the longing of the residents of one to go to the other. The earth and moon are depicted as both directly and indirectly illuminated by the sun, which proclaims 'to all things I give light, heat and motion'. The figure on the lower left, next to the title banner, is Copernicus, who holds a diagram of his world system and asks 'what if it were thus?', much as Tycho does on the frontispiece of Kepler's* Rudolphine Tables. *Galileo stands on the right of the banner, holding his telescope. He points at a bird and says 'here are his eyes', perhaps suggesting that the telescope extends human eyesight such that it rivals even birds of prey, or such that it brings us closer to the heavens. Galileo looks downward, however, perhaps due to his physical blindness in later life, or to what Wilkins considered his limited vision, as he never considered the possibility of life on other worlds. Kepler is depicted directly behind Galileo, adding 'and also his wings', perhaps suggesting that both observation and reasoned mathematics are necessary to reach the heavens.*

Descartes, *Principles of Philosophy*, 1644

The cosmological ideas of Descartes's The World *were ultimately rewritten and published in the 1644* Principles of Philosophy. *Here, Descartes also described the cosmos as filled with a fluid-like, penetrable matter that moved in swirling vortices, carrying planets around central suns. However, he protected himself from accusations like those levelled at Galileo by defining motion first, and then planetary motion specifically, such that the planets – including the earth – could all be said to be stationary, since they did not move relative to the fluid around them. However, they were carried by that fluid in its own motion around the sun. Descartes also carefully and explicitly insisted on the hypothetical nature of his work, though he claimed that 'it would scarcely be possible for the causes from which all phenomena can be clearly deduced not to be true'. His arguments in the book were primarily neither mathematical nor experimental, but rather based on reasoning from first principles.*

Second Part: On the principles of material things

xxiii. All variety of matter, or all diversity of its forms, depends on motion.

The matter that exists throughout the universe is one and the same, and is known only insofar as it is extended. And all the properties that we clearly perceive in it are reduced to this one: that it is dividable and movable according to its parts, and therefore capable of all those affections which we perceive and which are able to follow from the motion of its parts. For division by thought alone changes nothing; rather, all variation of matter, or all diversity of its forms, depends on motion. This seems to have been recognized everywhere even by the philosophers, who have said that nature is the principle of motion and rest. And by 'nature' they understood 'that through which all corporeal things become what we experience them to be'.

xxiv. What motion is, in the common sense.

Motion (specifically local, for no other kind makes sense to me, nor do I think that any other kind must be presumed in nature) as it is commonly understood, is nothing other than *an action by which some body moves from one place to another*. And just as we already explained above that the same thing can be said to change and not change place at the same time, so too the same thing can be said to be moved and not to be moved at the same time. Thus someone sitting on a ship that sets sail from the harbour thinks that he is indeed moved if he looks to the shore and considers it immovable; but he does not if he looks to the ship itself, as he always keeps the same position with respect to its parts. And as we commonly think that there is an action in every motion, but that there is in cessation of action in rest, then the man is more properly said to be at rest rather than to be moved, because he perceives no action in himself.

xxv. What motion is, in the proper sense.

But if we consider motion not in the common usage but rather in terms of the truth of the matter, and how it ought to be understood, so that some determinate nature can be attributed to it, we can say that it is the *translation of one part of matter, or of one body, from the vicinity of those bodies that are in immediate contact with it and are regarded as at rest to the vicinity of others*. By *one body* or *one part of matter* I understand all that is transferred at the same time, although it may consist of many parts that have different motions in themselves. And I say that it is the *translation*, not the force or action, which transfers, so that I can show that it is always in the thing that is movable, not in the mover, because these two are not usually distinguished accurately enough.

Third Part: On the Visible World

xv. The same appearances of the planets can be explained through various hypotheses.

When someone on one ship in the middle of a tranquil sea looks out at another in the distance that is changing its position with respect to it, he can often doubt to which of them the motion should be attributed. Likewise, the wanderings of the planets, seen from the earth, appear such that it cannot be ascertained from them alone to which bodies the motion should properly be attributed. And since they are very unequal and complex, it is not easy to explain them without choosing one of the various ways that those motions can be understood, and supposing that they are in accord with it. To this end there are three different hypotheses, invented by astronomers: namely, suppositions that are considered not as truths, but rather as suitable for explaining the phenomena.

xvi. The hypothesis of Ptolemy does not satisfy the appearances.

The first of these is Ptolemy's, which is now commonly rejected by all philosophers since it conflicts with many recent observations (especially the waxing and waning of light that is observed on Venus just as on the moon). Therefore, I omit it here.

xvii. The hypotheses of Copernicus and Tycho do not differ insofar as they are hypotheses.

The second is Copernicus's and the third is Tycho Brahe's. These two, insofar as they are only hypotheses, satisfy the phenomena in the same way, and there is no great difference between them, except that the hypothesis of Copernicus is somewhat simpler and clearer, so much so that Tycho would have had no reason to change it, except that he was trying to set forth not merely a hypothesis, but rather the actual truth of the matter.

xviii. Tycho claims to attribute less motion to the earth than Copernicus, but actually attributes more.

While Copernicus did not hesitate to attribute motion to the earth, Tycho wanted to amend this, believing it to be very absurd physically and contrary to the common sense of men. But because he did not sufficiently consider the true nature of motion, he only asserted in words that it was at rest, but in actuality he gave it more motion than Copernicus had.

xix. I deny the motion of the earth more carefully than Copernicus and more correctly than Tycho.

I differ from both of them in this alone: that I will remove all motion from the earth more correctly than Tycho and more diligently than Copernicus. I will propose here a hypothesis that seems to be the simplest of all, and the most suitable both for understanding the phenomena and for investigating their natural causes. I want this to be understood as only a hypothesis, however, and not as the truth of the matter.

xxv. The heavens carry with them everything contained within them.

Many seem to me to err in this: that, attributing fluidity to heavens, they imagine that space is completely empty, such that it does not resist the motions of other bodies, but also has no force to carry those same bodies with it. But there can be no such vacuum in the nature of things, and all fluids have this in common: that they do not resist the motion of other bodies only because they are themselves in motion, and this motion necessarily carries with it all the bodies that are contained in it – even if they are solid and at rest and hard – unless they are held back by an external cause, as is evident from what has been said previously.

xxvi. The earth is at rest in its heaven, but is nevertheless carried by it.

Since we see that the earth is not supported by pillars or suspended by cords, but is surrounded on every side by a very fluid heaven, we think that it is at rest and has no propensity to motion, since we observe none. But let us not think that this prevents it from being carried by that heaven, following its motions while itself unmoved. In this way a ship, driven by no winds or oars and tied to no anchors, rests in the middle of the sea, although perhaps a large mass of water, flowing undetectably on its course, bears the ship along with it.

xxvii. The same must be understood of all the planets.

And just as the other planets are like the earth in that they are opaque and reflect the rays of the sun, we are justified in thinking that they are similar to it in this as well: that each one is at rest in that region of the heaven where it is situated, and that every variation in position that is observed in them proceeds only from the fact that all the matter of the heaven that contains them is in motion.

xxviii. Properly speaking, the earth does not move, nor do any of the planets, although they are carried by the heaven.

Here let us remember what was said concerning the nature of the translation of one body from the vicinity of other bodies with which it has immediate contact and which are seen as at rest, to the vicinity of others. But often, according to common usage, every action in which a body moves from one place to another is called motion and this is the sense in which it can be said that something is moved and not moved at the same time, just as we can variously determine its place. From this it follows that no motion is properly found on earth or the other planets, because they are not transferred from the vicinity of the parts of heaven with which they have immediate contact, insofar as those parts are considered unmoved. For they would have to be separated from all these parts at once, and they are not: rather, because the matter of the heaven is fluid, first some particles of it and then others are removed from the planet that they touch, and that motion should be attributed to those particles, and not to the planet. The same must be understood of all the planets, just as the partial transfers of water and air which occur on the surface of the earth are not attributed to the earth itself, but rather to the parts of water and air that are transferred.

xxix. That no motion should be attributed to the earth, understood in its improper, common sense, but in this way it is rightly said that the planets are moved.

If we understand motion according to the common usage, it must be said that all the other planets are moved, and even the sun and the fixed stars. But it would be quite unsuitable to say the same thing of the earth. For people commonly determine the positions of the stars from parts of the earth seen as immobile. The stars they judge to be moved insofar as they depart from these determined positions. This is convenient for use in life, and therefore makes sense. Indeed, all of us from a young age have thought that the earth is not round but flat, and that it has the same up and down everywhere and the same poles of east, west, south and north, which we use to designate the positions of all other bodies. But if any philosopher, observing that the earth is a globe contained in a fluid and mobile heaven and that the sun and the fixed stars always keep the same positions among themselves, were to use them as immobile in order to determine his position, and thus affirms that the earth itself moves, he would be speaking without reason. For first, according to the philosophical sense, position should not be determined by very remote bodies like the fixed stars, but by those contiguous to the one that is said to be moved. And next, according to common usage, there is no reason to consider the fixed stars immovable rather than the earth.

xxx. All the planets are carried around the sun by the heaven.

With all our scruples about the motion of the earth put aside, let us suppose that the whole matter of the heaven in which the planets are situated turns constantly like a vortex in the centre of which is the sun, that the parts nearer to the sun are moving more rapidly than the ones further away, and that all the planets (among which is the earth) are always situated in the same parts of the celestial matter. From this alone, without any contrivances, all the phenomena of the planets can be very easily understood. For in those parts of rivers where the water twists on itself and forms a vortex, if some pieces of straw fall in the water we will see that they are carried along with it, and some also will be turned around their own centres, and those nearer to the centre are turned more quickly, and finally, although they always tend toward circular motion, they scarcely ever describe perfect circles, but rather vary somewhat in longitude and latitude. Thus we can imagine all the same

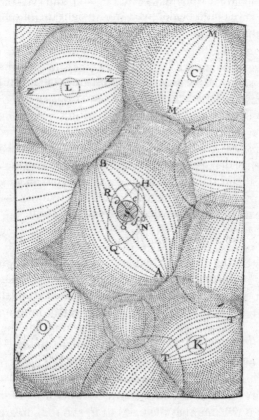

Vortices depicted in Descartes's *Principles of Philosophy*.

things about the planets without any difficulty, and all of their phenomena are explained through this one account.

xliii. It would scarcely be possible for the causes from which all phenomena can be clearly deduced not to be true.

And certainly if we use no principles except those which we have seen are clearly evident, and if we deduce nothing from them except their mathematical consequences, and meanwhile those things which we deduce from them accurately agree with all the phenomena of nature, we would seem to be doing an injustice to God if we suspected that the causes of things, found by us in this way, were false, as if he had produced us so imperfectly that we could be deceived by the right use of our reason.

xliv. Nevertheless, I want those things that I explain here to be regarded as hypotheses only.

Nevertheless, lest we seem too arrogant if we affirm that the genuine truth of things was discovered by us when philosophizing about such matters, I would rather leave the matter undecided, and put forth everything that I will write hereafter as a hypothesis only. And though one may think that they are false, I think that I have made something of great enough value with this work, if everything deduced from it agrees with experience. For thus we will secure as much utility in life from it as we would from knowledge of the truth itself.

xlv. I will also assume some things that are certain to be false.

Indeed, in order to better explain natural things, I will push their causes further back than I think they ever existed. For there is no doubt that the world was created from the beginning in all its perfection; thus the sun, earth, moon and stars existed then; and on the earth there were not only the seeds of plants, but the plants themselves; and Adam and Eve were not born as infants, but were made as adults. This is what the Christian faith teaches us, and natural reason also clearly convinces us: for when we consider the immense power of God, we cannot imagine that He ever did anything that was not complete on every level. But nevertheless, in order to understand the natures of plants or of men, it is far better to consider how they could have grown from seeds by degrees than to consider how they were created by God when the world was first created. Thus, if we can devise some principles that are very simple and easily understood, according to which all the things that we perceive in the visible world, including the stars and the earth, could have arisen from seeds – although we know well that they did not arise in this way – we will nevertheless by this means explain their natures far better than if we were to only describe them as they already are. And since I seem to have found such principles, I will explain them briefly here.

Giovanni Battista Riccioli, *New Almagest*, 1651

Giovanni Battista Riccioli (1598–1671) was an Italian Jesuit astronomer whose 1651 New Almagest *was a massive work of over 1,500 pages. A modern updating of Ptolemy's* Almagest, *it included all the subjects Riccioli deemed relevant to the astronomy of his day, gathered into two volumes and ten books. Book 9, itself several hundred pages long, focused on an analysis of the two rival world systems, geocentric and heliocentric (the former, to Riccioli, meant the Tychonic system, as the Ptolemaic had already been discredited). Riccioli offered 126 arguments concerning the earth's motion: forty-nine for and seventy-seven against. He offered answers to some of the arguments against each, but ultimately deemed some of the geocentric arguments against a moving earth unanswerable, and so argued that the Tychonic system was the most probable given the evidence available at the time. In particular, Riccioli felt that there should be some detectable evidence for a rotating earth, and yet such evidence was lacking. He also noted the lack of stellar parallax, and argued that if Copernicus were correct, stars would have to be massive – so much so that a single star in the Copernican system could be larger than the entire cosmos in the Ptolemaic. Like others before him, he deemed this absurd. The comprehensive nature of his book, which included all the latest astronomical discoveries to date, made it the standard reference book for astronomers of the time, even those who embraced the Copernican system.*

Part 2, Book 9, Section 4,
'On the System of the Earth's Motion'

Chapter 1

Now we face a controversy that is by far the most famous among astronomers, especially in this century. In order to solve this controversy, we must impose two supreme restraints and arbitrators on our intellect: reason and authority.

To briefly mention reason: nowhere did astronomy, which is a subdiscipline of mathematics and physics, clash more sharply with the rest of physics than on the occasion of this controversy. For Philolaus, Aristarchus and Copernicus armed themselves with arguments against the constancy and stability of the earth from every kind of ancient and modern phenomenon. Hence the Peripatetics, and indeed almost all physicists, gathered to defend the powers and particular properties of the elements. The former, so that their new hypotheses might end up victorious, also constructed a new physics based on the motion of the elements and the elementary bodies. The latter, by contrast, erected a new structure of very solid arguments, built on the most ancient foundation of nature, against the newcomers. Thus astronomers and physicists were no longer united, much like those storied gods and giants; instead, heaven itself seemed at war with the elements. Moreover, arguments were devised on both sides, sometimes with such subtlety and adorned with sophistries or paralogisms having such an appearance of truth, that even the most cautious could be deceived if he did not carefully and repeatedly examine them. So there is little wonder that, lightly grounded in the principles of astronomy but in some respects profoundly engaged with the sublimities of this science, the Copernican hypotheses did not penetrate deeply. But neither was their falsity discouraged by necessary arguments, but rather by arguments more trifling than were required. Thus its followers arose everywhere, and dared to sing of it triumphantly in Germany, England, France and Italy itself, and lay equally their ignorance of heavenly and terrestrial revolutions before the philosophers and theologians.

This would perhaps have been tolerable, had they not come into conflict with the authority of the Catholic Church in deciding controversies of this sort, particularly with respect to Sacred Scripture. Indeed, if we concede to Copernicans the licence to interpret the divine texts and evade Ecclesiastical decrees, then the matter will not be contained within the boundaries of astronomy alone, or even natural philosophy, but would be able to be extended to other more sacred doctrines by others, as it would evidently be permitted to deny the literal sense of the divine texts in this case without clear necessity. For there is no need, either from the principles of astronomy or the principles of physics, to depart from it, as I will shortly demonstrate, even though the hypothesis of the motion of the earth is preferred by some very clever and skilled men.

Before I attempt to do this, I think that the principal authors must be assessed, both those who have written in support of the Copernican suppositions and those who have written against, which I have thus far received.

Authors in favour of the Copernican hypothesis: Copernicus himself; Georg Joachim Rheticus in the *First Account* of the books on the revolutions of Copernicus, his teacher, with a preface and additions by Maestlin; Johannes Kepler in his *Prodromus, or Mysterium Cosmographicum*, with an appendix by Michael Maestlin, and in his Introduction to the *Commentaries on Mars*, and the *Epitome of Copernican Astronomy*; Christoph Rothmann in letters to Tycho; Galileo Galilei in four *Dialogues on the Great World Systems, Ptolemaic and Copernican*, printed first in Italy in 1632 and then afterwards in Latin in 1641 (although he declares there that he relates both sides without settling the matter, nevertheless if the arguments are followed to their conclusion, the Copernican hypothesis is clearly supported); William Gilbert, in the six books *On the Magnet*, although he asserts only the diurnal motion; Paolo Antonio Foscarini, the Carmelite, in the *Letter on the Mobility of the Earth and the Stability of the Sun*; Didacus Stunica in the *Commentaries on Job*; Ismaël Bullialdus in the *Astronomia*

Philolaica; Jacob Lansberger in the *Apologia for the Arguments of Philip Lansberg on the Diurnal and Annual Motion of the Earth*; Peter Heroginous in the *Cursus Mathematicus*, book 2; Petrus Gassendi in both his *Institutione Astronomica* book 3, and two letters, but chiefly the second concerning the impressed motion translated from the mover (though he keeps his intellect captive to the decrees of the Church); and Renatus Cartes [René Descartes] inclines toward this opinion in Part 3 of the *Principles of Philosophy*.

Authors who wrote against the hypothesis of the motion of the earth: Aristotle in the *De Caelo*; Ptolemy in the *Almagest*; Theon the Alexandrine; Regiomontanus in the *Epitome of the Almagest*; Alfraganus; Macrobius in the *Dream of Scipio*; Cleomedes; George Buchanan in the poem *On the Sphere*; Maurolycus in the dialogue on cosmography; Clavius in *On the Sphere*; Barocius in his *Cosmography*; Michael Neander in the *Elements of Astronomy*; Telesius in his *De Caelo*; Martinengus in his *Glossa Magna*; Justus Lipsius in the *Physiologiae*; Christoph Scheiner in his *Mathematical Disquisitions*; Tycho in his letters, and *Progymnasm.*; Alexander Tasso in his books *diversarum cont.*; Scipio Claramontius in the *Antiphilolaos* and *In Defence of AntiTycho against the Two Systems of Galileo*; Melchior Inchofer in the *Tractatus Syllepticus*; Libertus Fromundus in his *Vesta*; Jacob Ascarisius in the *Episcopus Vestanus*; Julius Caesar la Galla in *Concerning the Phenomena on the Orb of the Moon*; Adam Tanner, *De Caelo*; Bartholomeus Amicus, *De Caelo*; Antonio Rocco, *Exercitationes Philosophia*; Marin Mersenne *On Genesis*; Polacco in the *Anticopernico*; Athanasius Kircher *On the Magnet*; David Spinella in *Jove among the Ethiopians*; Johannes Pineda and Johannes Lorinus on *Ecclesiastes*; Bartholomaeus Mastrius and Bonaventure Bellutris on *De Caelo*; Johannes Ponscius on the *Physics*; Antonius Delphinus *On the Globes of the Heavens*; Johannes Elephantutius in *The Structures of the World*.

Chapter 35: Principal Conclusion

Considering only reason and intrinsic arguments, and putting aside all authority, it must absolutely be asserted that the hypothesis supposing the immobility or rest of the Earth is true, and the hypothesis that attributes either diurnal motion alone to the earth, or diurnal and annual motion, is false, as that hypothesis opposes physical and indeed physico-mathematical demonstrations.

Riccioli, Frontispiece,
New Almagest, 1651

The frontispiece for Riccioli's New Almagest *matched the book in its detailed contents. At its centre are two figures, Urania, the muse of astronomy, and Argus, the hundred-eyed giant. Argus holds a telescope with which he observes the sun through the eye on his knee; Urania holds a scale with which she measures two world systems, the Copernican on the left and the Tychonic on the right. The Tychonic system emerges as weightier and therefore to be preferred, perhaps because of the scriptural passage quoted by Urania: 'it shall not be moved for ever', about the earth. The Ptolemaic system lies discarded at the bottom right, in part because of the phases of Venus (upper right), while Ptolemy looks at the other two systems and says, 'I am enlightened while corrected.' A comet, the surface of the moon, and Jupiter with its moons are also depicted in the upper right corner, pointing to some of the developments in astronomy that Riccioli discussed in the book.*

Riccioli, Map of the Moon, from
New Almagest, 1651

Book four of the New Almagest *focused on the moon. Riccioli collaborated with the Jesuit Francesco Maria Grimaldi, who drew lunar maps. These maps followed other important recent lunar maps, like those of Michael van Langren in 1645, and of Johannes Hevelius in the 1647* Selenographia. *Riccioli gave new names to many of the locations on those maps, including the Sea of Tranquillity, still in use today. He also named craters for important astronomers in history, including – though he rejected heliocentric theory himself – Copernicus, Kepler and Galileo, as well as one after Tycho Brahe, and others after himself and Grimaldi. At the top of the map, Riccioli included the statement 'No men dwell on the moon', opposing beliefs like those of Wilkins who held that it was inhabited.*

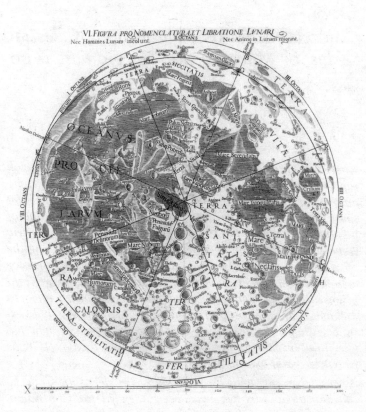

VI. FIGVRA PRO NOMENCLATVRA ET LIBRATIONE LVNARI.
Nec Homines Lunam incolunt.   Nec Anime in Lunam migrant.

Andreas Cellarius, Frontispiece, *Celestial Atlas, or Macrocosmic Harmonies*, 1660

Andreas Cellarius (c.1596–1665) was a German-Dutch cartographer best known for his lavishly illustrated celestial atlas, among the most beautiful ever produced. The atlas included twenty-nine colour maps: twenty-one devoted to the different possible cosmological systems and their implications, and the rest to star maps. The frontispiece below focuses on a group of astronomers gathered around a seated Urania, the muse of astronomy, who holds an armillary sphere and points up to the heavens. To her left is seated Tycho Brahe, who holds a compass pointed at a celestial globe. Next to him, behind Urania, stands Ptolemy, who points at a passage from his Almagest. *The turbaned figure behind him is Islamic astronomer al-Battani. Next to al-Battani stands King Alphonso the Wise of Castile; in what is perhaps an engraving error, he holds a model of the Copernican system, which may have been intended for Copernicus, seated next to him, who points to an armillary sphere while writing in a book. At the furthest right stands Philip Lansberge, a seventeenth-century Dutch astronomer famous for astronomical tables, pointing at the heliocentric system held up by the putti above him.*

John Milton, *Paradise Lost*, 1667

John Milton (1608–1674) was an English writer of poetry and prose best known for his biblical epic poem about the Fall of Man, Paradise Lost. *Milton had met Galileo, whose discoveries are referenced several times in the poem. Book 8 of* Paradise Lost *directly considered the question of cosmology and world systems. Adam asks the angel Raphael to explain the workings of the cosmos; he seems particularly puzzled by why the much larger sphere of the fixed stars would revolve around the earth, and wonders whether it might not be more economical for the smaller earth to move instead. Raphael responds by noting that nature is a book, but one that is difficult to read. He sounds, at times, sympathetic to the Copernican point of view, suggesting at one point that the earth might be like a star, reflecting light to the moon. But he does not explicitly endorse Copernican theory, instead reminding Adam that though humans can model the cosmos, their models may not be correct, and they ought to remain humble about matters beyond their own earthly abode.*

Book 8

The Angel ended, and in Adams Eare
So Charming left his voice, that he a while
Thought him still speaking, still stood fixt to hear;
Then as new wak't thus gratefully repli'd.
What thanks sufficient, or what recompence
Equal, have I to render thee, Divine
Hystorian, who thus largely hast allayd

The thirst I had of knowledge, and voutsaf't
This friendly condescention to relate
Things else by me unsearchable, now heard
With wonder, but delight, and, as is due,
With glorie attributed to the high
Creator! Something yet of doubt remaines,
Which onely thy solution can resolve.
When I behold this goodly Frame, this World
Of Heav'n and Earth consisting, and compute,
Their magnitudes; this Earth a spot, a graine,
An Atom, with the Firmament compar'd
And all her numberd Starrs, that seem to rowle
Spaces incomprehensible (for such
Their distance argues, and their swift return
Diurnal) meerly to officiate light
Round this opacous Earth, this punctual spot,
One day and night; in all their vast survey
Useless besides; reasoning I oft admire,
How Nature wise and frugal could commit
Such disproportions, with superfluous hand
So many nobler Bodies to create,
Greater so manifold, to this one use,
For aught appeers, and on their Orbs impose
Such restless revolution day by day
Repeated; while the sedentarie Earth,
That better might with farr less compass move,
Serv'd by more noble then her self, attaines
Her end without least motion, and receaves,
As Tribute, such a sumless journey brought
Of incorporeal speed, her warmth and light;
Speed, to describe whose swiftness Number failes.
And Raphael now, to Adam's doubt propos'd,
Benevolent and facil thus repli'd.
To ask or search I blame thee not, for Heav'n
Is as the Book of God before thee set,
Wherein to read his wondrous Works, and learne
His Seasons, Hours, or Dayes, or Months, or Yeares:
This to attain, whether Heav'n move or Earth,

Imports not, if thou reck'n right; the rest
From Man or Angel the great Architect
Did wisely to conceal, and not divulge
His secrets to be scann'd by them who ought
Rather admire; or, if they list to try
Conjecture, he his Fabric of the Heav'ns
Hath left to their disputes, perhaps to move
His laughter at thir quaint Opinions wide
Hereafter, when they come to model Heav'n
And calculate the Starrs, how they will wield
The mightie frame, how build, unbuild, contrive
To save appeerances, how gird the Sphear
With Centric and Eccentric scribl'd o're,
Cycle and Epicycle, Orb in Orb:
Alreadie by thy reasoning this I guess,
Who art to lead thy ofspring, and supposest
That bodies bright and greater should not serve
The less not bright, nor Heav'n such journies run,
Earth sitting still, when she alone receaves
The benefit: consider first, that Great
Or Bright inferrs not Excellence: the Earth
Though, in comparison of Heav'n, so small,
Nor glistering, may of solid good containe
More plenty than the Sun that barren shines,
Whose vertue on it self workes no effect,
But in the fruitful Earth; there first receavd
His beams, unactive else, their vigour find.
Yet not to Earth are those bright Luminaries
Officious, but to thee, Earth's habitant.
And for the Heav'ns wide Circuit, let it speak
The Makers high magnificence, who built
So spacious, and his Line stretcht out so farr;
That Man may know he dwells not in his own;
An Edifice too large for him to fill,
Lodg'd in a small partition, and the rest
Ordain'd for uses to his Lord best known.
The swiftness of those Circles attribute,
Though numberless, to his Omnipotence,

That to corporeal substances could adde
Speed almost Spiritual: mee thou thinkst not slow,
Who since the Morning hour set out from Heav'n
Where God resides, and ere mid-day arriv'd
In Eden; distance inexpressible
By Numbers that have name. But this I urge,
Admitting Motion in the Heav'ns, to shew
Invalid that which thee to doubt it mov'd;
Not that I so affirm, though so it seem
To thee who hast thy dwelling here on Earth.
God, to remove his wayes from human sense,
Plac'd Heav'n from Earth so farr, that earthly sight,
If it presume, might erre in things too high,
And no advantage gaine. What if the Sun
Be Centre to the World, and other Starrs,
By his attractive vertue and their own
Incited, dance about him various rounds?
Their wandring course now high, now low, then hid,
Progressive, retrograde, or standing still,
In six thou seest; and what if sev'nth to these
The Planet Earth, so stedfast though she seem,
Insensibly three different Motions move?
Which else to several Spheres thou must ascribe,
Mov'd contrarie with thwart obliquities,
Or save the Sun his labour, and that swift
Nocturnal and Diurnal rhomb suppos'd,
Invisible else above all Starrs, the Wheele
Of Day and Night; which needs not thy beleefe,
If Earth industrious of her self fetch Day
Travelling East, and with her part averse
From the Sun's beam meet Night, her other part
Still luminous by his ray. What if that light,
Sent from her through the wide transpicuous aire,
To the terrestrial Moon be as a Starr
Enlightning her by Day, as she by Night
This Earth? Reciprocal, if Land be there,
Fields and Inhabitants: Her spots thou seest
As Clouds, and Clouds may rain, and Rain produce

Fruits in her soft'nd Soile, for some to eate
Allotted there; and other Suns perhaps,
With their attendant Moons, thou wilt descrie
Communicating Male and Female Light;
Which two great Sexes animate the World,
Stor'd in each Orb perhaps with some that live.
For such vast room in Nature unpossest
By living Soule, desert and desolate,
Onely to shine, yet scarce to contribute
Each Orb a glimps of Light, conveyd so farr
Down to this habitable, which returnes
Light back to them, is obvious to dispute.
But whether thus these things, or whether not,
But whether the Sun, predominant in Heav'n,
Rise on the Earth, or Earth rise on the Sun,
Hee from the East his flaming rode begin,
Or Shee from West her silent course advance,
With inoffensive pace that spinning sleeps
On her soft Axle, while she paces Eev'n,
And beares thee soft with the smooth Air along;
Sollicit not thy thoughts with matters hid;
Leave them to God above, him serve and feare!
Of other Creatures, as him pleases best,
Wherever plac't, let him dispose; joy thou
In what he gives to thee, this Paradise
And thy faire Eve; Heav'n is for thee too high
To know what passes there; be lowlie wise:
Think onely what concernes thee and thy being;
Dream not of other Worlds, what Creatures there
Live, in what state, condition or degree;
Contented that thus farr hath been reveal'd
Not of Earth onely, but of highest Heav'n.

To whom thus Adam, cleerd of doubt, repli'd.
How fully hast thou satisfi'd me, pure
Intelligence of Heav'n, Angel serene!
And, freed from intricacies, taught to live
The easiest way; nor with perplexing thoughts

To interrupt the sweet of Life, from which
God hath bid dwell farr off all anxious cares,
And not molest us; unless we our selves
Seek them with wandring thoughts, and notions vain.
But apt the Mind or Fancy is to roave
Uncheckt, and of her roaving is no end;
Till warn'd, or by experience taught, she learne,
That not to know at large of things remote
From use, obscure and subtle, but to know
That which before us lies in daily life,
Is the prime Wisdom: what is more, is fume,
Or emptiness, or fond impertinence:
And renders us, in things that most concerne,
Unpractis'd, unprepar'd, and still to seek.
Therefore from this high pitch let us descend
A lower flight, and speak of things at hand
Useful; whence, haply, mention may arise
Of somthing not unseasonable to ask,
By sufferance, and thy wonted favour, deign'd.

Johannes Hevelius, *Celestial Machine*, 1673

Johannes Hevelius (1611–1687) was a Polish astronomer most famous for charting the surface of the moon, and for the instruments in his observatory in Danzig (now Gdańsk) where he devoted himself to precise astronomical observations. He did so alongside his astronomer wife, Elisabeth Hevelius. His 1673 Celestial Machine *focused on the instruments of his observatory and the observations obtained there; in the image below, Johannes and Elisabeth are both pictured using one of those instruments, a sextant requiring two observers. This image, along with the others in the book, are among the very first in which we see female astronomers at work. Tragically, a fire burned down Hevelius's house and observatory just as the copies of* Celestial Machine *had been returned to Hevelius by the printer; only a few copies survive.*

Robert Hooke, *An Attempt to Prove the Motion of the Earth by Observations*, 1674

The English polymath Robert Hooke (1635–1703) was curator of experiments at the Royal Society and assistant to Robert Boyle, and is best known for his microscopic observations described in the 1665 Micrographia. *In a work of 1674 supporting the Copernican system, Hooke claimed that with a self-improved telescope he could offer the definitive proof of the earth's motion that had so far been missing: stellar parallax. He described his observations of Gamma Draconis, a star which could be observed directly above London – therefore avoiding problematic atmospheric refraction – and argued that he had identified a change of position due to parallax. In fact, his equipment was not sufficiently accurate to produce reliable results, and successful detection of stellar parallax would not be achieved until well over a century later. This work is also notable because in it, Hooke briefly mentions a new 'system of the world' that blended Cartesian mechanical ideas with the magnetic theories of William Gilbert, which allowed for action at a distance. He thus articulated an early version of gravitational attraction, though he could not prove it mathematically. These ideas would be crucial in Newton's own development of gravitational theory in the* Principia.

Whether the Earth move or stand still has been a Problem, that since *Copernicus* revived it, has much exercised the Wits of our best modern Astronomers and Philosophers, amongst which notwithstanding there has not been any one who has found out a certain manifestation either of the one or the other Doctrine. The more knowing and judicious have for many plausible reasons adhered to the *Copernican* Hypothesis: But the generality of others, either out of ignorance or prejudice, have rejected it as a most extravagant opinion. To those indeed who understand not the grounds and principles of Astronomy, the prejudice of common converse does make it seem so absurd, that a man shall as soon persuade them that the Sun does not shine, as that it does not move; and as easily move the Earth as make them believe that it does so already. For such Persons I cannot suppose that they should understand the cogency of the Reasons here presented, drawn from the following observations of *Parallax*, much less therefore can I expect their belief and assent thereunto; to them I have only this to say, 'Tis not here my business to instruct them in the first principles of Astronomy, there being already Introductions enough for that purpose: But rather to furnish the Learned with an *experimentum crucis* to determine between the *Tychonick* and *Copernican* Hypotheses.

That which hath hitherto continued the dispute has been the plausibleness of some Arguments alleged by the one and the other party, with such who have been by nature or education prejudiced to this or that way. For to one that has been conversant only with illiterate persons, or such as understand not the principles of Astronomy and Geometry, and have had no true notion of the vastness of the Universe, and the exceeding minuteness of the Globe of the Earth in comparison therewith, who have confined their imaginations and fancies only within the compass and pale of their own walk and prospect, who can scarce imagine that the Earth is globous, but rather like some of old, imagine it to be a round plain covered with the Sky as with a Hemisphere, and the Sun, Moon and Stars to be holes through it by which the Light of Heaven comes down; that suppose themselves in the centre of this plain, and that the Sky does touch that plain round the edges, supported in part by the Mountains; that suppose the Sun as big as a Sieve, and the Moon as a *Cheddar* Cheese, and hardly a mile off. That wonder why the Sun, Moon and Stars do not fall down like Hailstones; and that will be martyr'd rather than grant that there may be Antipodes, believing it absolutely impossible, since they must necessarily fall down into the Abyss below them: For how can they go with their feet toward ours, and their heads downward, without making their brains addle. To one I say, thus prejudiced with these and a thousand other fancies and opinions more ridiculous and absurd to knowing men, who can ever imagine that the uniformity and harmony of the Celestial bodies and motions, should be an Argument prevalent to persuade that the Earth moves about the Sun: Whereas that Hypothesis which shows how to solve the appearances by the rest of the Earth and the motion of the Heavens seems generally so plausible that none of these can resist it.

Now though it may be said, 'Tis not only those but great Geometricians, Astronomers and Philosophers have also adhered to that side, yet generally the reason is the very same. For most of those, when young, have been imbued with principles as gross and rude as those of the Vulgar, especially as to the frame and fabrick of the World, which leave so deep an impression upon the fancy, that they are not without great pain and trouble obliterated: Others, as a further confirmation in their childish opinion, have been instructed in the *Ptolomaick* or *Tichonick* System, and by the Authority of their Tutors, over-awed into a belief, if not a veneration thereof: Whence for the most part such persons will not endure to hear Arguments against it, and if they do, 'tis only to find Answers to confute them.

On the other side, some out of a contradicting nature to their Tutors; others, by as great a prejudice of institution; and some few others upon better reasoned grounds, from the proportion and harmony of the World, cannot but embrace the *Copernican* Arguments, as demonstrations that the Earth moves, and that the Sun and Stars stand still.

I confess there is somewhat of reason on both sides, but there is also something of prejudice even on that side that seems the most rational. For by way of objection, what way of demonstration have we that the frame and constitution of the World is so harmonious according to our notion of its harmony, as we suppose? Is there not a possibility that the things may be otherwise? nay, is there not something of probability? may not the Sun move as *Ticho* supposes, and the Planets make their Revolutions about it whilst the Earth stands still, and by its magnetism attracts the Sun, and so keeps him moving about it, whilst at the same time Mercury and Venus move about the Sun, after the same manner as Saturn and Jupiter move about the Sun whilst the Satellites move about them? especially since it is not demonstrated without much art and difficulty, and taking many things for granted which are hard to be proved, that there is any body in the Universe more considerable then the Earth we tread on. Is there not much reason for the Hypothesis of *Ticho* at least, when he with all the accurateness that he arrived to with his vast Instruments, or *Riccioli*, who pretends much to out-strip him, were not able to find any sensible Parallax of the Earth's Orb among the fixt Stars, especially if the observations upon which they ground their assertions, were made to the accurateness of some few Seconds? What then, though we have a Chimera or Idea of perfection and harmony in that Hypothesis we pitch upon, may there not be a much greater harmony and proportion in the constitution itself which we know not, though it be quite differing from what we fancy?

Probable Arguments might thus have been urged both on the one and the other side to the World's end; but there never was nor could have been any determination of the Controversy, without some positive observation for determining whether there were a Parallax or no of the Orb of the Earth; This *Ticho* and *Riccioli* affirm in the Negative, that there is none at all: But I do affirm there is no one that can either prove that there is, or that there is not any Parallax of that Orb amongst the fixt Stars from the Suppellex of observations yet made either by *Ticho*, *Riccioli*, or any other Writer that I have yet met with from the beginning of writing to this day. For all Observators having hitherto made use of the naked eye for determining the exact place of the object, and the eye being unable to distinguish any angle less than a minute, and an observation requisite to determine this requiring a much greater exactness than to a minute, it doth necessarily follow that this *experimentum crucis* was not in their power, whatever either *Ticho* or *Riccioli* have said to the contrary, and would thence overthrow the *Copernican* System, and establish their own. We are not therefore wholly to acquiesce in their determination, since if we examine more nicely into the observations made by them, together with their Instruments and ways of using them, we shall find that their performances thereby were far otherwise than what they would seem to make us believe.

The Controversy therefore notwithstanding all that has been said either by the one or by the other Party, remains yet undetermined, Whether the Earth move above the Sun, or the Sun about the Earth; and all the Arguments alleged either on this or that side, are but probabilities at best, and admit not of a necessary and positive conclusion. Nor is there indeed any other means left for human industry to determine it, save this one which I have endeavoured to make; and the unquestionable certainty thereof is a most undeniable Argument of the truth of the *Copernican* System; and the want thereof has been the principal Argument that has hitherto somewhat detained me from declaring absolutely for that Hypothesis, for though it does in every particular almost seem to solve the appearances more naturally and easily, and to afford an exceeding harmonious constitution of the great bodies of the World compared one with another, as to their magnitudes, motions and distances, yet this objection was always very plausible to most men, that it is affirmed by such as have written more particularly of this subject, that there never was any sensible Parallax discovered by the best observations of this supposed annual motion of the Earth about the Sun as its centre, though moved in an Orb whose Diameter is by the greatest number of Astronomers reckoned between 11 and 12 hundred Diameters of the Earth: Though some others make it between 3 and 4 thousands; others between 7 and 8; and others between 14 and 15 thousands; and I am apt to believe it may be yet much more, each Diameter of the Earth being supposed to be between 7 and 8 thousand English miles, and consequently the whole being reduced into miles, if we reckon with the most, amounting to 120 millions of English miles. It cannot, I confess, but seem very uncouth and strange to such as have been used to confine the World with less dimensions, that this annual Orb of the Earth of so vast a magnitude, should have no sensible Parallax amongst the fixt Stars, and therefore 'twas in vain to endeavour to answer that objection. For it is unreasonable to expect that the fancies of most men should be so far strained beyond their narrow dimensions, as to make them believe the extent of the Universe so immensely great as they must have granted it to be, supposing no Parallax could have been found.

The Inquisitive Jesuit *Riccioli* has taken great pains by 77 Arguments to overthrow the *Copernican* Hypothesis, and is therein so earnest and zealous, that though otherwise a very learned man and good Astronomer, he seems to believe his own Arguments; but all his other 76 Arguments might have been spared as to most men, if upon making observations as I have done, he could have proved there had been no sensible Parallax this way discoverable, as I believe this one Discovery will answer them, and 77 more, if so many can be thought of and produced against it. Though yet I confess had I fail'd in discovering a Parallax this way, as to my own thoughts and persuasion, the almost infinite extension of the Universe had not to me seem'd altogether so great an absurdity to be believed as the Generality do esteem it; for since 'tis confessedly granted on all hands the distance of the fixed Stars is merely hypothetical, and not founded on any other ground or reason but fancy and supposition, and that there never was hitherto any Parallax observed, nor any other considerable Argument to prove the distances supposed by such as have been most curious and inquisitive in that particular, I see no Argument drawn from the nature of the thing that can have any necessary force in it to determine that the said distance cannot be more than this or that, whatever it be that is assigned. For the same God that did make this World that we would thus limit and bound, could as easily make it millions of millions of times bigger, as of that quantity we imagine; and all the other appearances except this of Parallax would be the very same that now they are. To me indeed the Universe seems to be vastly bigger than 'tis hitherto asserted by any Writer, when I consider the many differing magnitudes of the fixed Stars, and the continual increase of their number according as they are looked after with better and longer Telescopes. And could we certainly determine and measure their Diameters, and distinguish what part of their appearing magnitude were to be attributed to their bulk, and what to their brightness, I am apt to believe we should make another distribution of their magnitudes, thae what is already made by *Ptolomy*, *Ticho*, *Kepler*, *Bayer*, *Clavius*, *Grienbergerus*, *Piff*, *Hevelius* and others.

For supposing all the fixed Stars as so many Suns, and each of them to have a Sphere of activity or expansion proportionate to their solidity and activity, and a bigger and brighter bodied Star to have a proportionate bigger space or expansion belonging to it, we should from the knowledge of their Diameters and brightnesses be better able to judge of their distances, and consequently assign diverse of them other magnitudes than those already stated: Especially since we now find by observations, that of those which are accounted single Stars, diverse prove a congeries of many Stars, though from their near appearing to each other, the naked eye cannot distinguish them; Such as those Stars which are called *Nebulous*, and those in *Orion* Sword, and that in the head of *Aries*, and a multitude of others the Telescope does now detect. And possibly we may find that those twenty magnitudes of Stars now discovered by a fifteen-foot Glass, may be found to increase the magnitude of the Semidiameter of the visible World, forty times bigger than the *Copernicans* now suppose it between the Sun and the fixed Stars, and consequently sixty-four thousand times in bulk. And if a Telescope of double or treble the goodness of one of fifteen should discover double or treble the said number of magnitudes, would it not be an Argument of doubling or trebling the former Diameter, and of increasing the bulk eight or twenty-seven times. Especially if their apparent Diameters shall be found reciprocal to their Distances (for the determination of which I did make some observations, and design to complete with what speed I am able).

But to digress no further, This grand objection of the *Anticopernicans*, which to most men seem'd so plausible, that it was in vain to oppose it, though, I say, it kept me from declaring absolutely for the *Copernican* Hypothesis, yet I never found any absurdity or impossibility that followed thereupon: And I always suspected that though some great Astronomers had asserted that there was no Parallax to be found by their observations, though made with great accurateness, there might yet be a possibility that they might be mistaken; which made me always look upon it as an inquiry well worth examining: first,

Whether the ways they had already attempted were not subject and liable to great errors and uncertainties: and secondly, Whether there might not be some other ways found out which should be free from all the exceptions the former were encumbered with, and be so far advanced beyond the former in certainty and accurateness, as that from the diligent and curious use thereof, not only all the objections against the former might be removed, but all other whatsoever that were material to prove the ineffectualness thereof for this purpose.

I began therefore first to examine into the matter as it had already been performed by those who had asserted no sensible Parallax of the annual Orb of the Earth, and quickly found that (whatever they asserted) they could never determine whether there were any or no Parallax of this annual Orb; especially if it were less than a minute, which *Kepler* and *Riccioli* hypothetically affirm it to be: The former making it about twenty-four Seconds, and the latter about ten. For though *Ticho*, a man of unquestionable truth in his assertions, affirm it possible to observe with large Instruments, conveniently mounted and furnished with sights contrived by himself (and now the common ones for Astronomical Instruments) to the accurateness of ten Seconds; and though *Riccioli* and his ingenious and accurate Companion *Grimaldi* affirm it possible to make observations by their way, with the naked edge to the accurateness of five Seconds; Yet *Kepler* did affirm, and that justly, that 'twas impossible to be sure to a less Angle then twelve Seconds: And I from my own experience do find it exceeding difficult by any of the common sights yet used to be sure to a minute. I quickly concluded therefore that all their endeavours must have hitherto been ineffectual to this purpose, and that they had not been less imposed on themselves, than they had deceived others by their mistaken observations. And this mistake I found proceeded from diverse inconveniencies their ways of observations were liable to.

Having therefore examined the ways and Instruments for all manner of Astronomical observations hitherto made use of, and considered of the inconveniencies and imperfections of them; and having also duly weighed the great accurateness and certainty that this observation necessarily required: I did next contrive a way of making observations that might be free from all the former inconveniencies and exceptions, and as near as might be, fortified against any other that could be invented or raised against it. This way then was to observe by the passing of some considerable Star near the Zenith of *Gresham College*, whether it did not at one time of the year pass nearer to it, and at another further from it: for if the Earth did move in an Orb about the Sun, and that this Orb had any sensible Parallax amongst the fixed Stars; this must necessarily happen, especially to those fixed Stars which were nearest the Pole of the Ecliptick. And that this is so, anyone may plainly perceive.

Having thus resolved upon the way, and prepared the Instruments fit for the observation, I began to observe the Transits of the bright Star in the head of *Draco*. 'Tis manifest then by the observations of *July* the Sixth and Ninth: and that of the One and twentieth of *October*, that there is a sensible parallax of the Earth's Orb to the fixed Star in the head of *Draco*, and consequently a confirmation of the *Copernican* System against the *Ptolomaick* and *Tichonick*.

I should have here described some Clocks and Time-keepers of great use, nay absolute necessity in these and many other Astronomical observations, but that I reserve them for some attempts that are hereafter to follow, about the various ways I have tried, not without good success of improving Clocks and Watches, and adapting them for various uses, as for accurating Astronomy, completing the Tables of the fixt Stars to Seconds, discovery of Longitude, regulating Navigation and Geography, detecting the properties and effects of motions for promoting secret and swift conveyance and correspondence, and many other considerable scrutinies of nature: And shall only for the present hint that I have in some of my foregoing observations discovered some new Motions even in the Earth it self, which perhaps were not dreamed of before, which I shall hereafter more at large describe, when further trials have more fully confirmed and completed these beginnings. At which time also I shall explain a System of the World differing in many particulars from any yet known, answering in all things to the common Rules of Mechanical Motions: This depends upon three Suppositions. First, That all Celestial

Bodies whatsoever, have an attraction or gravitating power toward their own Centres, whereby they attract not only their own parts, and keep them from flying from them, as we may observe the Earth to do, but that they do also attract all the other Celestial Bodies that are within the sphere of their activity; and consequently that not only the Sun and Moon have an influence upon the body and motion of the Earth, and the Earth upon them, but that also Mercury, Venus, Mars, Saturn and Jupiter, by their attractive powers, have a considerable influence upon its motion as in the same manner the corresponding attractive power of the Earth has a considerable influence upon every one of their motions also. The second supposition is this, That all bodies whatsoever that are put into a direct and simple motion, will so continue to move forward in a straight line, till they are by some other effectual powers deflected and bent into a Motion, describing a Circle, Ellipsis, or some other more compounded Curve Line. The third supposition is, That these attractive powers are so much the more powerful in operating, by how much the nearer the body wrought upon is to their own Centres. Now what these several degrees are I have not yet experimentally verified; but it is a notion, which if fully prosecuted as it ought to be, will mightily assist the Astronomer to reduce all the Celestial Motions to a certain rule, which I doubt will never be done true without it. He that understands the nature of the Circular Pendulum and Circular Motion, will easily understand the whole ground of this Principle, and will know where to find direction in Nature for the true stating thereof. This I only hint at present to such as have ability and opportunity of prosecuting this Inquiry, and are not wanting of Industry for observing and calculating, wishing heartily such may be found, having my self many other things in hand which I would first complete, and therefore cannot so well attend it. But this I durst promise the Undertaker, that he will find all the great Motions of the World to be influenced by this Principle, and that the true understanding thereof will be the true perfection of Astronomy.

Bernard Le Bovier de Fontenelle, *Conversations on the Plurality of Worlds*, 1686

Bernard Le Bovier de Fontenelle (1657–1757) was a French writer of popular science. His most famous work, Conversations on the Plurality of Worlds, *was written in French and explicitly directed toward the general reader. The book consists of a series of conversations that take place over six evenings between Fontenelle and a marquise, making the book unusual in its inclusion of a serious female interlocutor in a discussion of science. Noteworthy also for its humorous and engaging style, the book endorsed the heliocentric system and considered the possibility that the stars were suns, each with their systems of planets and possible inhabitants. The book was very popular and regularly published in France and was translated into multiple other languages.*

First evening

One evening after supper we went to take a walk in the park. It was deliciously cool, which was recompense for the very hot day that we had endured. The moon had risen perhaps an hour before, and its rays, coming to us between the branches of the trees, made an agreeable mixture of bright white against the dark background. There was not a cloud to obscure the smallest star, and all appeared a pure and dazzling gold, heightened further by the blue background to which they were attached. This spectacle put me in a trance, which might have continued for a long time without the Marquise: but in the presence of such a lovely woman I could not allow myself to be lost to the moon and the stars.

'Don't you find,' I said, 'that the day itself is less beautiful than a beautiful night? Perhaps it is because the day appears too uniform, with only the sun and the blue sky, while the night reveals all these stars scattered about haphazardly and arranged in a thousand different shapes, which leads to a certain pleasurable disorder of thoughts.'

'I have always felt just that!' she said. 'I love the stars, and I would gladly scold the sun that hides them from us.'

'Ah!' I cried. 'I cannot forgive it for making me lose sight of all those worlds.'

'What are you calling worlds?' she asked, turning to look at me.

'Alas,' I replied, 'I must admit with some chagrin that I suppose that each star could well be a world. I wouldn't swear that it was true, but I believe it because it pleases me to do so. It is an idea that delights me and has strangely taken hold of my mind. And I believe that this feeling has a necessary relationship to the truth.'

'Well,' she went on, 'if it makes you so happy, share it with me. I will believe whatever you please about the stars, if I find pleasure in it.'

'Ah, Madame,' I answered quite frankly. 'It is not the sort of pleasure that you would feel at a comedy by Molière; it is one that relates to reason, and only gives pleasure to the mind.'

'What then,' she returned, 'do you believe that I am incapable of pleasures related to reason? I will show you how wrong you are right now. Teach me about your stars!'

'No,' I replied. 'Let me not be chided for having spoken of philosophers in a wood at ten o'clock in the evening to the most delightful person I know. Look elsewhere for philosophers.'

While I defended myself like this for a while longer, I had to give in. I made her promise, at least – for the preservation of my honour – that she would keep it a secret. And when I was unable to delay further and ready to speak, I didn't know where to begin. To a person like her, who knew nothing of physics, it would necessarily take time to prove that the earth could be a planet, and that other planets could be so many earths, and all the stars could be worlds. I repeatedly told her that it would be better to talk about trifles, as all reasonable people would have done in our place. In the end, however, to give her a general idea of philosophy, here is the reasoning into which I threw myself:

'All philosophy,' I told her, 'is founded on only two things: a curious mind and bad eyes. If you had better then could clearly see whether or not the stars are worlds, and if, on the other hand, you were less curious, you wouldn't care to know it, which would amount to the same thing. But we want to know more than we see, and that is the difficulty. Even if we could see the things that appear to us really well, we would know a lot, but things would still be quite different from how they appear. Thus true philosophers spend their lives not believing what they see, and trying to guess at what they do not see, and this is not, it seems to me, something to envy. With respect to this I always imagine that Nature is a spectacle much like an opera. From where you are at the opera, you don't see the theatre quite as it is. It has been arranged to look agreeable at a distance, and the wheels and counterweights that produce all the movements are hidden from your view. So you don't bother to guess at how it all works. Perhaps some machinist hidden below the stage, puzzled by a sight that seems extraordinary to him, will vow to unravel how it was accomplished. As you can see, this machinist is like the philosophers. But what makes it harder for the philosophers is that in the machines that Nature presents to our eyes, the wires are very well hidden: so well hidden that we have been guessing for a long time about what causes the movements of the universe. Thus imagine all the Sages at the opera: the Pythagorases, Platos, Aristotles – all those whose names are constantly chattered about. Let us suppose that they saw the flight of Phaeton, carried away by the winds, but could not discover the strings, and didn't know how the back of the theatre was laid out. One would say: "Some hidden virtue makes Phaeton light." Another: "Phaeton is composed of certain numbers that make him rise." The next: "Phaeton has a certain affinity for the top of the theatre, and is not at rest when he is not there." Another: "Phaeton was not made to fly, but would rather fly than leave the top of the theatre empty." And there would be a hundred other daydreams, so much that I am surprised that the reputation of antiquity has not been entirely ruined. In the end, Descartes and some other moderns would come, and say "Phaeton rises because he is pulled by ropes, because a weight heavier than he descends." Thus we no longer believe that a body moves if it is not pushed by another body or somehow pulled by wires; we no longer believe that it goes up or down except when affected by a counterweight or spring. Whoever wants to see Nature as she is need only look at what is behind the theatre at the opera.'

'By this account,' said the Marquise, 'nature has become very mechanical?'

'So mechanical,' I replied, 'that I fear we will soon be ashamed of it. We want the universe to be on a large scale only what a watch is on a small scale, such that everything in it is conducted by regular movements which depend on the arrangement of the parts. Admit the truth: have you not held a more sublime idea of the universe, and given it more honour than it deserved? I have seen many people who esteemed it less since they have come to know it.'

'And I,' she replied, 'esteem it much more since I know that it is like a watch. It's astonishing that the order of nature, admirable as it is, revolves only around such simple things.'

'I don't know,' I answered, 'who has given you such healthy ideas, but truthfully they are uncommon. Most people still hold in their minds a false idea of marvels under a shroud of obscurity. They only admire nature because they believe it to be a kind of magic that they don't understand, and as soon as they do understand it, they no longer admire it. But Madame,' I continued, 'you are so ready to accept everything I want to tell you that I think I need only draw the curtain and show you the world.

'From the earth, where we are, what we see as most distant is the blue sky, that great vault where the stars are attached like nails. They are called fixed because they seem to have only the movement of their heaven, which carries them with it from east to west. Between the earth and this last vault of the heavens are suspended at different heights the sun, the moon and five other stars, called planets: Mercury, Venus, Mars, Jupiter and Saturn.

These planets, not being attached to the same sky, and having unequal movements, appear differently with respect to each other, and relate to each other in different ways, whereas the fixed stars are always in the same position with respect to each other. The Chariot, for example, which you see is formed of the seven stars, has always been made as it is now, and always will be so; but the moon is sometimes close to the sun and sometimes far away from it, and the same is true for the other planets. This is how things appeared to the ancient Chaldean shepherds, whose great leisure produced the first observations which were the foundation of astronomy.'

'When someone had recognized this arrangement of the heavens,' she asked, 'what was the question next?'

'It was a question,' I continued, 'of divining how all the parts of the universe should be arranged, and this is what the scholars call building a system. But as I am going to explain to you the first of the systems, please first note that we are naturally like a certain Athenian fool you have heard of, who entertained the fantasy that all the vessels that landed at the Port of Piraeus belonged to him. Our foolishness is likewise to believe that all of nature, without exception, is intended for our use, and when we ask our philosophers what the use is of so prodigious a number of fixed stars, when a small part alone would be enough to do what they all do, they answer coldly that they serve to gratify our sight. By the same principle we necessarily imagined that the earth would be at rest at the centre of the universe, while all the celestial bodies, which were made for it, would take the trouble to revolve around it and illuminate it. Therefore, the moon was placed above the earth; and above the moon, Mercury; then Venus; the sun; Mars; Jupiter; and Saturn. Above all this was placed the heaven of the fixed stars. The earth was placed precisely in the middle of the circles described by these planets, and these circles were larger the more distant they were from the earth; consequently, the further planets took more time to complete their courses, which is indeed true.'

'But I don't know,' interrupted the Marquise, 'why you don't approve of this order in the universe; it seems to me as clear as it is intelligible, and for my part I declare to you that I am satisfied with it.'

'I can congratulate myself,' I replied, 'that I sweetened this system for you. If I had given it to you as it was conceived by Ptolemy, its author, or by those who worked on it after him, it would throw you into a horrible terror. For the movements of the planets are not so regular, and sometimes they go faster, sometimes more slowly; sometimes in one direction, sometimes in another; sometimes they are very far from the earth, and sometimes closer. The Ancients imagined I know not how many circles interwoven differently with one another, by which they saved all these oddities. The embarrassment of all these circles was so great that in a time when nothing better was yet known, a king of Aragon – a great mathematician, but apparently not very devout – said that if God had called him to His council when He made the world, he would have given Him better advice. The thought is too libertine, but it is rather pleasant to realize that the system itself is what led to his sin, because it was too confused. The good advice that this king wished to give no doubt had to do with the suppression of all these circles, which had confused the celestial movements. Apparently, it also pertained to the two or three superfluous heavens that had been placed beyond the fixed stars. These philosophers – to explain a kind of movement of the celestial bodies – constructed a heaven made of crystal, beyond the last heaven that we see, which impresses motion on the lower heavens. Whenever they had news of another motion, they immediately erected another crystal heaven; for in the end, crystal skies didn't cost them anything.'

'And why were the heavens only made of crystal?' said the Marquise. 'Would not some other material have been as good?'

'No,' I replied. 'The light had to pass through them, while they still had to be solid. This was necessary to Aristotle, who found that solidity was part of what made them noble in nature, and since he said it, no one doubted it. But we have since seen comets, which were higher than people formerly believed, which burst through the crystal of the heavens as they passed and broke the whole universe. To resolve the matter, they made the heavens into a fluid material like air. Finally, it is beyond doubt due to the observations of the last centuries that Venus and Mercury revolve around the sun, and not around the earth, and that the old system is absolutely unsustainable at this point. I am therefore going to propose another which would absolve the king of Aragon from giving advice, for it is charmingly simple, which alone would make it preferable.'

'It would seem,' interrupted the Marquise, 'that your philosophy is a kind of auction, where those who offer to do things for the least expense prevail over the rest.'

'That's true,' I continued, 'and it is only thereby that one can grasp the plan on which nature has done her work. She is extraordinarily frugal; whatever she can do in a way that will cost her a little less, even if that less is close to nothing, she will surely do it in only that way. Nonetheless this frugality is matched by a startling magnificence that shines in all that she has done. The magnificence is in the design, and the frugality in the execution. Nothing is better than a great design that is executed at little cost. We ourselves often reverse this in our ideas. We assign thrift to nature's design, and magnificence to her execution. We assign her a little design, which she executes with ten times greater expense than is necessary: this is absolutely ridiculous.'

'I would be very glad,' she said, 'if your description of nature imitates this aspect of it very closely: this manner of speaking will profit my imagination, as I will then have less trouble understanding what you tell me.'

'I will not make things more needlessly difficult for you,' I replied. 'Imagine a German named Copernicus, who takes hold of all these different circles, and the whole solid heavens that have been imagined by antiquity, destroying the former and tearing the latter to pieces. Seized with the noble fury of astronomy, he snatches the earth and sends it far from the centre of the universe, where it had been placed. In this centre he places the sun, to whom this honour was really due. The planets no longer revolve around the earth, and no longer enclose it in the middle of the circles they describe. If they illuminate us, it is only by chance, as they meet us on their way. Everything currently revolves around the sun, as does the earth itself, and to punish it for the long rest it had granted itself, Copernicus burdens it as much as he can with all the movements it had given to the planets and the heavens. Finally, of all the celestial equipage that used to accompany and surround the earth only the moon remains to it, and still revolves around it.'

'Wait,' said the Marquise, 'you got excited and explained things so pompously that I don't think I heard them. The sun is at the centre of the universe, and it is motionless there: what follows after it?'

'Mercury,' I replied. 'It revolves around the sun, so that the sun is at the centre of the circle that Mercury describes. Above Mercury is Venus, which similarly revolves around the sun. Next comes the earth which, being higher than Mercury and Venus, describes a greater circle around the sun and these planets. Finally follow Mars, Jupiter and Saturn, in the order I've named them to you; and you can see that Saturn must describe the greatest circle of all around the sun, so that it takes more time than any other planet to make its revolution.'

'You forget the moon,' she interrupted.

'I will find her again,' I said. 'The moon revolves around the earth but does not abandon it; as the earth advances in its circle around the sun, the moon always follows it and revolves around it. And if it revolves around the sun, it is only so as not to leave the earth.'

'I understand you,' she replied, 'and I love the moon for sticking with us when all the other planets are leaving us. Admit that if your German could have caused us to lose it, he would have done so willingly; for I see that in his approach he had bad intentions for the earth.'

'I am grateful to him,' I replied, 'for reducing the vanity of men, who had placed themselves in the best part of the universe, and I am pleased to see the earth now in the crowd of planets.'

'What?' she replied. 'Do you really believe that the vanity of men extends to astronomy? Do you think you have humbled me by teaching me that the earth revolves around the sun? I swear to you that I don't think less of myself.'

'By God, Madame,' I replied, 'I know very well that people are less jealous of their rank in the universe than of the one they believe they should hold in society, and that the precedence of two planets will never be as great an affair as that of two ambassadors. However, the same inclination that makes one want to have the most honourable place in a ceremony makes a philosopher put himself at the centre of a world system, if he can. He rejoices when everything is done for him, and supposes, perhaps without realizing it, any principle that flatters him, and his heart will turn a matter of pure speculation to his interest.'

'Frankly,' she replied, 'that is a calumny that you have invented against humankind. By this account we should never have accepted the Copernican system, since it is so humiliating.'

'Copernicus himself,' I returned, 'was very distrustful of the success of his opinion. He did not want to publish it for a long time. Finally he resolved to do so, at the request of very important people; but the very day that they brought him the first printed copy of his book, do you know what he did? He died. He didn't want the responsibility of all the contradictions he foresaw, and skilfully removed himself from the affair.'

'Listen,' said the Marquise. 'We must do justice to everyone. It is certainly difficult to imagine that we revolve around the sun, for we never change our individual places, and always end up in the morning where we went to bed at night. I think I see, by your attitude, that you are going to tell me that as the whole Earth moves—'

'Certainly,' I interrupted. 'It's the same as if you fell asleep in a boat going down the river – when you woke up you would find yourself in the same place and in the same situation with respect to all the parts of the boat.'

'Yes,' she replied, 'but there is a difference: I would find when I woke that the shore had changed, and that would make me see that my boat had changed its place. But it is not the same with the earth – I find everything just as I have left it.'

'No Madame,' I answered. 'No. The shore has changed too. You know that beyond all the circles of the planets are the fixed stars: that is our shore. I am on the earth, and the earth describes a great circle around the sun. I look at the centre of the circle and see the sun there. If it didn't block the stars, then, if I directed my gaze in a straight line beyond the sun, I would necessarily see it correspond to some fixed stars; but at night I easily see what stars it corresponds to during the day, and it is exactly the same thing. If the earth did not change its place on its circle, I would always see the sun correspond to the same fixed stars, but as soon as the earth changes its position, I must see it correspond to other stars. This is the shore that changes every day, and as the earth makes its circle in a year around the sun, I see the sun, in the space of that year, corresponding successively to various fixed stars that make up a circle. This circle is called the Zodiac.'

'Let's talk about the sun,' she said. 'I understand well how we can imagine that it follows the circle that we ourselves make. But this circuit is only completed in a year, and the sun goes over our head every day – how is that?'

'Have you noticed,' I replied, 'that a ball rolling down a path has two motions? It goes toward the end of the lane, and at the same time it turns on itself several times, so that the top parts of the ball descend downwards and the bottom parts rise up. The earth does the same thing. In the time that it advances on the circle that it describes in a year around the sun, it turns on itself in twenty-four hours; thus in twenty-four hours each part of the earth loses the sun and recovers it. As we turn toward the side where the sun is, it seems to rise; and as we begin to move away from it, continuing to turn, it seems to set.'

'It's quite pleasant,' she said, 'that the earth takes everything upon itself and the sun does nothing. And when the moon and the other planets and the fixed stars seem to go around our heads in twenty-four hours, is that also imagined?'

'Pure imagination,' I replied, 'which comes from the same cause. The planets make their circles around the sun at unequal times according to their unequal distances, and the one that we see today corresponding to a certain point of the zodiac, we see at the same time tomorrow corresponding to another point, both because it has advanced on its circle and because we have advanced on ours. We move and the other planets move too, but more or less quickly than we do; this puts us at different points of view with respect to them, and makes oddities appear in their courses, which I don't need to speak about with you now. It is enough for you to know that what seems irregular in the planets comes only from the various ways in which our motion makes us encounter them, and that basically they are all very regular.'

'I can agree that they are so,' said the Marquise, 'but I would prefer that their regularity cost the earth less. It is a great bother to demand that so heavy and solid a mass be so agile.'

'But,' I replied, 'would you rather have the sun and all the other stars, which are very large bodies, make an immense tour around the earth, over an infinite number of leagues, in twenty-four hours? Because all this must happen if the earth does not turn on itself in twenty-four hours.'

'Oh!' replied the Marquise. 'The sun and the stars are all fire, so movement costs them nothing; but the earth hardly seems portable.'

'And would you believe,' I replied, 'if you did not have the experience, that a big ship, mounted with a hundred and fifty cannons and loaded with more than three thousand men, along with a very large quantity of goods, was something portable? And yet only a small puff of wind is needed to make it move on the water, because the water is liquid, and being easily divided it offers little resistance to the movement of the ship. So too the earth, massive as it is, is easily carried in the middle of the celestial matter, which is a thousand times more fluid than water, and which fills this whole great space where the planets swim. And where would the earth have to be fastened to resist the movement of this celestial matter so that it wouldn't be carried away by it? It would be as if a small ball of wood could resist the current of a river.'

'But,' she replied again, 'how does the earth, with all its weight, support itself on your celestial matter, which, since it is so fluid, must be very light?'

'It is not the case,' I replied, 'that what is fluid is lighter. What do you have to say of our great ship, which, with all its weight, is lighter than water, since it floats on it?'

'But I have a serious difficulty,' she said. 'If the earth rotates, then we change air at every moment, and we are always breathing that of another country.'

'Not at all, Madame,' I replied. 'The air that surrounds the earth extends only up to a certain height, perhaps up to twenty leagues; it follows us, and turns with us. Have you ever seen the work of a silkworm, and those shells that these little animals fashion with so much art to imprison themselves in? They are made of a very tight silk, but are covered with some very light and loose down. This is how the earth, which is quite solid, is covered from its surface up to a certain height – with a kind of down, which is the air – and the whole silkworm shell rotates at the same time. Beyond the air is celestial matter, incomparably purer, subtler and even more agitated than it is.'

'You present the earth to me using a very lowly comparison,' said the Marquise. 'But it is on this silkworm shell that such great works are accomplished, such great wars are waged, and such great agitation takes hold everywhere.'

'Yes,' I answered, 'and through all that time Nature is unaware of all these little particular movements, and carries us all together in one general movement, as though playing with a little ball.'

'It seems to me,' she replied, 'that it is ridiculous to be on something that turns and torment oneself so much about it. But the real problem is that we are not sure that we are moving: for after all, to tell you the truth, I am suspicious about all the care you are taking to explain why we don't notice the motion of the earth. Is it possible that it does not leave any noticeable little mark by which such motion can be recognized?'

'The most natural movements,' I replied, 'and the most ordinary, are those which make themselves the least felt – and that is true even in morality. The movement of self-love is so natural to us that most often we do not feel it, and we believe that we act according to other principles.'

'Ah!' she said. 'You are moralizing when it comes to physics, and that's making me yawn. Let us retire, for that is enough for the first time. Tomorrow we'll come back here, you with your systems, and I with my ignorance.'

Returning to the château, to exhaust the matter of the systems, I told her that there was a third one invented by Tycho Brahe who, absolutely invested in the earth's immobility, placed it at the centre of the world and made the sun revolve around it, and the other planets revolve around the sun: for since the new discoveries there was no way to make the planets revolve around the earth. But the Marquise, who has a keen and prompt discernment, judged that it was too affected to exempt the earth from revolving around the sun while not exempting so many other large bodies; that it was no longer appropriate for the sun to revolve around the earth, since all the other planets revolved around it; and that this system was at most only appropriate for supporting the immobility of the earth when one was already convinced of it, but would not persuade someone who wasn't. Finally, it was resolved that we would stick to the system of Copernicus, which is more uniform and agreeable, and is not mixed with prejudice. Ultimately, its simplicity persuaded us, and its boldness pleased us.

Second Evening

The next morning, as soon as it was possible for someone to enter the Marquise's apartment, I sent for news of her, and asked if she had been able to sleep while turning. She replied that she was already quite accustomed to this turning of the earth, and had passed the night as quietly as Copernicus himself would have. Some time later people came to her house and stayed there until evening, according to the tiresome country custom. We were still very much obliged to them, for the custom of the country also gives them the right to extend their visit until the following day, if they so desire, but they had the decency not to do so. So the Marquise and I found ourselves free in the evening. We went into the park again, and our conversation turned immediately to our systems. She had understood them so well that she did not want to speak of them a second time and wanted me instead to teach her something new.

'Well then,' I said to her, 'since the sun, which is now immobile, has ceased to be a planet, and the earth, which moves around it, has now become one, you will not be so surprised to hear that the moon is an earth like this one, and that apparently it is inhabited.'

'But I have never heard of an inhabited moon,' she replied, 'except as a foolish daydream!'

'Maybe this is one too,' I replied. 'I only take sides in these things as one does in civil wars, where the uncertainty of what may happen means that one always maintains intelligence of the opposite party and keeps channels open even with one's enemies. As for me, although I believe that the moon is inhabited, I continue to live civilly with those who do not believe it, and I always hold myself in a position where I could take their side honourably, if they had the upper hand. But until they have some considerable advantage over us, here is what made me incline toward an inhabited Moon.

'Suppose there had never been any communication between Paris and Saint-Denis, and that a resident of Paris, who had never left his city, saw Saint-Denis from afar from the towers of Notre-Dame. Ask him if he believes that Saint-Denis is inhabited like Paris: he will boldly answer no, for he will say that he sees the inhabitants of Paris but not those of Saint-Denis, nor has he heard of them. There will be someone who will note to him that it is true that one cannot see inhabitants of Saint-Denis from the towers of Notre-Dame, but the reason is the distance; that everything one can see in Saint-Denis strongly resembles Paris; that Saint-Denis has steeples, houses, walls; and thus that it might well resemble Paris in being inhabited. None of this will mean anything to our Parisian, who will always persist in maintaining that Saint-Denis is not inhabited, since he sees no one there. Our Saint-Denis is the moon, and each of us is a Parisian who has never left his city.'

'Ah!' interrupted the Marquise. 'You do us wrong, for we are not as stupid as your Parisian. Since he sees that Saint-Denis is made just like Paris, he must have lost his reason not to believe it inhabited. But the moon is not made like the earth at all.'

'Take care, Madame,' I replied, 'because if the moon does resemble the earth in everything, you would then be obliged to believe that it is inhabited.'

'I admit,' she replied, 'that there would be no way to avoid it, and I see that you have an air of confidence that already frightens me. The two movements of the earth, which I had never suspected, make me timid about everything else. And yet, would it be possible for the earth to be luminous like the moon? For that would be necessary in order for them to be truly alike.'

'Alas, Madame,' I replied, 'being luminous is not as great a thing as you think. It is only the sun that has this notable quality. It is luminous by itself, and by virtue of the particular nature that it has, but the planets are luminous only because they are illuminated by it. It sends its light to the moon, which sends it back to us; and the earth must also send back the light from the sun to the moon, for it is no further from the earth to the moon than from the moon to the earth.'

'But,' said the Marquise, 'is the earth as fit as the moon to reflect the light of the sun?'

'I see you are still one for the moon,' I replied, 'and cannot shake off the remnant of your esteem for it. But light is composed of small balls that bounce off what is solid and move in another direction, but pass in a straight line through whatever gives them an opening, like air or glass. So what makes the moon shine on us is that it is a hard and solid body, which bounces these little balls back at us. And I believe you will not dispute that the earth has this same hardness and this same solidity. Admire, then, what is simply a difference of advantageous placement. Because the moon is so far away from us, we only see it as a luminous body, and we don't know that it is a big, earth-like mass. By contrast, because, to the earth's misfortune, we see it so closely, it seems to us just a large mass, fit only to provide food for animals, and we do not notice that it is luminous, since we cannot put ourselves at some distance from it.'

'It is the same for us,' said the Marquise. 'We are dazzled by the status of those above us, and we do not see that beneath everything we all are very much alike.'

'It's the same thing,' I replied. 'We want to judge everything, and we are always at a bad point of view. We want to judge ourselves, but we are too close; we want to judge others, but we are too far. Only one who was between the moon and the earth would be able to see them well, but that person would simply be a spectator of the world, not an inhabitant.'

'I shall never be at peace with the injustice we do to the earth,' she said, 'and with the excessively favourable concern we have for the moon, if you do not assure me that the people of the moon do not know their advantages any better than we know ours, and that they take our earth for a star, without knowing that their dwelling is one too.'

'I guarantee it to you,' I replied. 'We seem to them to perform our function quite regularly as a star. It is true that they do not see us making a circle around them, but that doesn't matter. Here is how it is: the half of the moon that was turned toward us at the beginning of the world has always been turned toward us since; she only ever shows us the same eyes, mouth and face that our imagination composes on the basis of the spots that she shows us. If the other, opposite half presented itself to us, other, differently arranged spots would no doubt lead us to imagine some other figure. It's not that the moon doesn't rotate in the time that it moves around the earth – that is to say, in a month. Rather, when she makes part of her rotation and might hide from us a cheek of this imagined face, for example, she makes precisely the same part of her circle around the earth, putting herself in a new point of view and still showing us the same cheek. Thus, the moon rotates with regard to the sun and the other stars, but not with regard to us. They all seem to her to rise and set in the space of fifteen days, but earth seems to always hang in the same place in the sky. This apparent immobility hardly suits a body that should pass for a star, but the moon is not perfect either. She has a kind of trembling that causes a small corner of her face to hide sometimes, and a small corner of the opposite side to appear. But on my honour, she does not fail to attribute this trembling to us, and to imagine that we move in the sky like a pendulum that comes and goes.'

'All these planets,' said the Marquise, 'are just like us, always throwing back on others what is in ourselves. The earth says: "It is not I who turns, it is the sun." The moon says: "It is not I who trembles, it is the earth." There are many errors everywhere.'

'I would not advise you to undertake to reform anything here,' I replied. 'It would be much better that you finish convincing yourself of the complete resemblance of the earth and the moon. And remember, the Europeans were not in America until after six thousand years. It took them that long to perfect navigation to the point of being able to cross the ocean. The people of the moon may already know how to make small trips through the air; perhaps at the moment they are practising, and when they are more skilful and experienced, we'll see them. God knows, that would be a surprise!'

'You are unbearable,' she said, 'to push me to the limit with reasoning as hollow as that!'

'If you are mad at me,' I replied, 'I know how to strengthen it. Notice that the world expands little by little. The ancients were quite sure that the torrid and frigid zones could not be inhabited because of the excessive heat or cold. And in the time of the Romans, the general map of the world was hardly more extensive than the map of their empire, which had grandeur on the one hand, and marked a great deal of ignorance on the other. However, they found men without fail in very hot and very cold countries, and thus the world already grew larger. Then it was believed that the ocean covered the whole earth except for what was already known, and that there were no Antipodes, for no one had heard of them, and how could they have their feet up and heads down? But after this excellent reasoning the Antipodes were discovered. And so a new reform of the map and a new half of the world. Hear me well, Madame: these Antipodes that we found against all expectation should teach us to be restrained in our judgements. Perhaps the world will complete its development for us, and then we will know even the moon. We are not there yet, because all the earth isn't yet discovered, and it seems that this must be done in order. When we know our own home well, we will be allowed to know that of our neighbours, the people of the moon.'

Fourth Evening

'The sun is not a body of the same type as the earth, or any of the planets. He is the source of all the light that the planets only send to each other after receiving it from him. They can exchange it, so to speak, among themselves, but they cannot produce it. He alone draws this precious substance from himself and emits it with force on all sides. The light reflects when it meets something solid, and spreads from one planet to another; vast, long streaks of light which intersect, cross each other and intertwine in a thousand different ways, and form wonderful tissues of the richest material in the world. And so the sun is placed in the centre, which is the most convenient place from which it can distribute it equally and animate everything by its heat. The sun is therefore quite a particular body: but what kind of body? We find it difficult to say. It had always been believed that it was a very pure fire, but we overturned that belief at the beginning of this century, when we saw spots on its surface. As new planets had been discovered a short time before – which I will tell you about – and as those filled the minds of everyone in philosophy, and finally as the new planets became quite the fashion, we immediately judged that these spots were planets that moved around the sun, and that they necessarily hid some part of it from us, and turned their dark half toward us.

'It turned out they weren't planets, but clouds, smoke, or foam rising from the sun. They are sometimes great in number, sometimes small, and sometimes they all disappear. Sometimes several of them come together, sometimes they separate; sometimes they are lighter, sometimes darker. Sometimes we see lots of them, and at other times, equally long, none appear. It seems that the sun is a liquid matter – some say molten gold – which boils incessantly, and produces impurities that the force of its movement throws back on to its surface. There they are consumed, and then others are produced. Imagine what strange bodies these are, when some are as big as the earth. Judge by this what the quantity of this molten gold is, or the extent of this great sea of light and fire. Others say that through a telescope the spots appear full of mountains that vomit fire and are like millions of Mount Etnas. But some say that these mountains are just a vision caused by a fault in the telescope. But what can we trust, if we must distrust the very telescopes to which we owe our awareness of so many new objects? Finally, whatever the sun may be, it does not seem in any way fit for habitation. It's a pity, however, for what a lovely place it would be to live! We would be at the centre of everything, and would see all the planets revolving regularly around us, instead of see-ing in their courses an infinity of oddities that only appear because we are not in the proper place to judge well, namely, the centre of their movements. Isn't that pitiful? There is only one place in the world where the study of the stars would be extremely easy, and precisely in that place there's no one.

'But do you want to continue our voyage through the worlds? We have arrived at the centre, which is always the lowest place in everything that is round. It is now necessary to retrace our steps and go back up. We will find Mercury, Venus, the earth, the moon, all the planets we have visited. Next is Mars. Mars has nothing curious that I know of; its days are not quite an hour longer than ours, but its year is equal to two of ours. It is smaller than the earth, and it sees the sun a little smaller and less bright than we see it. In the end, it is not really worth stop-ping at Mars. But how pretty is Jupiter with its four moons or satellites! They are four small planets that revolve around it as our moon revolves around us.'

'But,' interrupted the Marquise, 'why are there planets that revolve around other planets no better than they are? Seriously, it would seem to me more regular and uniform if all the planets, both large and small, had only the same motion around the sun.'

'Ah, Madame,' I replied, 'if you knew what Descartes's vor-tices are – those whirlwinds whose name is so terrible, even while the idea is so pleasing – you would not speak as you do.'

'Even if my head were to start spinning,' she said, laughing, 'it would be nice to know what vortices are. Finish driving me mad – I no longer spare myself or hold myself back from phil-osophy. Let the world say what it will, as we give ourselves over to vortices.'

'I have not known you so transported,' I replied. 'It's a pity that it was inspired only by vortices. What is called a vortex is a mass of matter whose parts are detached from each other and all move in the same direction. At the same time, they are each allowed to have small particular movements, provided that they always follow the general movement. Thus a vortex of wind is an infinity of small parts of air, which turn around all together and envelop what they encounter. You know that the planets are carried in the celestial matter, which is prodigiously subtle and agitated. This whole great mass of celestial matter, which lies from the sun to the fixed stars, turns in circles and carries the planets with it, and makes them all turn in the same direction around the sun, which occupies the centre, but in longer or shorter times depending on whether they are more or less distant from it. Even the sun turns on itself, because it is precisely in the middle of all this celestial matter. You will notice in passing as well that if the earth were in the place of the sun, it could also do no less than turn on itself.

'This is the great vortex of which the sun is like the master, but at the same time the planets comprise small individual vortices in imitation of the sun's vortex. Each of them, while revolving around the sun, does not stop turning on itself, and also causes some quantity of the celestial matter (which is always ready to follow whatever movements we give it if they do not divert it from its general motion) to revolve around it in the same direction. This is the particular vortex of the planet, which pushes it as far as the strength of the planet's motion extends. If some smaller planet falls into this little vortex dominated by the larger one, then it is carried away and necessarily forced to revolve around it, while all of them – the large planet, the small planet and the vortex that contains them – revolve still around the sun. This is how, at the beginning of the world, we came to be followed by the moon, because it was within the range of our vortex and completely at our disposal. Jupiter, whom I had begun to speak to you about, was more fortunate or more powerful than we were. There were four small planets in its vicinity, and it subjugated all four of them. And we, who are a principal planet – do you know what would have happened had we been close to it? It is ninety times bigger than we are, and would have easily swallowed us up in its vortex. Then we would have been only a moon dependent on it, instead of having one of our own. So it is true that a chance situation alone often decides one's whole fortune.'

'If the earth is so small compared to Jupiter,' she said, 'does Jupiter see us? I'm afraid we are strangers to him.'

'Honestly, I believe this to be so,' I replied. 'Jupiter would have to see the earth ninety times smaller than we see it. It is too little for him to see her. Here is the only thing we can hope for: that there will be astronomers on Jupiter who, after having taken great pains to construct excellent telescopes, and after having chosen the best nights for observation, will ultimately discover in the heavens a very small planet they have never seen. First, the *Journal des Savants* of that country will speak of it, and the people of Jupiter will either understand nothing about it or only laugh at it. The philosophers, whose opinions it destroys, adopt a plan of believing nothing about it, and only a few very reasonable people want to inquire further. They observe again, and see the little planet again; they are sure it is not an illusion, and even begin to suspect that it moves around the sun. They find, after a thousand observations, that this movement lasts one year. And finally, thanks to all the pains that their scholars have taken, they know on Jupiter that our earth is a world. The curious rush to see it at the end of a telescope, and can barely discern it.'

Fifth Evening

The Marquise felt very impatient to know what would become of the fixed stars. 'Will they be inhabited like the planets or won't they be?' she asked me. 'And what do we ultimately do with them?'

'You might soon guess, if you wanted to,' I replied. 'The fixed stars cannot be less distant from the earth than fifty million leagues – and if you annoy an astronomer, he will put them even further away. The distance from the sun to the furthest planet is nothing compared to the distance from the sun to the fixed stars, which is beyond the bother of counting it. Their light, as you see, is quite brilliant. If they received it from the sun, it would have to be very feeble after such a journey of fifty million leagues, and they would need to reflect it back to us at this same distance, which would weaken it still more. It would be impossible for a light that had undergone reflection and journeyed fifty million leagues twice to have the strength and vivacity that the fixed stars do. So they must be luminous in themselves, and all of them, in a word, so many suns.'

'Am I mistaken,' exclaimed the marquise, 'or do I see where you want to lead me? Are you going to say to me: "The fixed stars are so many suns; our sun is the centre of a vortex that revolves around it; why should each fixed star not also be the centre of a vortex that moves around it? Our sun has planets that it illuminates; why shouldn't each fixed star also have planets that it illuminates?"'

'Let me only answer to you,' I told her, 'what Phaedra said to Oenone: "You have said it!"'

'But,' she replied, 'the universe is now so big that I lose myself in it. I no longer know where I am – I'm nothing. What, will everything be divided into vortices thrown confusedly among each other? Each star will be the centre of a vortex perhaps as large as the one we're in? All this immense space which includes our sun and planets will only be a small part of the universe? As many similar spaces as fixed stars? This confuses me and disturbs me. In fact, it terrifies me.'

'And I,' I answered, 'am put at ease. When the sky was only this blue vault where the stars were nailed, the universe seemed small and narrow to me – I felt myself stifled there. Now that we have given infinitely more breadth and depth to this vault by dividing it into thousands and thousands of vortices, it seems to me that I breathe with more freedom, that the air is greater, and certainly that the universe has a new magnificence. Nature has spared nothing in producing it; she has made a profusion of treasures quite worthy of her. Nothing is so beautiful to imagine as this prodigious number of vortices, with a sun at the middle of each around which planets revolve. The inhabitants of a planet of one of these infinite vortices see on all sides the suns of the vortices by which they are surrounded, but they cannot see the planets that have only a weak light borrowed from their sun which does not extend beyond their world.'

'You offer me,' she said, 'a kind of perspective so long that no eyes can reach the end of it. I clearly see the inhabitants of the earth; then you make me see those of the moon, and of the other planets of our vortex pretty clearly, but less so than those of the earth. After them come the inhabitants of the planets of the other vortices. I confess to you that they are completely obscured, and that no matter how much effort I make to see them, I can hardly distinguish them. And indeed, are they not almost annihilated by the very expression that you need to use when speaking of them? You must call them the inhabitants of one of the planets of one of the vortices, whose number is infinite. Yet the very same expression is true of we ourselves – admit that you would hardly know how to disentangle us any longer from the midst of so many worlds! For my part, I'm beginning to see the earth as so frightfully small that after this I don't think I'll be bothered about anything anymore. It seems clear that if we have such a great desire for self-aggrandizement, if we make plan after plan, if we go to so much trouble, it is because we do not know about vortices. I will claim that my laziness stems from my new insights, and when someone reproaches me for my indolence, I will answer: "Ah! If only you knew what the fixed stars are!"

'But do you know that by multiplying worlds so liberally, you've created a real difficulty for me?' she said. 'The vortices whose suns we see touch the vortex where we are: how can so many globes touch a single one? I want to imagine it, but I cannot.'

'They interlock with one another,' I replied, 'somewhat similarly to the gears of a watch, and mutually aid each other's movements. It is true, however, that they also act against each other. Each world, it is said, is like a balloon that would expand if we let it, but it is immediately repelled by the neighbouring worlds, and it deflates, after which it begins to inflate and expand again, and so on. Some say that the fixed stars twinkle and don't always shine because their vortices perpetually push ours and are perpetually repelled by it.'

'I really like all these ideas,' said the Marquise. 'I like these balloons that inflate and deflate at each moment, and these worlds that always fight each other, and above all I like to see how this fight creates an exchange of light, which seems to be the only exchange they can have.'

'No, no,' I replied. 'It's not the only one. Neighbouring worlds sometimes send visitors, and do so quite magnificently: they are comets.'

'We who never leave our vortex,' she replied, 'lead rather boring lives. If the inhabitants of a comet are witty enough to predict the time of their passage to our world, then those who have already made the journey might announce to the others in advance what they'll see. "You will soon discover a planet that has a large ring around it," they'd perhaps say, speaking of Saturn. "You will see another one that has four little ones following it." Perhaps there are even people whose job it is to observe the moment when they enter our world, and who immediately cry out "New sun! New sun!" like those sailors who cry "Land! Land!"'

'You therefore wouldn't think,' I said, 'of pitying the inhabitants of a comet. But I hope at least you will pity those who live in a vortex in which the sun has just been extinguished, and who remain in eternal night.'

'What!' she cried. 'Suns are extinguished?'

'Yes, without doubt,' I replied. 'The Ancients saw fixed stars in the heavens that we no longer see there. These suns have lost their light, which certainly means great desolation throughout that vortex, and a general mortality on all the planets: for what can be done without a sun?'

'How can a sun be darkened and extinguished,' said the Marquise, 'when it itself is a source of light?'

'With all the ease in the world, according to Descartes,' I answered. 'He supposes that the spots of our sun, being foam or mist, can thicken and join one to the other. Finally they will form a crust around the sun which will continue to grow, and farewell to the sun. We've already had a narrow escape, they say. The sun was very pale for entire years – for example, during those that followed Caesar's death. It was the crust beginning to form. The force of the sun broke and dissipated it, but if it had continued, we would have been lost.'

'You make me tremble,' she said. 'Now that I know the consequences of the pallor of the sun, I believe that instead of going to my mirror in the mornings to see if I'm pale, I will go look in the sky to see if the sun is pale!'

'Ah, Madame,' I replied. 'Don't worry – it takes time to ruin a world.'

'And yet,' she said, 'does it not only take time?'

'I admit it,' I said. 'All this immense mass of matter that makes up the universe is in perpetual motion, from which none of its parts is entirely exempt. As soon as there is no motion somewhere, you can be sure that change must occur – slowly or quickly, but always in a time proportionate to the effect. The Ancients were happy to imagine that the celestial bodies were of an unchanging nature, because they had not seen them change. Had they the time to ascertain this by experience? Compared to us, the Ancients were children.'

'Honestly,' said the Marquise, 'I now find the worlds, the heavens and the celestial bodies so subject to change that I am completely returned to myself.'

'If you believe, me, let's return even more fully,' I replied, 'and say no more about it. It's just as well, for you've arrived at the last vault of heaven, and to tell you if there are still stars beyond it would require more skill than I possess. Put more worlds there or don't – it's up to you. Those great invisible countries, which can be there or not based on what you wish, are properly the realm of the philosophers. It is enough for me to have led your mind as far as your eyes can see.'

'Well!' she exclaimed. 'I have the whole system of the universe in my head! Am I a scholar?'

'Yes,' I replied. 'You are well enough on your way, and you have the advantage of being able to disbelieve anything I have told you whenever the mood strikes you. I only ask of you, as a reward for my pains, that whenever you look at the sun, the sky, or the stars, you think of me.'

Edmond Halley, Opening Poem to Isaac Newton, *Mathematical Principles of Natural Philosophy (Principia)*, 1687*

Edmond Halley (1656–1742) was an English astronomer, most famous for using Newton's laws of motion to compute the period of the comet that would be named for him. Before that, Halley was one of the central figures responsible for the publication of Newton's great work, the Principia, *which he not only convinced Newton to make public but also edited and published at his own expense. At the start of the book, he included a Latin poem he had written celebrating Newton and his achievements, which served as a model for the many later works in praise of Newton.*

Ode on This Splendid Ornament of Our Time and Our Nation, the Mathematical-Physical Treatise by the Eminent Isaac Newton

Behold the pattern of the heavens, and the balances of the divine
    structure;
Behold Jove's calculation and the laws
That the creator of all things, while he was setting the beginnings of
    the world would not violate.
Behold the foundations he gave to his works.
Heaven has been conquered and its innermost secrets are revealed;
The force that turns the outermost orbs around is no longer hidden.

---

* Selections taken from *The Principia: The Authoritative Translation: Mathematical Principles of Natural Philosophy*, by Isaac Newton, translated by I. Bernard Cohen and Anne Whitman (University of California Press, 1999).

The Sun sitting on his throne commands all things
To tend downward toward himself, and does not allow the chariots
    of the heavenly bodies to move
Through the immense void in a straight path, but hastens them along
In unmoving circles around himself as centre.
Now we know what curved path the frightful comets have;
No longer do we marvel at the appearances of a bearded star.
From this treatise we learn at last why silvery Phoebe moves at an
    unequal pace,
Why, till now, she has refused to be bridled by the numbers of any
    astronomer,
Why the nodes regress, and why the upper apsides move forward.
We learn also the magnitude of the forces with which wandering
    Cynthia
Impels the ebbing sea, while its weary waves leave the seaweed far
    behind
And the sea bares the sands that sailors fear, and alternately beat
    high up on the shores.
The things that so often vexed the minds of the ancient philosophers
And fruitlessly disturb the schools with noisy debate
We see right before our eyes, since mathematics drives away the cloud.
Error and doubt no longer encumber us with mist;
For the keenness of a sublime intelligence has made it possible for us
    to enter
The dwellings of the gods above and to climb the heights of heaven.
Mortals arise, put aside earthly cares,
And from this treatise discern the power of a mind sprung from
    heaven,
Far removed from the life of beasts.
He who commanded us by written tablets to abstain from murder,
Thefts, adultery, and the crime of bearing false witness,
Or he who taught nomadic peoples to build walled cities, or he who
    enriched the nations with the gift of Ceres,
Or he who pressed from the grape a solace for cares,
Or he who with a reed from the Nile showed how to join together
Pictured sounds and to set spoken words before the eyes,
Exalted the human lot less, inasmuch as he was concerned with only
    a few comforts of a wretched life,

And thus did less than our author for the condition of mankind.
But we are now admitted to the banquets of the gods;
We may deal with the laws of heaven above; and we now have
The secret keys to unlock the obscure earth; and we know the
  immovable order of the world
And the things that were concealed from the generations of the past.
O you who rejoice in feeding on the nectar of the gods in heaven,
Join me in singing the praises of NEWTON, who reveals all this,
Who opens the treasure chest of hidden truth,
NEWTON, dear to the Muses,
The one in whose pure heart Phoebus Apollo dwells and whose mind
  he has filled with all of his divine power;
No closer to the gods can any mortal rise.

                                                                Edm. Halley

Isaac Newton, *Mathematical Principles of Natural Philosophy (Principia)*, 1687*

Isaac Newton (1642–1727) was an English astronomer and mathematician who invented calculus and developed the theory of universal gravitation, while also devoting a great deal of his time to optics, alchemy, theology and his work as Master of the Mint. Newton's Mathematical Principles of Natural Philosophy, *popularly referred to as the* Principia, *was in part an updating of Descartes's* Principles of Philosophy, *by making natural philosophy not only mechanical but also mathematical, as well as eliminating Descartes's vortices in favour of action at a distance. By combining the terrestrial physics of Galileo with the celestial physics of Kepler, and by offering a theory of gravitation that would explain the orbits of the planets in the heliocentric system just as it explained the falling of objects on earth, Newton's work tied together different threads from the previous century and offered something new: a complete system of the world, and a new way of doing science. His work explained why planets obeyed Kepler's three laws and proved those laws mathematically. The* Principia *was published in three books, with the first two focusing on mathematics (its proofs are all geometric, though it was written after Newton invented calculus), and the third applying that mathematics to the movements of bodies in the solar system.*

* Selections taken from *The Principia: The Authoritative Translation: Mathematical Principles of Natural Philosophy*, by Isaac Newton, translated by I. Bernard Cohen and Anne Whitman (University of California Press, 1999).

Author's Preface to the Reader

Since the Ancients considered *mechanics* to be of the greatest importance in the investigation of nature and science and since the moderns – rejecting substantial forms and occult qualities – have undertaken to reduce the phenomena of nature to mathematical laws, it has seemed best in this treatise to concentrate on *mathematics* as it relates to natural philosophy. Therefore, our present work sets forth mathematical principles of natural philosophy. For the basic problem of philosophy seems to be to discover the forces of nature from the phenomena of motions and then to demonstrate the other phenomena from these forces. It is to these ends that the general propositions in books 1 and 2 are directed, while in book 3 our explanation of the system of the world illustrates these propositions. For in book 3, by means of propositions demonstrated mathematically in books 1 and 2, we derive from celestial phenomena the gravitational forces by which bodies tend toward the sun and toward the individual planets. Then the motions of the planets, the comets, the moon and the sea are deduced from these forces by propositions that are also mathematical. If only we could derive the other phenomena of nature from mechanical principles by the same kind of reasoning! For many things lead me to have a suspicion that all phenomena may depend on certain forces by which the particles of bodies, by causes not yet known, either are impelled toward one another and cohere in regular figures, or are repelled from one another and recede. Since these forces are unknown, philosophers have hitherto made trial of nature in vain. But I hope that the principles set down here will shed some light on either this mode of philosophizing or some truer one.

Axioms, or the Laws of Motion

Law 1: *Every body perseveres in its state of being at rest or of moving uniformly straight forward, except insofar as it is compelled to change its state by forces impressed.*

Law 2: *A change in motion is proportional to the motive force impressed and takes place along the straight line in which that force is impressed.*

Law 3: *To any action there is always an opposite and equal reaction; in other words, the actions of two bodies upon each other are always equal and always opposite in direction.*

Scholium

The principles I have set forth are accepted by mathematicians and confirmed by experiments of many kinds. By means of the first two laws and the first two corollaries Galileo found that the descent of heavy bodies is the squared ratio of the time and the motion of projectiles occurs in a parabola, as experiment confirms, except insofar as these motions are somewhat retarded by the resistance of air. When a body falls, uniform gravity, by acting equally in individual equal particles of time, impresses equal forces upon that body and generates equal velocities; and in the total time it impresses a total force and generates a total velocity proportional to the time. And the spaces described in proportional times are as the velocities and the times jointly, that is, in the squared ratio of the times. And when a body is projected upward, uniform gravity impresses forces and takes away velocities proportional to the times; and the times of ascending to the greatest heights are as the velocities to be taken away, and these heights are as the velocities and the times jointly, or as the squares of the velocities. And when a body is projected along any straight line, its motion arising from the projection is compounded with the motion arising from gravity.

Book 3: *The System of the World*

In the preceding books I have presented principles of philosophy that are not, however, philosophical but strictly mathematical – that is, those on which the study of philosophy can be based. These principles are the laws and conditions of motions and of forces, which especially relate to philosophy. But in order to prevent these principles from seeming sterile, I have illustrated them with some philosophical scholiums, treating topics that are general and that seem to be the most fundamental for philosophy, such as the density and resistance of bodies, spaces void of bodies, and the motion of light and sounds. It still remains for us to exhibit the system of the world from these same principles. On this subject I composed an earlier version of book 3 in popular form, so that it might be more widely read. But those who have not sufficiently grasped the principles set down here will certainly not perceive the force of the conclusions, nor will they lay aside the preconceptions to which they have become accustomed over many years; and therefore, to avoid lengthy disputations, I have translated the substance of the earlier version into propositions in a mathematical style, so that they may be read only by those who have first mastered the principles. But since in books 1 and 2 a great number of propositions occur which might be too time-consuming even for readers who are proficient in mathematics, I am unwilling to advise anyone to study every one of these propositions. It will be sufficient to read with care the Definitions, the Laws of Motion and the first three sections of book 1, and then turn to this book 3 on the system of the world, consulting at will the other propositions of books 1 and 2 which are referred to here.

Rules for the Study of Natural Philosophy

RULE 1: *No more causes of natural things should be admitted than are both true and sufficient to explain their phenomena.*

As the philosophers say: Nature does nothing in vain, and more causes are in vain when fewer suffice. For nature is simple and does not indulge in the luxury of superfluous causes.

RULE 2: *Therefore, the causes assigned to natural effects of the same kind must be, so far as possible, the same.*

Examples are the cause of respiration in man and beast, or of the falling of stones in Europe and America, or of the light of a kitchen fire and the sun, or of the reflection of light on our earth and the planets.

RULE 3: *Those qualities of bodies that cannot be intended and remitted [i.e., qualities that cannot be increased and diminished] and that belong to all bodies on which experiments can be made should be taken as qualities of all bodies universally.*

For the qualities of bodies can be known only through experiments; and therefore qualities that square with experiments universally are to be regarded as universal qualities; and qualities that cannot be diminished cannot be taken away from bodies. Certainly, idle fancies ought not to be fabricated recklessly against the evidence of experiments, nor should we depart from the analogy of nature, since nature is always simple and ever consonant with itself. The extension of bodies is known to us only through our senses, and yet there are bodies beyond the range of these senses; but because extension is found in all sensible bodies, it is ascribed to all bodies universally. We know by experience that some bodies are hard. Moreover, because the hardness of the whole arises from the hardness of its parts, we justly infer from this not only the hardness of the undivided particles of bodies that are accessible to our senses, but also of all other bodies. That all bodies are impenetrable we gather not by reason but by our senses. We find those bodies that we handle to be impenetrable, and hence we conclude that impenetrability is a property of all bodies universally. That all bodies are movable and persevere in motion or in rest by means of certain forces (which we call forces of inertia) we infer from finding these properties in the bodies that we have seen. The extension, hardness, impenetrability, mobility and force of inertia of the whole arise from the extension, hardness, impenetrability, mobility and force of inertia of each of the parts; and thus we conclude that every one of the least parts of all bodies is extended, hard, impenetrable, movable and endowed with a force of inertia. And this is the foundation of all natural philosophy. Further, from phenomena we know that the divided, contiguous parts of bodies can be separated from one another, and from mathematics it is certain that the undivided parts can be distinguished into smaller parts by our reason. But it is uncertain whether those parts which have been distinguished in this way and not yet divided can actually be divided and separated from one another by the forces of nature. But if it were established by even a single experiment that in the breaking of a hard and solid body, any undivided particle underwent division, we should conclude by the force of this third rule not only that divided parts are separable but also that undivided parts can be divided indefinitely.

Finally, if it is universally established by experiments and astronomical observations that all bodies on or near the earth gravitate [lit. are heavy] toward the earth, and do so in proportion to the quantity of matter in each body, and that the moon gravitates [is heavy] toward the earth in proportion to the quantity of its matter, and that our sea in turn gravitates [is heavy] toward the moon, and that all planets gravitate [are heavy] toward one another, and that there is a similar gravity [heaviness] of comets toward the sun, it will have to be concluded by this third rule that all bodies gravitate toward one another. Indeed, the argument from phenomena will be even stronger for universal gravity than for the impenetrability of bodies, for which, of course, we have not a single experiment, and not even an observation, in the case of the heavenly bodies. Yet I am by no means affirming that gravity is essential to bodies. By inherent force I mean only the force of inertia. This is immutable. Gravity is diminished as bodies recede from the earth.

RULE 4: *In experimental philosophy, propositions gathered from phenomena by induction should be considered either exactly or very nearly true notwithstanding any contrary hypotheses, until yet other phenomena make such propositions either more exact or liable to exceptions.*

This rule should be followed so that arguments based on induction may not be nullified by hypotheses.

Phenomena

PHENOMENON 1: *The circumjovial planets [or satellites of Jupiter], by radii drawn to the centre of Jupiter, describe areas proportional to the times, and their periodic times – the fixed stars being at rest – are as the 3:2 powers of their distances from that centre.*

PHENOMENON 2: *The circumsaturnian planets [or satellites of Saturn], by radii drawn to the centre of Saturn, describe areas proportional to the times, and their periodic times – the fixed stars being at rest – are as the 3:2 powers of their distances from that centre.*

PHENOMENON 3: *The orbits of the five primary planets – Mercury, Venus, Mars, Jupiter and Saturn – encircle the sun.*

That Mercury and Venus revolve about the sun is proved by their exhibiting phases like the moon's. When these planets are shining with a full face, they are situated beyond the sun; when half full, to one side of the sun; when horned, on this side of the sun; and they sometimes pass across the sun's disc like spots. Because Mars also shows a full face when near conjunction with the sun, and appears gibbous in the quadratures, it is certain that Mars goes around the sun. The same thing is proved also with respect to Jupiter and Saturn from their phases being always full; and in these two planets, it is manifest from the shadows that their satellites project upon them that they shine with light borrowed from the sun.

PHENOMENON 4: *The periodic times of the five primary planets and of either the sun about the earth or the earth about the sun – the fixed stars being at rest – are as the 3:2 powers of their mean distances from the sun.*

This proportion, which was found by Kepler, is accepted by everyone. In fact, the periodic times are the same, and the dimensions of the orbits are the same, whether the sun revolves about the earth, or the earth about the sun. There is universal agreement among astronomers concerning the measure of the periodic times.

PHENOMENON 5: *The primary planets, by radii drawn to the earth, describe areas in no way proportional to the times but, by radii drawn to the sun, traverse areas proportional to the times.*

For with respect to the earth they sometimes have a progressive [direct or forward] motion, they sometimes are stationary, and sometimes they even have a retrograde motion; but with respect to the sun they move always forward, and they do so with a motion that is almost uniform – but, nevertheless, a little more swiftly in their perihelia and more slowly in their aphelia, in such a way that the description of areas is uniform. This is a proposition very well known to astronomers and is especially provable in the case of Jupiter by the eclipses of its satellites; by means of these eclipses we have said that the heliocentric longitudes of this planet and its distances from the sun are determined.

PHENOMENON 6: *The moon, by a radius drawn to the centre of the earth, describes areas proportional to the times.*

This is evident from a comparison of the apparent motion of the moon with its apparent diameter. Actually, the motion of the moon is somewhat perturbed by the force of the sun, but in these phenomena I pay no attention to minute errors that are negligible.

Propositions

PROPOSITION 5, THEOREM 5: *The circumjovial planets [or satellites of Jupiter] gravitate toward Jupiter, the circumsaturnian planets [or satellites of Saturn] gravitate toward Saturn, and the circumsolar [or primary] planets gravitate toward the sun, and by the force of their gravity they are always drawn back from rectilinear motions and kept in curvilinear orbits.*

For the revolutions of the circumjovial planets about Jupiter, of the circumsaturnian planets about Saturn, and of Mercury and Venus and the other circumsolar planets about the sun are phenomena of the same kind as the revolution of the moon about the earth, and therefore (by rule 2) depend on causes of the same kind, especially since it has been proved that the forces on which those revolutions depend are directed toward the centres of Jupiter, Saturn and the sun, and decrease according to the same ratio and law (in receding from Jupiter, Saturn and the sun) as the force of gravity (in receding from the earth).

COROLLARY 1. Therefore, there is gravity toward all planets universally. For no one doubts that Venus, Mercury and the rest [of the planets, primary and secondary] are bodies of the same kind as Jupiter and Saturn. And since, by the third law of motion, every attraction is mutual, Jupiter will gravitate toward all its satellites, Saturn toward its satellites, and the earth will gravitate toward the moon, and the sun toward all the primary planets.

COROLLARY 2. The gravity that is directed toward every planet is inversely as the square of the distance of places from the centre of the planet.

COROLLARY 3. All the planets are heavy toward one another by corols. 1 and 2. And hence Jupiter and Saturn near conjunction, by attracting each other, sensibly perturb each other's motions, the sun perturbs the lunar motions, and the sun and moon perturb our sea, as will be explained in what follows.

Scholium: Hitherto we have called 'centripetal' that force by which celestial bodies are kept in their orbits. It is now established that this force is gravity, and therefore we shall call it gravity from now on. For the cause of the centripetal force by which the moon is kept in its orbit ought to be extended to all the planets, by rules 1, 2 and 4.

PROPOSITION 6, THEOREM 6: *All bodies gravitate toward each of the planets, and at any given distance from the centre of any one planet the weight of any body whatever toward that planet is proportional to the quantity of matter which the body contains.*

Others have long since observed that the falling of all heavy bodies toward the earth (at least on making an adjustment for the inequality of the retardation that arises from the very slight resistance of the air) takes place in equal times, and it is possible to discern that equality of the times, to a very high degree of accuracy, by using pendulums. Now, there is no doubt that the nature of gravity toward the planets is the same as toward the earth. For imagine our terrestrial bodies to be raised as far as the orbit of the moon and, together with the moon, deprived of all motion, to be released so as to fall to the earth simultaneously; and by what has already been shown, it is certain that in equal times these falling terrestrial bodies will describe the same spaces as the moon, and therefore that they are to the quantity of matter in the moon as their own weights are to its weight. Further, since the satellites of Jupiter revolve in times that are as the 3:2 power of their distances from the centre of Jupiter, their accelerative gravities toward Jupiter will be inversely as the squares of the distances from the centre of Jupiter, and, therefore, at equal distances from Jupiter their accelerative gravities would come out equal. Accordingly, in equal times in falling from equal heights [toward Jupiter] they would describe equal spaces, just as happens with heavy bodies on this earth of ours. And by the same argument the circumsolar [or primary] planets, let fall from equal distances from the sun, would describe equal spaces in equal times in their descent to the sun.

COROLLARY 5. The force of gravity is of a different kind from the magnetic force. For magnetic attraction is not proportional to the [quantity of] matter attracted. Some bodies are attracted [by a magnet] more [than in proportion to their quantity of matter], and others less, while most bodies are not attracted [by a magnet at all]. And the magnetic force in one and the same body can be intended and remitted [i.e., increased and decreased] and is sometimes far greater in proportion to the quantity of matter than the force of gravity; and this force, in receding from the magnet, decreases not as the square but almost as the cube of the distance, as far as I have been able to tell from certain rough observations.

PROPOSITION 7, THEOREM 7: *Gravity exists in all bodies universally and is proportional to the quantity of matter in each.*

We have already proved that all planets are heavy [or gravitate] toward one another and also that the gravity toward any one planet, taken by itself, is inversely as the square of the distance of places from the centre of the planet. And it follows (by book 1, prop. 69 and its corollaries) that the gravity toward all the planets is proportional to the matter in them. Further, since all the parts of any planet A are heavy [or gravitate] toward any planet B, and since the gravity of each part is to the gravity of the whole as the matter of that part to the matter of the whole, and since to every action (by the third law of motion) there is an equal reaction, it follows that planet B will gravitate in turn toward all the parts of planet A, and its gravity toward any one part will be to its gravity toward the whole of the planet as the matter of that part to the matter of the whole. Q.E.D.

HYPOTHESIS 1: *The centre of the system of the world is at rest.*

No one doubts this, although some argue that the earth, others that the sun, is at rest in the centre of the system. Let us see what follows from this hypothesis.

PROPOSITION 11, THEOREM 11: *The common centre of gravity of the earth, the sun and all the planets is at rest.*

For that centre (by corol. 4 of the Laws) either will be at rest or will move uniformly straight forward. But if that centre always moves forward, the centre of the universe will also move, contrary to the hypothesis.

PROPOSITION 12, THEOREM 12: *The sun is engaged in continual motion but never recedes far from the common centre of gravity of all the planets.*

For since (by prop. 8, corol. 2) the matter in the sun is to the matter in Jupiter as 1,067 to 1, and the distance of Jupiter from the sun is to the semidiameter of the sun in a slightly greater ratio, the common centre of gravity of Jupiter and the sun will fall upon a point a little outside the surface of the sun. By the same argument, since the matter in the sun is to the matter in Saturn as 3,021 to 1, and the distance of Saturn from the sun is to the semidiameter of the sun in a slightly smaller ratio, the common centre of gravity of Saturn and the sun will fall upon a point a little within the surface of the sun. And continuing the same kind of calculation, if the earth and all the planets were to lie on one side of the sun, the distance of the common centre of gravity of them all from the centre of the sun would scarcely be a whole diameter of the sun. In other cases the distance between those two centres is always less. And therefore, since that centre of gravity is continually at rest, the sun will move in one direction or another, according to the various configurations of the planets, but will never recede far from that centre.

COROLLARY. Hence the common centre of gravity of the earth, the sun and all the planets is to be considered the centre of the universe. For since the earth, sun and all the planets gravitate toward one another and therefore, in proportion to the force of the gravity of each of them, are constantly put in motion according to the laws of motion, it is clear that their mobile centres cannot be considered the centre of the universe, which is at rest. If that body toward which all bodies gravitate most had to be placed in the centre (as is the commonly held opinion), that privilege would have to be conceded to the sun. But since the sun itself moves, an immobile point will have to be chosen for that centre from which the centre of the sun moves away as little as possible and from which it would move away still less, supposing that the sun were denser and larger, in which case it would move less.

PROPOSITION 13, THEOREM 13: *The planets move in ellipses that have a focus in the centre of the sun, and by radii drawn to that centre they describe areas proportional to the times.*

We have already discussed these motions from the phenomena. Now that the principles of motions have been found, we deduce the celestial motions from these principles *a priori*. Since the weights of the planets toward the sun are inversely as the squares of the distances from the centre of the sun, it follows (from book 1, props. 1 and 11, and prop. 13, corol. 1) that if the sun were at rest and the remaining planets did not act upon one another, their orbits would be elliptical, having the sun in their common focus, and they would describe areas proportional to the times. The actions of the planets upon one another, however, are so very small that they can be ignored, and they perturb the motions of the planets in ellipses about the mobile sun less (by book 1, prop. 66) than if those motions were being performed about the sun at rest.

PROPOSITION 14, THEOREM 14: *The aphelia and nodes of the [planetary] orbits are at rest.*

COROLLARY 1. The fixed stars also are at rest, because they maintain given positions with respect to the aphelia and nodes.

COROLLARY 2. And so, since the fixed stars have no sensible parallax arising from the annual motion of the earth, their forces will produce no sensible effects in the region of our system, because of the immense distance of these bodies from us. Indeed, the fixed stars, being equally dispersed in all parts of the heavens, by their contrary attractions annul their mutual forces, by book 1, prop. 70.

Newton, Parabolic Path of a Comet from *Mathematical Principles of Natural Philosophy (Principia)*, 1687

As Tycho's comet had been central to his own theories so many years earlier, comets were central to Newton's development of the Principia *and its arguments. In 1680, a large comet passing earth caught the attention of astronomers, who still did not know how to describe their orbital paths. Kepler had believed they moved along straight lines, though Hevelius had later suggested that they moved on parabolic paths around the sun. However, it still wasn't clear how to calculate these paths, whether the comets came singly or in pairs, whether they ever returned to these positions again, or what precisely caused their motion in the first place. Newton carefully observed the 1680 comet and engaged in a lively correspondence with the royal astronomer, John Flamsteed, about it. As he was working on the ideas of the* Principia *at the same time, Newton considered whether comets could follow the same inverse square law that applied to the planets, and ultimately proved that they did. He described how to calculate the path of the comet of 1680 within the* Principia, *'this being a problem of very great difficulty', and showed that it was parabolic. Comets were therefore a central aspect of showing how the theory of gravitation was truly universal.*

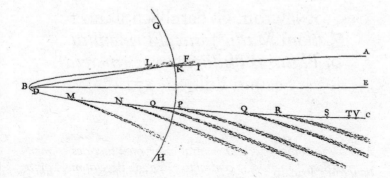

Newton, General Scholium
from *Mathematical Principles of Natural Philosophy (Principia)*, Second Edition, 1713*

Newton's Principia *evoked spirited controversy, particularly given its claims that gravity was a force that operated across a distance without direct contact, against the requirements of Cartesian mechanical philosophy. In the second and third editions of the* Principia, *Newton published an appendix called a 'General Scholium', in which he responded to some of the criticism he received. He began by attacking Descartes's theory of vortices, claiming that evidence of orbital motion, especially of comets, was not consistent with the system Descartes had devised. He then moved to theology, arguing that the system he described in the* Principia *relied on the design and dominion of God, and that considerations of God based on natural phenomena are 'certainly a part of natural philosophy'. Finally, he responded to criticism about the force of gravity and its seemingly occult nature, arguing that it clearly existed and could be described mathematically, even if he had not yet been able to figure out exactly what it was or how it worked. He famously argued that 'I do not feign hypotheses', and that proper science, or 'experimental philosophy', operated by induction from phenomena rather than hypothetical speculation.*

* Selections taken from *The Principia: The Authoritative Translation: Mathematical Principles of Natural Philosophy*, by Isaac Newton, translated by I. Bernard Cohen and Anne Whitman (University of California Press, 1999).

General Scholium

The hypothesis of vortices is beset with many difficulties. If, by a radius drawn to the sun, each and every planet is to describe areas proportional to the time, the periodic times of the parts of the vortex must be as the squares of the distances from the sun. If the periodic times of the planets are to be as the 3:2 powers of the distances from the sun, the periodic times of the parts of the vortex must be as the 3:2 powers of the distances. If the smaller vortices revolving about Saturn, Jupiter and the other planets are to be preserved and are to float without agitation in the vortex of the sun, the periodic times of the parts of the solar vortex must be the same. The axial revolutions [i.e., rotations] of the sun and planets, which would have to agree with the motions of their vortices, differ from all these proportions. The motions of comets are extremely regular, observe the same laws as the motions of planets, and cannot be explained by vortices. Comets go with very eccentric motions into all parts of the heavens, which cannot happen unless vortices are eliminated.

The only resistance which projectiles encounter in our air is from the air. With the air removed, as it is in Boyle's vacuum, resistance ceases, since a tenuous feather and solid gold fall with equal velocity in such a vacuum. And the case is the same for the celestial spaces, which are above the atmosphere of the earth. All bodies must move very freely in these spaces, and therefore planets and comets must revolve continually in orbits given in kind and in position, according to the laws set forth above. They will indeed persevere in their orbits by the laws of gravity, but they certainly could not originally have acquired the regular position of the orbits by these laws.

The six primary planets revolve about the sun in circles concentric with the sun, with the same direction of motion, and very nearly in the same plane. Ten moons revolve about the earth, Jupiter and Saturn in concentric circles, with the same direction of motion, very nearly in the planes of the orbits of the planets. And all these regular motions do not have their origin in mechanical causes, since comets go freely in very eccentric orbits and into all parts of the heavens. And with this kind of motion the comets pass very swiftly and very easily through the orbits of the planets; and in their aphelia, where they are slower and spend a longer time, they are at the greatest possible distance from one another, so as to attract one another as little as possible.

This most elegant system of the sun, planets and comets could not have arisen without the design and dominion of an intelligent and powerful being. And if the fixed stars are the centres of similar systems, they will all be constructed according to a similar design and subject to the dominion of *One*, especially since the light of the fixed stars is of the same nature as the light of the sun, and all the systems send light into all the others. And so that the systems of the fixed stars will not fall upon one another as a result of their gravity, He has placed them at immense distances from one another.

He rules all things, not as the world soul but as the lord of all. And because of His dominion He is called Lord God *Pantokrator*. He is eternal and infinite, omnipotent and omniscient, that is, He endures from eternity to eternity, and He is present from infinity to infinity; He rules all things, and He knows all things that happen or can happen. He is not eternity and infinity, but eternal and infinite; He is not duration and space, but He endures and is present. He endures always and is present everywhere, and by existing always and everywhere He constitutes duration and space. Since each and every particle of space is always, and each and every indivisible moment of duration is everywhere, certainly the maker and lord of all things will not be never or nowhere.

Every sentient soul, at different times and in different organs of senses and motions, is the same indivisible person. There are parts that are successive in duration and coexistent in space, but neither of these exist in the person of man or in his thinking principle, and much less in the thinking substance of God. Every man, insofar as he is a thing that has senses, is one and the same man throughout his lifetime in each and every organ of his senses. God is one and the same God always and everywhere. He is omnipresent not only virtually but also substantially, for action requires substance [lit. for active power [*virtus*] cannot subsist without substance]. In Him all things are contained and move, but He does not act on them nor they on Him. God experiences nothing from the motions of bodies; the bodies feel no resistance from God's omnipresence.

This concludes the discussion of God, and to treat of God from phenomena is certainly a part of natural philosophy.

Thus far I have explained the phenomena of the heavens and of our sea by the force of gravity, but I have not yet assigned a cause to gravity. Indeed, this force arises from some cause that penetrates as far as the centres of the sun and planets without any diminution of its power to act, and that acts not in proportion to the quantity of the surfaces of the particles on which it acts (as mechanical causes are wont to do) but in proportion to the quantity of solid matter, and whose action is extended everywhere to immense distances, always decreasing as the squares of the distances. Gravity toward the sun is compounded of the gravities toward the individual particles of the sun, and at increasing distances from the sun decreases exactly as the squares of the distances as far out as the orbit of Saturn, as is manifest from the fact that the aphelia of the planets are at rest, and even as far as the furthest aphelia of the comets, provided that those aphelia are at rest. I have not as yet been able to deduce from phenomena the reason for these properties of gravity, and I do not feign hypotheses. For whatever is not deduced from the phenomena must be called a hypothesis; and hypotheses, whether metaphysical or physical, or based on occult qualities, or mechanical, have no place in experimental philosophy. In this experimental philosophy, propositions are deduced from the phenomena and are made general by induction. The impenetrability, mobility and impetus of bodies, and the laws of motion and the law of gravity have been found by this method. And it is enough that gravity really exists and acts according to the laws that we have set forth and is sufficient to explain all the motions of the heavenly bodies and of our sea.

Newton, *A Treatise of the System of the World*, 1728

Newton had initially planned for his Principia *to have two books, with the second applying the results of the first to the 'system of the world'. The second book was first written so that it was more popularly accessible, focused on description rather than mathematics. Newton ultimately expanded the planned first book into two, and what became the third book he rewrote in a denser and more mathematical style, with some additions, especially the material on comets. After Newton's death in 1727, the original, more accessible second book was translated into English (it is not known by whom) with some minor updating and published under the title* A Treatise of the System of the World. *The volume was quite popular and helped spread Newton's ideas to a broader audience.*

Translators' Preface

In the beginning of the third book of the *Mathematical Principles of Natural Philosophy* of the late illustrious Sir Isaac Newton, he acquaints us 'that he had once composed that book in a popular method that it might be generally read; but that afterwards to avoid the disputes which might arise, if those who were unacquainted with the principles laid down in the preceding books, and full of the prejudices which many years had made natural to them, should take it in hand, he put the substance of that book into propositions in the mathematical way, that it might be read by those only who had studied the principles beforehand'. The prejudices that Great Man apprehended, when the principles whereby he explained the appearances of Nature were new, are now so well dissipated by the universal applause and approbation they have been justly attended with, that his reason for suppressing that performance being entirely ceased, we think the same may be very acceptable to the World. And to render it the more useful to our countrymen, we choose to give it in a careful translation rather than in the original Latin.

By the imperfection of our sense it comes to pass, that the appearances of the heavens and heavenly bodies are entirely different from what they really are in themselves. The Sun, the Moon, the other Planets and the fixed Stars, appear to the eye only like small circles or like points of light, at equal distances from us; sticking to the top of a blue hemispherical vault, not very far from the Earth, which it seems to touch on every side near the horizon. And this blue vault turning upon its axis, these bodies being carried round with it, rise and set alternately, and so present themselves in order to the several parts of the Earth.

But our reason assisted by geometry opens to us a very different scene. It shows us an immense ocean of celestial space spreading itself on all hands from our Earth without possibility that any limits should be set to it. That the several heavenly bodies that seem to our eye so small, and equally distant from us, are in reality of prodigious magnitudes, and placed at vast and very different intervals from the Earth and each other, in that boundless abyss. These things were long ago known and consented to by all; but the exact form in which these bodies and the Earth are situate with respect to each other in that immensity of space, was not well settled until about sixty years ago; at which time Astronomers generally agreed in the following System. That the Sun, though it seems to move, is indeed fixed and quiescent in a certain given part of the celestial regions. That at a great distance from thence that Planet we call Mercury goes round him in an orbit either circular or very nearly so, in the centre of which the Sun is. That at a greater distance the Planet Venus revolves about the Sun in like manner. That next to Venus the Earth goes round the Sun, turning at the same time upon its axis, and attended by the Moon; which revolves about the Earth as the Earth does about the Sun. That the other Planets, Mars, Jupiter and Saturn, move also round the Sun in like orbits with the others; the two last being observed to have Satellites or Moons revolving about them, as our Earth has its Moon. All this is represented in the common figures of the Solar or *Copernican* System.

But Astronomers did not stop here. They not only found the order in which the Planets move and the situation of their orbits; but the sagacious Kepler made also the following great discoveries concerning the laws of their motions. The first, that all the Planets, in moving round the Sun, describe by radii drawn to the Sun, areas proportional to the times; and, secondly, that the cubes of their distances increase in the same proportion as the squares of their periodical times, or that the distances and periodical times are in a sesquialteral ratio.

This was the state of the astronomy of the Planets, when Sir Isaac Newton wrote. As to the Comets, they were generally thought to move uniformly in right lines through the celestial spaces, and not to describe any orbit recurring into itself, or at all incurvated; and it was a prevailing opinion, that at the time those bodies were visible, their motions were performed in the regions beyond Saturn.

There remained one thing further to be known, which was, the cause which retains the Planets in those regular motions which they are found to revolve with. For since projected bodies, impelled by a single force, move by their own nature uniformly along the right lines in which the forces that impel them are directed, without any possibility of deviating from those rectilinear paths; it follows that bodies, like the Planets, that move in curve lines, must be acted on by more forces than one, or by some compound force, drawing them out of the right lines they naturally tend to describe. The question then was, of what nature that force is which is diffused through so vast an extent as that of the Planetary System, and which seems to be simple and uniform, by its uniform effects above-mentioned, of the proportionality of the areas and times; and the sesquialteral ratio of the distances and periods, said above to be collected from the observations.

To this end the Author did not frame hypotheses as other

philosophers used to do, but set himself to examine the phenomena themselves, by mathematical reasoning. And first of
all, he found that bodies describing areas proportional to the
times by radii drawn to any given point are impelled by two
forces; by one of which they continually endeavour to fly off in
the tangents of the orbits they describe, and by the other, which
continually impels them to the given point, they are withheld
from doing so, and retained in those curvilinear paths.

Since then the Planets by radii drawn to the Sun describe
areas proportional to the times, it follows that the compound
force which keeps them in their regular motions according to
that law, is compounded of two forces; one of which impels
them according to the tangents of their orbits, and the other
impels them toward the Sun; the first of which we may call the
Projectile, the other the Centripetal Force.

Now by this discovery we have made a great advance, having found, First, that this phenomenon of the motions of the
Planets may be accounted for by two such forces; and, Secondly, that the centripetal force is directed to one fixed and
given place, the Sun's centre; and not to any uncertain and
wandering places; which would have made our future speculations very intricate, if it had proved so. What the cause is of this
force, we do not yet pretend to determine; but our business is,
since such a force is found to exist, to search into the properties
and proportions of that force, before we think of inquiring into
the cause of it.

The next question is then what proportion two centripetal
forces actually bear to one another in two different Planets. For
forces being measurable quantities, are capable of being compared together as to proportion; which proportion may be
expressed by numbers; in the same manner as the proportions of
any two lines, weights, etc. And since these two forces actually
exist, and operate upon the Planets, they must of course bear
some certain proportion to each other, if that could be found
out. They must be either equal in all the Planets, or greater in the
nearest, and less in the more remote, or vice versa.

The investigation of these proportions is a mathematical
problem; which the Author solves by finding that if several bodies

revolve about a centre in such orbits as the Planets, and the times of their periods be to each other in a sesquialteral ratio, their centripetal forces will be reciprocally as the squares of the distances from the centre. Whence it follows that since the Planets move about the Sun in periods having that ratio to each other, their centripetal forces toward the Sun are reciprocally as the squares of their distances from the Sun.

There is therefore a virtue diffused throughout the whole Planetary System by which the Planets are obliged to tend on all sides toward the Sun placed in the common centre or focus of their orbits; and whose effect is always in the reciprocal duplicate ratio of the distances from the Sun.

This is a discovery for which we are beholden to the penetrating genius of our great Author. This virtue we will name an attractive virtue; though the effect thereof is more probably caused by an impulse from without the attracted body, rather than by any thing emitted from that which it is said to attract. But since the names of things are arbitrary, and after this caution premised, this name cannot mislead us, we use it for the convenience of brevity.

All the Planets may be considered as centres and sources of this attractive force which spreads itself about them on all sides, attracting toward them from all quarters, all bodies that are not hindered by a more forcible attraction to some other body. This is further confirmed by the revolution of secondary Planets about some of the primary (as about the Earth, Jupiter and Saturn), according to the same laws, which the primary observe about the Sun, as to the proportionality of times and areas, and the sesquialteral ratio of the distances and periods; and the same thing appears by the sensible effects of the Moon upon the Earth. All which demonstrate that an attractive virtue is propagated from those bodies, of the very same kind with that which is propagated from the Sun.

The sum therefore of these new inventions is contained in this great and universal truth, and its corollaries. That all bodies through the whole solar System (and therefore throughout the whole universe, as we have all the reason that can be to suppose) attract, or (which is the term used most generally) gravitate toward, each other with forces that are as their quantities of matter directly, and as the squares of their distances from each other reciprocally.

On this great foundation the Author builds two important discoveries, entirely new; the first relating to the Moon and the tides; the second relating to the motions of the Comets, and the figure of their orbits.

Some persons will probably be ready to inquire what is the cause of this hidden virtue of gravity which is here attributed to the heavenly bodies. To this the only answer is, that this cause is as yet one of Nature's secrets: and perhaps it will ever remain so. 'Tis only plain from mathematical reasoning and undoubted observation, that such a virtue exists and operates in the System of the Planets, and that according to the laws above mentioned. Till a better explication be found we may very well resolve it into the Will of the All-wise and Almighty Creator, who when He formed and disposed these bodies, by this invisible bond *made them fast for ever and ever: and gave them a law which shall not be broken.*

Of the System of the World

That the matter of the heavens is fluid

It was the ancient opinion of not a few, in the earliest ages of philosophy, that the fixed stars stood immovable in the highest parts of the world; that under the fixed stars the planets were carried about the sun; that the earth, as one of the planets, described an annual course about the sun, while by a diurnal motion it was in the meantime revolved about its own axis; and that the sun, as the common fire which served to warm the whole, was fixed in the centre of the universe.

This was the philosophy taught of old by Philolaus, Aristarchus of Samos, Plato in his riper years, and the whole sect of the Pythagoreans: and this was the judgement of Anaximander, more ancient still; and of that wise king of the Romans, Numa Pompilius, who, as a symbol of the figure of the world with the sun in the centre, erected a round temple in honour of Vesta, and ordained perpetual fire to be kept in the middle of it.

The Egyptians were early observers of the heavens; and from them, probably, this philosophy was spread abroad among other nations; for from them it was, and the nations about them, that the Greeks, a people more addicted to the study of philology than of Nature, derived their first, as well as soundest, notions of philosophy; and in the Vestal ceremonies we may yet trace the ancient spirit of the Egyptians; for it was their way to deliver their mysteries, that is, their philosophy of things above the common way of thinking, under the veil of religious rites and hieroglyphic symbols.

It is not to be denied that Anaxagoras, Democritus and others, did now and then start up, who would have it that the earth possessed the centre of the world, and that the stars were revolved toward the west about the earth quiescent in the centre, some at a swifter, others at a slower rate.

However, it was agreed on both sides that the motions of the celestial bodies were performed in spaces altogether free and void of resistance. The whim of solid orbs was of a later date, introduced by Eudoxus, Callippus and Aristotle; when the ancient philosophy began to decline, and to give place to the new prevailing fictions of the Greeks.

But, above all things, the phenomena of comets can by no means tolerate the idea of solid orbits. The Chaldeans, the most learned astronomers of their time, looked upon the comets (which of ancient times before had been numbered among the celestial bodies) as a particular sort of planets, which, describing eccentric orbits, presented themselves to view only by turns, once in a revolution, when they descended into the lower parts of their orbits.

And as it was the unavoidable consequence of the hypothesis of solid orbits, while it prevailed, that the comets should be thrust into spaces below the moon; so, when later observations of astronomers restored the comets to their ancient places in the higher heavens, these celestial spaces were necessarily cleared of the incumbrance of solid orbits, which by these observations were broke into pieces and discarded for ever.

The principle of circular motion in free spaces

Whence it was that the Planets came to be retained within any certain bounds in these free spaces, and to be drawn off from the rectilinear courses, which, left to themselves, they should have pursued, into regular revolutions in curvilinear orbits, are questions which we do not know how the ancients explained. And probably it was to give some sort of satisfaction to this difficulty that solid orbs had been introduced.

The later philosophers pretend to account for it either by the action of certain vortices, as Kepler and Descartes; or by some other principle of impulse or attraction, as Borelli, Hooke and others of our nation; for, from the laws of motion, it is most certain that these effects must proceed from the action of some force or other.

But our purpose is only to trace out the quantity and properties of this force from the phenomena, and to apply what we discover in some simple cases as principles, by which, in a mathematical way, we may estimate the effects thereof in more involved cases; for it would be endless and impossible to bring every particular to direct and immediate observation.

We said, in a mathematical way, to avoid all questions about the nature or quality of this force, which we would not be understood to determine by any hypothesis; and therefore call it by the general name of a centripetal force, as it is a force which is directed toward some centre; and as it regards more particularly a body in that centre, we call it circumsolar, circum-terrestrial, circumjovial; and so in respect of other central bodies.

The effects of centripetal forces

That by means of centripetal forces the planets may be retained in certain orbits, we may easily understand, if we consider the motions of projectiles for a stone that is projected is by the pressure of its own weight forced out of the rectilinear path, which by the initial projection alone it should have pursued, and made to describe a curved line in the air; and through that crooked way is at last brought down to the ground; and the greater the velocity is with which it is projected, the further it goes before it falls to the earth. We may therefore suppose the velocity to be so increased, that it would describe an arc of 1, 5, 10, 1,000 miles before it arrived at the earth, till at last, exceeding the limits of the earth, it should pass into space without touching it.

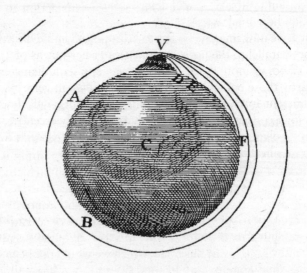

Let AFB represent the surface of the earth, C its centre, VD, VE, VF the curved lines which a body would describe, if projected in a horizontal direction from the top of a high mountain successively with more and more velocity, and, because the celestial motions are scarcely retarded by the little or no resistance of the spaces in which they are performed, to keep up the parity of cases, let us suppose either that there is no air about the earth, or at least that it is endowed with little or no power of resisting; and for the same reason that the body projected with a less velocity describes the lesser arc VD, and with a greater velocity the greater arc VE, and, augmenting the velocity, it goes further and further to F and G, if the velocity was still more and more augmented, it would reach at last quite beyond the circumference of the earth, and return to the mountain from which it was projected.

And since the areas which by this motion it describes by a radius drawn to the centre of the earth are proportional to the times in which they are described, its velocity, when it returns to the mountain, will be no less than it was at first; and, retaining the same velocity, it will describe the same curve over and over, by the same law.

But if we now imagine bodies to be projected in the directions of lines parallel to the horizon from greater heights, as of 5, 10, 100, 1,000, or more miles, or rather as many semi-diameters of the earth, those bodies, according to their different velocity, and the different force of gravity in different heights, will describe arcs either concentric with the earth, or variously eccentric, and go on revolving through the heavens in those orbits just as the planets do in their orbits.

The certainty of the argument

As when a stone is projected obliquely, that is, any way but in the perpendicular direction, the continual deflection thereof toward the earth from the right line in which it was projected is a proof of its gravitation to the earth, no less certain than its direct descent when suffered to fall freely from rest; so the deviation of bodies moving in free spaces from rectilinear paths, and continual deflection therefrom toward any place, is a sure indication of the existence of some force which from all quarters impels those bodies toward that place.

And as, from the supposed existence of gravity, it necessarily follows that all bodies about the earth must press downwards, and therefore must either descend directly to the earth, if they are let fall from rest, or at least continually deviate from right lines toward the earth, if they are projected obliquely; so, from the supposed existence of a force directed to any centre, it will follow, by the like necessity, that all bodies upon which this force acts must either descend directly to that centre, or at least deviate continually toward it from right lines, if otherwise they should have moved obliquely in these right lines.

And from the motions given we may infer the forces, or from the forces given we may determine the motions, by mathematical reasoning.

What follows from the supposed diurnal motion of the stars

If the earth is supposed to stand still, and the fixed stars to be revolved in free spaces in the space of twenty-four hours, it is certain the forces by which the fixed stars are held in their orbits are not directed to the earth, but to the centres of those orbits, that is, of the several parallel circles, which the fixed stars, declining to one side and the other from the equator, describe daily; also that by radii drawn to the centres of the orbits the fixed stars describe areas exactly proportional to the times of description. Then, because the periodic times are equal, it follows that the centripetal forces are as the radii of the several orbits, and that they will continually revolve in the same orbits. And the like consequences may be drawn from the supposed diurnal motion of the planets.

The incongruous consequences of this supposition

That forces should be directed to no body on which they physically depend, but to innumerable imaginary points in the axis of the earth, is an hypothesis too incongruous. It is more incongruous still that those forces should increase exactly in proportion of the distances from this axis; for this is an indication of an increase to immensity, or rather to infinity; whereas the forces of natural things commonly decrease in receding from the fountain from which they flow. But, what is yet more absurd, neither are the areas described by the same star proportional to the times, nor are its revolutions performed in the same orbit; for as the star recedes from the neighbouring pole, both areas and orbits increase; and from the increase of the area it is demonstrated that the forces are not directed to the axis of the earth. And this difficulty arises from the two-fold motion that is observed in the fixed stars, one diurnal round the axis of the earth, the other exceedingly slow round the axis of the ecliptic. And the explication thereof requires a composition of forces so involved and so variable, that it is hardly to be reconciled with any physical theory.

Frontispiece, Andrew Motte's English Translation of Book 1 of Newton's *Mathematical Principles of Natural Philosophy*, 1730

Newton's work continued to evoke a great deal of interest, excitement and controversy after his death. There were great debates about how to interpret his legacy, particularly about the idea of gravity as an attractive force operating at a distance, and about the relationship between the experimental and the mathematical components of the model for science that he had articulated. Yet there also slowly emerged a popular vision of Newton as the archetypal modern scientist, or even as an almost godlike genius – a vision that would persist in myths about Newton through the centuries. We can see this sort of veneration of Newton here, in the frontispiece to the very first English translation of the Principia *by Andrew Motte. Newton appears seated in heaven, receiving instruction from the muse of geometry. Below him, the clouds part, revealing a heliocentric system operating according to the laws he has articulated. The lines at the top are taken from the prefatory poem to the* Principia *by Halley. The image has much in common with the epitaph written for Newton by the poet Alexander Pope: 'Nature and Nature's laws lay hid in night: God said, Let Newton be! and all was light.'*

Voltaire, Frontispiece, *Elements of the Philosophy of Newton*, 1738

François-Marie Arouet (1694–1778), better known as Voltaire, was a French Enlightenment writer and an apostle of Newton's, who did his utmost to convert the French public away from the theories of Descartes and toward those of Newton. His French Elements of the Philosophy of Newton *conveyed those theories in popular language and helped spread both the ideas and the legend of Newton. On the frontispiece, we see Voltaire sitting at his desk. The manuscript on which he is working is illuminated by light reflected in a mirror, which originates behind parted clouds. At the opening of the clouds from which the light emerges sits Newton in a godlike pose. He gazes at the woman holding the mirror, who may represent the Marquise Émilie du Châtelet, who translated the* Principia *into French and published her own works on physics; she was also a collaborator of Voltaire's, with whom she was in a relationship for many years.*

James Ferguson, *Astronomy, Explained Upon Sir Isaac Newton's Principles, and Made Easy to Those Who Have Not Studied Mathematics*, 1756

James Ferguson (1710–1776) was a Scottish astronomer who was a successful writer, speaker and inventor of astronomical instruments, despite having no formal education or training. He was one of the most popular scientific lecturers of his day, and this work promoting and simplifying the astronomy of Newton was an immediate success, earning him election to the Royal Society of London. When the first edition of the Encyclopaedia Britannica *was published, in three volumes between 1768 and 1771, its entry on 'Astronomy' was essentially a reprint of Ferguson's work, including many of its plates. Thus, a casual reader in England in 1768 who wished to understand the consensus about astronomy at that time would have read the words below. Three of the book's thirteen plates, which were also included in the* Encyclopaedia, *are reproduced below.*

CHAP. I. Of Astronomy in General

The general use of astronomy.

1. Of all the sciences cultivated by mankind, Astronomy is acknowledged to be, and undoubtedly is, the most sublime, the most interesting and the most useful. For, by knowledge derived from this science, not only the bulk of the Earth is discovered, the situation and extent of the countries and kingdoms upon it ascertained, trade and commerce carried on to the remotest

parts of the world, and the various products of several countries distributed for the health, comfort and conveniency of its
inhabitants; but our very faculties are enlarged with the grandeur of the ideas it conveys, our minds exalted above the low
contracted prejudices of the vulgar, and our understandings
clearly convinced, and affected with the conviction, of the
existence, wisdom, power, goodness and superintendency of
the SUPREME BEING! So that without an hyperbole, 'An
undevout Astronomer is mad.'

2. From this branch of knowledge we also learn by what means
or laws the Almighty carries on, and continues the admirable
harmony, order and connexion observable throughout the planetary system; and are led by very powerful arguments to form
the pleasing deduction, that minds capable of such deep
researches not only derive their origin from that adorable Being,
but are also incited to aspire after a more perfect knowledge of
His nature, and a stricter conformity to His will.

The Earth but a point as seen from the Sun.

3. By Astronomy we discover that the Earth is at so great a distance from the Sun, that if seen from thence it would appear no
bigger than a point; although its circumference is known to be
25,020 miles. Yet that distance is so small, compared with the
distance of the Fixed Stars, that if the Orbit in which the Earth
moves round the Sun were solid, and seen from the nearest
Star, it would likewise appear no bigger than a point, although
it is at least 162 millions of miles in diameter. For the Earth in
going round the Sun is 162 millions of miles nearer to some of
the Stars at one time of the year than at another; and yet their
apparent magnitudes, situations and distances from one
another still remain the same; and a telescope which magnifies
above 200 times does not sensibly magnify them: which proves
them to be at least 400 thousand times further from us than we
are from the Sun.

The Stars are Suns.

4. It is not to be imagined that all the Stars are placed in one
concave surface, so as to be equally distant from us; but that

they are scattered at immense distances from one another through unlimited space. So that there may be as great a distance between any two neighbouring Stars, as between our Sun and those which are nearest to him. Therefore an Observer, who is nearest any fixed Star, will look upon it alone as a real Sun; and consider the rest as so many shining points, placed at equal distances from him in the Firmament.

And innumerable.

5. By the help of telescopes we discover thousands of Stars which are invisible to the naked eye; and the better our glasses are, still the more become visible: so that we can set no limits either to their number or their distances. The celebrated HUY-GENS carries his thoughts so far, as to believe it not impossible that there may be Stars at such inconceivable distances, that their light has not yet reached the Earth since its creation; although the velocity of light be a million of times greater than the velocity of a cannon bullet, as shall be demonstrated afterwards: and, as Mr ADDISON very justly observes, this thought is far from being extravagant, when we consider that the Universe is the work of infinite power, prompted by infinite goodness; having an infinite space to exert itself in; so that our imaginations can set no bounds to it.

Why the Sun appears bigger than the Stars.

6. The Sun appears very bright and large in comparison of the Fixed Stars, because we keep constantly near the Sun, in comparison of our immense distance from the Stars. For, a spectator, placed as near to any Star as we are to the Sun, would see that Star a body as large and bright as the Sun appears to us: and a spectator, as far distant from the Sun as we are from the Stars, would see the Sun as small as we see a Star, divested of all its circumvolving Planets; and would reckon it one of the Stars in numbering them.

The Stars are not enlightened by the Sun.

7. The Stars, being at such immense distances from the Sun, cannot possibly receive from him so strong a light as they seem to

have; nor any brightness sufficient to make them visible to us. For the Sun's rays must be so scattered and dissipated before they reach such remote objects, that they can never be transmitted back to our eyes, so as to render these objects visible by reflection. The Stars therefore shine with their own native and unborrowed lustre, as the Sun does; and since each particular Star, as well as the Sun, is confined to a particular portion of space, 'tis plain that the Stars are of the same nature with the Sun.

They are probably surrounded by Planets.

8. It is no ways probable that the Almighty, who always acts with infinite wisdom and does nothing in vain, should create so many glorious Suns, fit for so many important purposes, and place them at such distances from one another, without proper objects near enough to be benefited by their influences. Whoever imagines they were created only to give a faint glimmering light to the inhabitants of this Globe, must have a very superficial knowledge of Astronomy, and a mean opinion of the Divine Wisdom: since, by an infinitely less exertion of creating power, the Deity could have given our Earth much more light by one single additional Moon.

9. Instead then of one Sun and one World only in the Universe, as the unskilful in Astronomy imagine, *that* Science discovers to us such an inconceivable number of Suns, Systems and Worlds, dispersed through boundless Space, that if our Sun, with all the Planets, Moons and Comets belonging to it were annihilated, they would be no more missed out of the Creation than a grain of sand from the seashore. The space they possess being comparatively so small, that it would scarce be a sensible blank in the Universe; although Saturn, the outermost of our planets, revolves about the Sun in an Orbit of 4,884 millions of miles in circumference, and some of our Comets make excursions upwards of ten thousand millions of miles beyond Saturn's Orbit; and yet, at that amazing distance, they are incomparably nearer to the Sun than to any of the Stars; as is evident from their keeping clear of the attractive Power of all the Stars, and returning periodically by virtue of the Sun's attraction.

The stellar Planets may be habitable.

10. From what we know of our own System it may be reasonably concluded that all the rest are with equal wisdom contrived, situated and provided with accommodations for rational inhabitants. Let us therefore take a survey of the System to which we belong; the only one accessible to us; and from thence we shall be the better enabled to judge of the nature and end of the other Systems of the Universe. For although there is almost an infinite variety in all the parts of the Creation which we have opportunities of examining; yet there is a general analogy running through and connecting all the parts into one scheme, one design, one whole!

As our Solar Planets are.

11. And then, to an attentive considerer, it will appear highly probable, that the Planets of our System, together with their attendants called Satellites or Moons, are much of the same nature with our Earth, and destined for the like purposes. For, they are solid opaque Globes, capable of supporting animals and vegetables. Some of them are bigger, some less, and some much about the size of our Earth. They all circulate round the Sun, as the Earth does, in a shorter or longer time according to their respective distances from him: and have, where it would not be inconvenient, regular returns of summer and winter, spring and autumn. They have warmer and colder climates, as the various productions of our Earth require: and, in such as afford a possibility of discovering it, we observe a regular motion round their Axes like that of our Earth, causing an alternate return of day and night; which is necessary for labour, rest and vegetation, and that all parts of their surfaces may be exposed to the rays of the Sun.

The furthest from the Sun have most Moons to enlighten their nights.

12. Such of the Planets as are furthest from the Sun, and therefore enjoy least of his light, have that deficiency made up by several Moons, which constantly accompany, and revolve about them, as our Moon revolves about the Earth. The remotest Planet has, over and above, a broad Ring encompassing it;

which like a lucid Zone in the Heavens reflects the Sun's light very copiously on that Planet: so that if the remoter Planets have the Sun's light fainter by day than we, they have an addition made to it morning and evening by one or more of their Moons, and a greater quantity of light in the night-time.

Our Moon mountainous like the Earth.

13. On the surface of the Moon, because it is nearer us than any other of the celestial Bodies are, we discover a nearer resemblance of our Earth. For, by the assistance of telescopes we observe the Moon to be full of high mountains, large valleys and deep cavities. These similarities leave us no room to doubt but that all the Planets and Moons in the System are designed as commodious habitations for creatures endowed with capacities of knowing and adoring their beneficent Creator.

14. Since the Fixed Stars are prodigious spheres of fire, like our Sun, and at inconceivable distances from one another, as well as from us, it is reasonable to conclude they are made for the same purposes that the Sun is; each to bestow light, heat and vegetation on a certain number of inhabited Planets, kept by gravitation within the sphere of its activity.

Numberless Suns and Worlds.

15. What an august! what an amazing conception, if human imagination can conceive it, does this give of the works of the Creator! Thousands of thousands of Suns, multiplied without end, and ranged all around us, at immense distances from each other, attended by ten thousand times ten thousand Worlds, all in rapid motion, yet calm, regular and harmonious, invariably keeping the paths prescribed them; and these Worlds peopled with myriads of intelligent beings, formed for endless progression in perfection and felicity.

16. If so much power, wisdom, goodness and magnificence is displayed in the material Creation, which is the least considerable part of the Universe, how great, how wise, how good must HE be, who made and governs the Whole!

CHAP. III. The Copernican
System Demonstrated to Be True

Of matter and motion.

99. Matter is of itself inactive, and indifferent to motion or
rest. A body at rest can never put itself in motion; a body in
motion can never stop nor move slower of itself. Hence, when
we see a body in motion we conclude some other substance
must have given it that motion; when we see a body fall from
motion to rest we conclude some other body or cause stoped it.
100. All motion is naturally rectilineal. A bullet thrown by the
hand, or discharged from a cannon, would continue to move in
the same direction it received at first, if no other power diverted
its course. Therefore, when we see a body moving in a curve of
whatever kind, we conclude it must be acted upon by two powers at least: one to put it in motion, and another drawing it off
from the rectilineal course which it would otherwise have continued to move in.

Gravity demonstrable.

101. The power by which bodies fall toward the Earth is
called *Gravity* or *Attraction*. By this power in the Earth it is,
that all bodies, on whatever side, fall in lines perpendicular to
its surface. On opposite parts of the Earth bodies fall in opposite directions, all toward the centre where the force of gravity
is as it were accumulated. By this power constantly acting on
bodies near the Earth they are kept from leaving it altogether;
and those on its surface are kept thereto on all sides, so that
they cannot fall from it. Bodies thrown with any obliquity are
drawn by this power from a straight line into a curve, until they
fall to the Ground: the greater the force by which they are
thrown, the greater is the distance they are carried before they
fall. If we suppose a body carried several miles above the Earth,
and there projected in an horizontal direction, with so great a
velocity that it would move more than a semidiameter of the
Earth, in the time it would take to fall to the Earth by gravity;

in that case, if there were no resisting medium in the way, the body would not fall to the Earth at all; but continue to circulate round the Earth, keeping always the same path, and returning to the point from whence it was projected, with the same velocity as at first.

Projectile force demonstrable.

102. We find the Moon moves round the Earth in an Orbit nearly circular. The Moon therefore must be acted on by two powers or forces; one which would cause her to move in a right line, another bending her motion from that line into a curve. This attractive power must be seated in the Earth; for there is no other body within the Moon's Orbit to draw her. The attractive power of the Earth therefore extends to the Moon; and, in combination with her projectile force, causes her to move round the Earth in the same manner as the circulating body above supposed.

The Sun and Planets attract each other.

103. The Moons of Jupiter and Saturn are observed to move round their primary Planets: therefore there is such a power as gravity in these Planets. All the Planets move round the Sun, and respect it for their centre of motion: therefore the Sun must be endowed with attracting force, as well as the Earth and Planets. The like may be proved of the Comets. So that all the bodies or matter in the Solar System are possessed of this power; and perhaps so is all matter whatsoever.

104. As the Sun attracts the Planets with their Satellites, and the Earth the Moon, so the Planets and Satellites re-attract the Sun, and the Moon the Earth: action and reaction being always equal. This is also confirmed by observation; for the Moon raises tides in the ocean, the Satellites and Planets disturb one another's motions.

105. Every particle of matter being possessed of an attracting power, the effect of the whole must be in proportion to the number of attracting particles: that is, to the quantity of matter in the body. This is demonstrated from experiments on pendulums: for, if they are of equal lengths, whatever their weights

be, they always vibrate in equal times. Now, if one be double the weight of another, the force of gravity or attraction must be double to make it oscillate with the same celerity: if one is thrice the weight or quantity of matter of another, it requires thrice the force of gravity to make it move with the same celerity. Hence it is certain, that the power of gravity is always proportional to the quantity of matter in bodies, whatever their bulks or figures are.

106. Gravity also, like all other virtues or emanations issuing from a centre, decreases as the square of the distance increases: that is, a body at twice the distance attracts another with only a fourth part of the force; at four times the distance, with a sixteenth part of the force. This too is confirmed from observation, by comparing the distance which the Moon falls in a minute from a right line touching her Orbit, with the space which bodies near the Earth fall in the same time: and also by comparing the forces which retain Jupiter's Moons in their Orbits. This will be more fully explained in the seventh Chapter.

Gravitation and projection exemplified.

107. The mutual attraction of bodies may be exemplified by a boat and a ship on the Water, tied by a rope. Let a man either in the ship or boat pull the rope (it is the same in effect at which end he pulls, for the rope will be equally stretched throughout), the ship and boat will be drawn towards one another; but with this difference, that the boat will move as much faster than the ship as the ship is heavier than the boat. Suppose the boat as heavy as the ship, and they will draw one another equally (setting aside the greater resistance of the Water on the bigger body) and meet in the middle of the first distance between them. If the ship is a thousand or ten thousand times heavier than the boat, the boat will be drawn a thousand or ten thousand times faster than the ship; and meet proportionably nearer the place from which the ship set out. Now, whilst one man pulls the rope, endeavouring to bring the ship and boat together, let another man, in the boat, endeavour to row her off sidewise, or at right Angles to the rope; and the former, instead of being able to draw the boat to the ship, will find it enough for him to keep the boat from going

further off; whilst the latter, endeavouring to row off the boat in a straight line, will, by means of the other's pulling it toward the ship, row the boat round the ship at the rope's length from her. Here, the power employed to draw the ship and boat to one another represents the mutual attraction of the Sun and Planets, by which the Planets would fall freely toward the Sun with a quick motion; and would also in falling attract the Sun toward them. And the power employed to row off the boat represents the projectile force impressed on the Planets at right Angles, or nearly so, to the Sun's attraction; by which means the Planets move round the Sun, and are kept from falling to it. On the other hand, if it be attempted to make a heavy ship go round a light boat, they will meet sooner than the ship can get round; or the ship will drag the boat after it.

The absurdity of supposing the Earth at rest.

108. Let the above principles be applied to the Sun and Earth; and they will evince, beyond a possibility of doubt, that the Sun, not the Earth, is the centre of the System; and that the Earth moves round the Sun as the other Planets do.

For, if the Sun moves about the Earth, the Earth's attractive power must draw the Sun toward it from the line of projection so, as to bend its motion into a curve; and the Earth being at least 169 thousand times lighter than the Sun, by being so much less as to its quantity of matter, must move 169 thousand times faster toward the Sun than the Sun does toward the Earth; and consequently would fall to the Sun in a short time if it had not a very strong projectile motion to carry it off. The Earth therefore, as well as every other Planet in the System, must have a rectilineal impulse to prevent its falling into the Sun. To say, that gravitation retains all the other Planets in their Orbits without affecting the Earth, which is placed between the Orbits of Mars and Venus, is as absurd as to suppose that six cannon bullets might be projected upwards to different heights in the Air, and that five of them should fall down to the ground; but the sixth, which is neither the highest nor the lowest, should remain suspended in the Air without falling; and the Earth move round about it.

109. There is no such thing in nature as a heavy body moving round a light one as its centre of motion. A pebble fastened to a millstone by a string, may by an easy impulse be made to circulate round the millstone: but no impulse can make a millstone circulate round a loose pebble, for the heaviest would undoubtedly carry the lightest along with it wherever it goes.

110. The Sun is so immensely bigger and heavier than the Earth, that if he was moved out of his place, not only the Earth, but all the other Planets if they were united into one mass, would be carried along with the Sun as the pebble would be with the millstone.

The harmony of the celestial motions

The absurdity of supposing the Stars and Planets to move round the Earth.

111. By considering the law of gravitation, which takes place throughout the Solar System, in another light, it will be evident that the Earth moves round the Sun in a year; and not the Sun round the Earth. It has been shewn that the power of gravity decreases as the square of the distance increases: and from this it follows with mathematical certainty, that when two or more bodies move round another as their centre of motion, the squares of their periodic times will be to one another in the same proportion as the cubes of their distances from the central body. This holds precisely with regard to the Planets round the Sun, and the Satellites round the Planets; the relative distances of all which, are well known. But, if we suppose the Sun to move round the Earth, and compare its period with the Moon's by the above rule, it will be found that the Sun would take no less than 173,510 days to move round the Earth, in which case our year would be 475 times as long as it now is. To this we may add, that the aspects of increase and decrease of the Planets, the times of their seeming to stand still, and to move direct and retrograde, answer precisely to the Earth's motion; but not at all to the Sun's without introducing the most absurd and monstrous suppositions, which would destroy all harmony, order and simplicity in

the System. Moreover, if the Earth is supposed to stand still, and the Stars to revolve in free spaces about the Earth in twenty-four hours, it is certain that the forces by which the Stars revolve in their Orbits are not directed to the Earth, but to the centres of the several Orbits: that is, of the several parallel Circles which the Stars on different sides of the Equator describe every day: and the like inferences may be drawn from the supposed diurnal motion of the Planets, since they are never in the Equinoctial but twice, in their courses with regard to the starry Heavens. But, that forces should be directed to no central body, on which they physically depend, but to innumerable imaginary points in the axe of the Earth produced to the Poles of the Heavens, is an hypothesis too absurd to be allowed of by any rational creature. And it is still more absurd to imagine that these forces should increase exactly in proportion to the distances from this axe; for this is an indication of an increase to infinity: whereas the force of attraction is found to decrease in receding from the fountain from whence it flows. But, the further that any Star is from the quiescent Pole the greater must be the Orbit which it describes; and yet it appears to go round in the same time as the nearest Star to the Pole does. And if we take into consideration the two-fold motion observed in the Stars, one diurnal round the Axis of the Earth in twenty-four hours, and the other round the Axis of the Ecliptic in 25,920 years, it would require an explication of such a perplexed composition of forces, as could by no means be reconciled with any physical Theory.

Objections against the Earth's motion answered.

112. There is but one objection of any weight that can be made to the Earth's motion round the Sun; which is, that in opposite points of the Earth's Orbit, its Axis which always keeps a parallel direction would point to different fixed Stars; which is not found to be fact. But this objection is easily removed by considering the immense distance of the Stars in respect of the diameter of the Earth's Orbit; the latter being no more than a point when compared to the former. If we lay a ruler on the side of a table, and along the edge of the ruler view the top of a spire at ten miles distance; then lay the ruler on the opposite

side of the table in a parallel situation to what it had before, and the spire will still appear along the edge of the ruler; because our eyes, even when assisted by the best instruments, are incapable of distinguishing so small a change.

113. Dr BRADLEY, our present Astronomer Royal, has found by a long series of the most accurate observations, that there is a small apparent motion of the fixed Stars, occasioned by the aberration of their light, and so exactly answering to an annual motion of the Earth, as evinces the same, even to a mathematical demonstration. Those who are qualified to read the Doctor's modest Account of this great discovery may consult the *Philosophical Transactions*, N° 406.

Why the Sun appears to change his place.

114. It is true that the Sun seems to change his place daily, so as to make a tour round the starry Heavens in a year. But whether the Earth or Sun moves, this appearance will be the same; for, when the Earth is in any part of the Heavens, the Sun will appear in the opposite. And therefore, this appearance can be no objection against the motion of the Earth.

115. It is well known to every person who has sailed on smooth Water, or been carried by a stream in a calm, that however fast the vessel goes he does not feel its progressive motion. The motion of the Earth is incomparably more smooth and uniform than that of a ship, or any machine made and moved by human art: and therefore it is not to be imagined that we can feel its motion.

The Earth's motion on its Axis demonstrated.

116. We find that the Sun, and those Planets on which there are visible spots, turn round their Axes: for the spots move regularly over their Discs. From hence we may reasonably conclude that the other Planets on which we see no spots, and the Earth which is likewise a Planet, have such rotations. But being incapable of leaving the Earth, and viewing it at a distance; and its rotation being smooth and uniform, we can neither see it move on its Axis as we do the Planets, nor feel ourselves affected by its motion. Yet there is one effect of such a motion which will

enable us to judge with certainty whether the Earth revolves on its Axis or not. All Globes which do not turn round their Axes will be perfect spheres, on account of the equality of the weight of bodies on their surfaces; especially of the fluid parts. But all Globes which turn on their Axes will be oblate spheroids; that is, their surfaces will be higher, or further from the centre, in the equatoreal than in the polar Regions: for, as the equatoreal parts move quickest, they will recede further from the Axis of motion, and enlarge the equatoreal diameter. That our Earth is really of this figure is demonstrable from the unequal vibrations of a pendulum, and the unequal lengths of degrees in different latitudes. Since then, the Earth is higher at the Equator than at the Poles, the sea, which naturally runs downward, or toward the places which are nearest the centre, would run toward the polar Regions, and leave the equatoreal parts dry, if the centrifugal force of these parts did not raise and carry the waters thither. The Earth's equatoreal diameter is 35 miles longer than its Axis.

All bodies heavier at the Poles than they would be at the Equator.

117. Bodies near the Poles are heavier than those toward the Equator, because they are nearer the Earth's centre, where the whole force of the Earth's attraction is accumulated. They are also heavier because their centrifugal force is less on account of their diurnal motion being slower. For both these reasons, bodies carried from the Poles toward the Equator, gradually lose of their weight. Experiments prove that a pendulum, which vibrates seconds near the Poles vibrates slower near the Equator, which shews that it is lighter or less attracted there. To make it oscillate in the same time, 'tis found necessary to diminish its length. By comparing the different lengths of pendulums swinging seconds at the Equator and at *London*, it is found that a pendulum must be $2^{169}/_{1000}$ lines shorter at the Equator than at the Poles. A line is a twelfth part of an inch.

How they might lose all their weight.

118. If the Earth turned round its Axis in 84 minutes 43 seconds, the centrifugal force would be equal to the power of gravity at the Equator; and all bodies there would entirely lose their weight. If the Earth revolved quicker they would all fly off, and leave it.

The Earth's motion cannot be felt.

119. One on the Earth can no more be sensible of its undisturbed motion on its Axis, than one in the cabin of a ship on smooth Water can be sensible of her motion when she turns gently and uniformly round. It is therefore no argument against the Earth's diurnal motion that we do not feel it: nor is the apparent revolutions of the celestial bodies every day a proof of the reality of these motions; for whether we or they revolve, the appearance is the very same. A person looking through the cabin windows of a ship as strongly fancies the objects on land to go round when the ship turns, as if they were actually in motion.

To the different Planets, the Heavens appear to turn round on different Axes.

120. If we could translate ourselves from Planet to Planet, we should still find that the Stars would appear of the same magnitudes, and at the same distances from each other, as they do to us here; because the width of the remotest Planet's Orbit bears no sensible proportion to the distance of the Stars. But then, the Heavens would seem to revolve about very different Axes; and consequently, those quiescent Points which are our Poles in the Heavens would seem to revolve about other points, which, though apparently in motion to us on Earth would be at rest as seen from any other Planet. Thus, the Axis of Venus, which lies almost at right Angles to the Axis of the Earth, would have its motionless Poles in two opposite points of the Heavens lying almost in our Equinoctial, where the motion appears quickest because it is performed in the greatest Circle. And the very Poles, which are at rest to us, have the quickest motion of all as seen from Venus. To Mars and Jupiter the Heavens appear to turn

round with very different velocities on the same Axis, whose Poles are about 23½ degrees from ours. Were we on Jupiter we should be at first amazed at the rapid motion of the Heavens; the Sun and Stars going round in 9 hours 56 minutes. Could we go from thence to Venus we should be as much surprised at the slowness of the heavenly motions: the Sun going but once round in 584 hours, and the Stars in 540. And could we go from Venus to the Moon we should see the Heavens turn round with a yet slower motion; the Sun in 708 hours, the Stars in 655. As it is impossible these various circumvolutions in such different times and on such different Axes can be real, so it is unreasonable to suppose the Heavens to revolve about our Earth more than it does about any other Planet. When we reflect on the vast distance of the fixed Stars, to which 162,000,000 of miles is but a point, we are filled with amazement at the immensity of their distance. But if we try to frame an idea of the extreme rapidity with which the Stars must move, if they move round the Earth in twenty-four hours, the thought becomes so much too big for our imagination, that we can no more conceive it than we do infinity or eternity. If the Sun was to go round the Earth in a day, he must travel upwards of 300,000 miles in a minute: but the Stars being at least 10,000 times as far as the Sun from us, those about the Equator must move 10,000 times as quick. And all this to serve no other purpose than what can be as fully and much more simply obtained by the Earth's turning round eastward as on an Axis, every twenty-four hours, causing thereby an apparent diurnal motion of the Sun westward, and bringing about the alternate returns of day and night.

121. As to the common objections against the Earth's motion on its Axis, they are all easily answered and set aside. That it may turn without being seen or felt to do so, has been already shewn. But some are apt to imagine that if the Earth turns eastward (as it certainly does if it turns at all) a ball fired perpendicularly upward in the air must fall considerably westward of the place it was projected from. This objection, which at first seems to have some weight, will be found to have none at all when we consider that the gun and ball partake of the Earth's motion; and therefore the ball being carried forward with the air as quick as the

Earth and air turn, must fall down again on the same place. A stone let fall from the top of a mainmast, if it meets with no obstacle, falls on the deck as near the foot of the mast when the ship sails as when it does not. And if an inverted bottle, full of liquor, be hung up to the ceiling of the cabin, and a small hole be made in the cork to let the liquor drop through to the floor, the drops will fall just as far forward on the floor when the ship sails as when it is at rest. And gnats or flies can as easily dance among one another in a moving cabin as in a fixed chamber. As for those Scripture expressions which seem to contradict the Earth's motion, this general answer may be made to them all, *viz.* 'tis plain from many instances that the Scriptures were never intended to instruct us in Philosophy or Astronomy; and there-fore, on those subjects, expressions are not always to be taken in the strictest sense; but for the most part as accommodated to the common apprehensions of mankind. Men of sense in all ages, when not treating of the sciences purposely, have followed this method: and it would be in vain to follow any other in address-ing ourselves to the vulgar, or bulk of any community. *Moses* calls the Moon A GREAT LUMINARY (as it is in the Hebrew) as well as the Sun: but the Moon is known to be an opaque body, and the smallest that Astronomers have observed in the Heavens and shines upon us not by any inherent light of its own, but by reflecting the light of the Sun. If *Moses* had known this, and told the *Israelites* so, they would have stared at him; and considered him rather as a madman than as a person commissioned by the Almighty to be their leader.

CHAP. VII. The Physical Causes of the Motions of the Planets. The Excentricities of Their Orbits. The Times in Which the Action of Gravity Would Bring Them to the Sun. Archimedes's Ideal Problem for Moving the Earth. The World Not Eternal.

Gravitation and Projection.

150. From the uniform projectile motion of bodies in straight lines, and the universal power of attraction, arises the curvilineal motions of all the Heavenly bodies. If the body A be projected along the right line ABX, in open Space, where it meets with no resistance, and is not drawn aside by any other power, it will for ever go on with the same velocity, and in the same direction. For, the force which moves it from A to B in any given time, will carry it from B to X in as much more time; and so on, there being nothing to obstruct or alter its motion. But if, when this projectile force has carried it, suppose to B, the body S begins to attract it, with a power duly adjusted, and perpendicular to its motion at B, it will then be drawn from the straight line ABX, and forced to revolve about S in the Circle BYTU. When the body A comes to U, or any other part of its Orbit, if the small body u, within the sphere of U's attraction, be projected as in the right line Z, with a force perpendicular to the attraction of U, then u will go round U in the Orbit W, and accompany it in its whole course round the body S. Here, S may represent the Sun, U the Earth, and u the Moon.

151. If a Planet at B gravitates, or is attracted, toward the Sun, so as to fall from B to y in the time that the projectile force would have carried it from B to X, it will describe the curve BY by the combined action of these two forces, in the same time that the projectile force singly would have carried it from B to X, or the gravitating power singly have caused it to descend from B to y; and these two forces being duly proportioned, and perpendicular to one another, the Planet obeying them both, will move in the circle BYTU.

152. But if, whilst the projectile force carries the Planet from B to b, the Sun's attraction (which constitutes the Planet's gravitation) should bring it down from B to I, the gravitating power would then be too strong for the projectile force; and would cause the Planet to describe the curve BC. When the Planet comes to C, the gravitating power (which always increases as the square of the distance from the Sun S diminishes) will be yet stronger for the projectile force; and by conspiring in some degree therewith, will accelerate the Planet's motion all the way from C to K; causing it to describe the arcs BC, CD, DE, EF, &c. all in equal times. Having its motion thus accelerated, it gains so much centrifugal force, or tendency to fly off at K in the line Kk, as overcomes the Sun's attraction: and the centrifugal force being too great to allow the Planet to be brought nearer the Sun, or even to move round him in the Circle Klmn, &c. it goes off, and ascends in the curve KLMN, &c. its motion decreasing as gradually from K to B as it increased from B to K, because the Sun's attraction acts now against the Planet's projectile motion just as much as it acted with it before. When the Planet has got round to B, its projectile force is as much diminished from its mean state about G or N, as it was augmented at K; and so, the Sun's attraction being more than sufficient to keep the Planet from going off at B, it describes the same Orbit over again, by virtue of the same forces or laws.

The Planets describe equal Areas in equal times.

153. A double projectile force will always balance a quadruple power of gravity. Let the Planet at B have twice as great an impulse from thence towards X, as it had before: that is, in the same length of time that it was projected from B to b, as in the last example, let it now be projected from B to c; and it will require four times as much gravity to retain it in its Orbit: that is, it must fall as far as from B to 4 in the time that the projectile force would carry it from B to c; otherwise it could not describe the curve BD, as is evident by the Figure. But, in as much time as the Planet moves from B to C in the higher part of its Orbit, it moves from I to K or from K to L in the lower part thereof; because, from the joint action of these two forces,

it must always describe equal areas in equal times, throughout its annual course. These Areas are represented by the triangles BSC, CSD, DSE, ESF, &c. whose contents are equal to one another, quite round the Figure.

A difficulty removed.

154. As the Planets approach nearer the Sun, and recede further from him, in every Revolution; there may be some difficulty in conceiving the reason why the power of gravity, when it once gets the better of the projectile force, does not bring the Planets nearer and nearer the Sun in every Revolution, till they fall upon and unite with him. Or why the projectile force, when it once gets the better of gravity, does not carry the Planets further and further from the Sun, till it removes them quite out of the sphere of his attraction, and causes them to go on in straight lines for ever afterward. But by considering the effects of these powers as described in the two last Articles, this difficulty will be removed. Suppose a Planet at B to be carried by the projectile force as far as from B to b, in the time that gravity would have brought it down from B to 1: by these two forces it will describe the curve BC. When the Planet comes down to K, it will be but half as far from the Sun S as it was at B; and therefore, by gravitating four times as strongly toward him, it would fall from K to V in the same length of time that it would have fallen from B to 1 in the higher part of its Orbit, that is, through four times as much space; but its projectile force is then so much increased at K, as would carry it from K to k in the same time; being double of what it was at B, and is therefore too strong for the tendency of the gravitating power, either to draw the Planet to the Sun, or cause it to go round him in the circle Klmn, &c. which would require its falling from K to w, through a greater space than gravity can draw it whilst the projectile force is such as would carry it from K to k: and therefore the Planet ascends in its Orbit KLMN, decreasing in its velocity for the cause already assigned.

The Planetary Orbits Elliptical

Their Excentricities.

155. The Orbits of all the Planets are Ellipses, very little different from Circles: but the Orbits of the Comets are very long Ellipses; the lower focus of them all being in the Sun. If we suppose the mean distance (or middle between the greatest and least) of every Planet and Comet from the Sun to be divided into 1,000 equal parts, the Excentricities of their Orbits, both in such parts and in *English* miles, will be as follows. Mercury's, 210 parts, or 6,720,000 miles; Venus's, 7 parts, or 413,000 miles; the Earth's, 17 parts, or 1,377,000 miles; Mars's, 93 parts, or 11,439,000 miles; Jupiter's, 48 parts, or 20,352,000 miles; Saturn's, 55 parts, or 42,735,000 miles. Of the nearest of the three forementioned Comets, 1,458,000 miles; of the middlemost, 2,025,000,000 miles; and of the outermost, 6,600,000,000.

Hard to determine what Gravity is.

160. The Sun and Planets mutually attract each other: the power by which they do so we call *Gravity*. But whether this power be mechanical or no, is very much disputed. We are certain that the Planets disturb one another's motions by it, and that it decreases according to the squares of the distances of the Sun and Planets; as light, which is known to be material, likewise does. Hence Gravity should seem to arise from the agency of some subtle matter pressing toward the Sun and Planets, and acting, like all mechanical causes, by contact. But on the other hand, when we consider that the degree or force of Gravity is exactly in proportion to the quantities of matter in those bodies, without any regard to their bulks or quantity of surface, acting as freely on their internal as external parts, it seems to surpass the power of mechanism; and to be either the immediate agency of the Deity, or effected by a law originally established and imprest on all matter by him. But some affirm that matter, being altogether inert, cannot be impressed with any Law, even by almighty Power: and that the Deity must therefore be constantly impelling the Planets toward the Sun, and moving them with

the same irregularities and disturbances which Gravity would cause, if it could be supposed to exist. But, if a man may venture to publish his own thoughts (and why should not one as well as another?), it seems to me no greater absurdity, to suppose the Deity capable of superadding a Law, or what Laws He pleases, to matter, than to suppose Him capable of giving it existence at first. The manner of both is equally inconceivable to us; but neither of them imply a contradiction in our ideas: and what implies no contradiction is within the power of Omnipotence. Do we not see that a human creature can prepare a bar of steel so as to make it attract needles and filings of iron; and that he can put a stop to that power or virtue, and again call it forth again as often as he pleases? To say that the workman infuses any new power into the bar, is saying too much; since the needle and filings, to which he has done nothing, re-attract the bar. And from this it appears that the power was originally imprest on the matter of which the bar, needle and filings are composed; but does not seem to act until the bar be properly prepared by the artificer: somewhat like a rope coiled up in a ship, which will never draw a boat or any other thing toward the ship, unless one end be tied to it, and the other end to that which is to be hauled up; and then it is no matter which end of the rope the sailors pull at, for the rope will be equally stretched throughout, and the ship and boat will move toward one another. To say that the Almighty has infused no such virtue or power into the materials which compose the bar, but that He waits till the operator be pleased to prepare it by due position and friction, and then, when the needle or filings are brought pretty near the bar, the Deity presses them toward it, and withdraws His hand whenever the workman either for use, curiosity or whim, does what appears to him to destroy the action of the bar, seems quite ridiculous and trifling; as it supposes God not only to be subservient to our inconstant wills, but also to do what would be below the dignity of any rational man to be employed about.

161. That the projectile force was at first given by the Deity is evident. For, since matter can never put itself into motion, and all bodies may be moved in any direction whatsoever; and yet all the Planets both primary and secondary move from west to

east, in planes nearly coincident; whilst the Comets move in all directions, and in planes so different from one another; these motions can be owing to no mechanical cause of necessity, but to the free choice and power of an intelligent Being.

162. Whatever Gravity be, 'tis plain that it acts every moment of time: for should its action cease, the projectile force would instantly carry off the Planets in straight lines from those parts of their Orbits where Gravity left them. But, the Planets being once put into motion, there is no occasion for any new project-ile force, unless they meet with some resistance in their Orbits; nor for any mending hand, unless they disturb one another too much by their mutual attractions.

The Planets Disturb One Another's Motion

The consequences thereof.

163. It is found that there are disturbances among the Planets in their motions, arising from their mutual attractions when they are in the same quarter of the Heavens; and that our years are not always precisely of the same length. Besides, there is reason to believe that the Moon is somewhat nearer the Earth now than she was formerly; her periodical month being shorter than it was in former ages. For, our Astronomical Tables, which in the pres-ent Age shew the times of Solar and Lunar Eclipses to great precision, do not answer so well for very ancient Eclipses. Hence it appears, that the Moon does not move in a medium void of all resistance; and therefore her projectile force being a little weak-ened, whilst there is nothing to diminish her gravity, she must be gradually approaching nearer the Earth, describing smaller and smaller Circles round it in every revolution, and finishing her Period sooner, although her absolute motion with regard to space be not so quick now as it was formerly: and therefore, she must come to the Earth at last; unless that Being, which gave her a sufficient projectile force at the beginning, adds a little more to it in due time. And, as all the Planets move in spaces full of ether and light, which are material substances, they too must meet with some resistance. And therefore, if their gravities are not

diminished, nor their projectile forces increased, they must necessarily approach nearer and nearer the Sun, and at length fall
upon and unite with him.

The World not eternal.

164. Here we have a strong philosophical argument against the
eternity of the World. For, had it existed from eternity, and been
left by the Deity to be governed by the combined actions of the
above forces or powers, generally called Laws, it had been at an
end long ago. And if it be left to them it must come to an end.
But we may be certain that it will last as long as was intended by
its Author, who ought no more to be found fault with for fram-
ing so perishable a work, than for making man mortal.

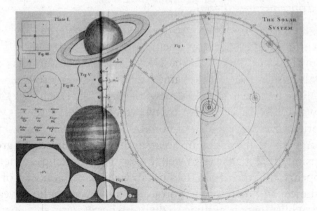

Plate I. The Solar System

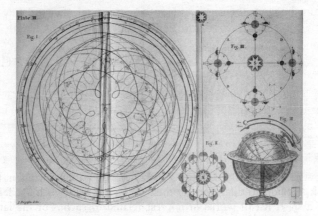

Plate III. The very irregular planetary motions,
as seen from the Earth

Plate V. The Seasons

Appendix: Other Translations and Works of Interest

Below you will find a list of many of the principal characters in this book, including translations of their relevant work (when available) and a short list of works of interest related to each. For the latter, given the broad intended audience of this book, I have focused on books when possible, rather than scholarly articles. At the end, I have also included a selection of books that focus on the Copernican Revolution more broadly.

Bellarmine, Robert and Foscarini, Paolo Antonio

TRANSLATIONS

Both Foscarini and Bellarmine's letters are translated in Blackwell, Richard J., *Galileo, Bellarmine, and the Bible* (University of Notre Dame Press, 1991).

WORKS OF INTEREST

Godman, Peter, *The Saint as Censor: Robert Bellarmine between Inquisition and Index* (Brill, 2000).

Brahe, Tycho

TRANSLATIONS

Tycho Brahe's Description of His Instruments and Scientific Work, trans. H. Ræder, E. Strömgren and B. Strömgren (Royal Danish Academy of Sciences and Letters, 1946).

Learned: Tico Brahæ His Astronomicall Coniectur of the New and Much Admired [Star] Which Appered in the Year 1572, trans. V. V. S., printed by B. A. and T. F. for Michaell [Sparke] and Samuell Nealand, 1632.

WORKS OF INTEREST

Christianson, J. R., *On Tycho's Island: Tycho Brahe and His Assistants 1570–1601* (Cambridge University Press, 2000).

Dreyer, J. L. E., *Tycho Brahe: A Picture of Scientific Life and Work in the Sixteenth Century* (Dover, 1963).

Mosley, Adam, *Bearing the Heavens: Tycho Brahe and the Astronomical Community of the Late Sixteenth Century* (Cambridge University Press, 2007).

Thoren, Victor E., *The Lord of Uraniborg: A Biography of Tycho Brahe* (Cambridge University Press, 1990).

Bruno, Giordano

TRANSLATIONS

The Ash Wednesday Supper, trans. Edward Gosselin and Lawrence Lerner (Shoe String Press, 1977).

The Ash Wednesday Supper, trans. Hilary Gatti (University of Toronto Press, 2018).

WORKS OF INTEREST

Gatti, Hilary, *Giordano Bruno and Renaissance Science* (Cornell University Press, 1999).

Michel, Paul-Henri, *The Cosmology of Giordano Bruno*, trans. R. E. W. Maddison (Cornell University Press, 1973).

Singer, Dorothea Waley, *Giordano Bruno: His Life and Thought – With Annotated Translation of His Work,* On the Infinite Universe and Worlds (Schuman, 1950).

Yates, Frances, *Giordano Bruno and the Hermetic Tradition* (University of Chicago Press, 1964).

Campanella, Tommaso

TRANSLATIONS

A Defense of Galileo, the Mathematician from Florence by Thomas Campanella, trans. Richard J. Blackwell (University of Notre Dame Press, 1994).

WORKS OF INTEREST

Headley, John M., *Tommaso Campanella and the Transformation of the World* (Princeton University Press, 1997).

Copernicus, Nicolaus

TRANSLATIONS

Three Copernican Treatises, trans. Edward Rosen, 2nd edn (Dover, 1959).

On the Revolutions, vol. 2 of *Complete Works*, trans. Edward Rosen (Johns Hopkins Press, 1978).

On the Revolutions of the Heavenly Spheres, trans. A. M. Duncan (Barnes & Noble, 1976).

WORKS OF INTEREST

Gingerich, Owen, *Copernicus: A Very Short Introduction* (Oxford University Press, 2016).

Gingerich, Owen, *The Book Nobody Read: Chasing the Revolutions of Nicolaus Copernicus* (Walker and Co., 2004).

Ravetz, Jerome R., *Astronomy and Cosmology in the Achievement of Nicolaus Copernicus* (Ossolineum, 1965).

Rosen, Edward, *Copernicus and the Scientific Revolution* (Krieger, 1984).

Sobel, Dava, *A More Perfect Heaven: How Copernicus Revolutionised the Cosmos* (Bloomsbury, 2011).

Swerdlow, N. M. and Neugebauer, O., *Mathematical Astronomy in Copernicus's* De Revolutionibus (Springer-Verlag, 1984).

Vollman, William T., *Uncentering the Earth: Copernicus and* The Revolutions of the Heavenly Spheres (W. W. Norton & Co., 2007).

Descartes, René

TRANSLATIONS

Le Monde, ou Traité de la lumière [*The World*], trans. Michael S. Mahoney (Abaris Books, 1979).

Principles of Philosophy, trans. Valentine R. Miller and Reese P. Miller (Reidel, 1983).

WORKS OF INTEREST

Garber, Daniel, *Descartes' Metaphysical Physics* (University of Chicago Press, 1992).

Gaukroger, Stephen, *Descartes: An Intellectual Biography* (Clarendon Press, 1995).

Fontenelle, Bernard Le Bovier de

TRANSLATIONS

Conversations on the Plurality of Worlds, trans. H. A. Hargreaves (University of California Press, 1990).
A Discovery of New Worlds from the French, trans. Aphra Behn (Printed for William Canning, 1688).

Galilei, Galileo

TRANSLATIONS

The Galileo Affair, trans. Maurice A. Finocchiaro (University of California Press, 1989).
The Trial of Galileo: Essential Documents, ed. and trans. Maurice A. Finocchiaro (Hackett Publishing Company, 2014).
Sidereus Nuncius, or the Sidereal Messenger, trans. Albert Van Helden (University of Chicago Press, 1989).
Dialogue Concerning the Two Chief World Systems, trans. Stillman Drake (University of California Press, 1953).
Galileo on the World Systems: A New Abridged Translation and Guide, trans. Maurice A. Finocchiaro (University of California Press, 1997).
Discoveries and Opinions of Galileo, trans. Stillman Drake (Doubleday, 1957).
The Controversy on the Comets of 1618, trans. Stillman Drake and C. D. O'Malley (University of Pennsylvania Press, 1960).

WORKS OF INTEREST

Biagioli, Mario, *Galileo, Courtier* (University of Chicago Press, 1993).
Drake, Stillman, *Galileo at Work: His Scientific Biography* (University of Chicago Press, 1978).
Drake, Stillman, *Galileo: A Very Short Introduction* (Oxford University Press, 2001).
Feldhay, Rivka, *Galileo and the Church: Political Inquisition or Critical Dialogue?* (Cambridge University Press, 1995).
Heilbron, J. L., *Galileo* (Oxford University Press, 2012).
Reeves, Eileen, *Painting the Heavens: Art and Science in the Age of Galileo* (Princeton University Press, 1997).
Wootton, David, *Galileo: Watcher of the Skies* (Yale University Press, 2013).

Gilbert, William

TRANSLATIONS

On the Magnet, trans. Silvanus P. Thompson (Chiswick Press, 1900).

WORKS OF INTEREST

Roller, Duane H. D., *The* De Magnete *of William Gilbert* (Menno Hertzberger, 1959).

Kepler, Johannes

TRANSLATIONS

Mysterium Cosmographicum: The Secret of the Universe, trans. A. M. Duncan (Abaris Books, 1981).

New Astronomy, trans. William H. Donahue (Cambridge University Press, 1992).

Kepler's Conversation with Galileo's Sidereal Messenger, trans. Edward Rosen (Johnson Reprint, 1965).

Somnium: The Dream, or Posthumous Work on Lunar Astronomy, trans. Edward Rosen (University of Wisconsin Press, 1967).

The Sidereal Messenger *of Galileo Galilei: And a Part of the Preface to Kepler's Dioptrics*, trans. Edward Stafford Carlos (Dawsons of Pall Mall, 1960).

Epitome of Copernican Astronomy [books IV and V]; The Harmonies of the World [Book V], trans. Charles Glenn Wallis. In *Great Books of the Western World*, vol. 16 (Encyclopaedia Britannica, 1955).

The Harmony of the World, trans. E. J. Aiton, A. M. Duncan and J. V. Field (American Philosophical Society, 1997).

WORKS OF INTEREST

Boner, Patrick, *Kepler's Cosmological Synthesis* (Brill, 2013).

Caspar, Max, *Kepler*, trans. C. Doris Hellman (Dover, 1993).

Chen-Morris, Raz, *Measuring Shadows: Kepler's Optics of Invisibility* (Penn State University Press, 2016).

Field, J. V., *Kepler's Geometrical Cosmology* (University of Chicago Press, 1988).

Jardine, Nicholas, *The Birth of History and Philosophy of Science: Kepler's* A Defence of Tycho against Ursus (Cambridge University Press, 1984).

Rosen, Edward, *Three Imperial Mathematicians: Kepler Trapped between Tycho Brahe and Ursus* (Abaris Books, 1986).

Rothman, Aviva, *The Pursuit of Harmony: Kepler on Cosmos, Confession, and Community* (University of Chicago Press, 2017).

Stephenson, Bruce, *Kepler's Physical Astronomy* (Springer-Verlag, 1987).

Voelkel, J. R., *The Composition of Kepler's* Astronomia Nova (Princeton University Press, 2001).

Locher, Johannes Georg

TRANSLATIONS

Mathematical Disquisitions: The Booklet of Theses Immortalized by Galileo, trans. Christopher M. Graney (University of Notre Dame Press, 2017).

Milton, John

WORKS OF INTEREST

Danielson, Dennis, Paradise Lost *and the Cosmological Revolution* (Cambridge University Press, 2014).

Newton, Isaac

TRANSLATIONS

The Principia: The Authoritative Translation – Mathematical Principles of Natural Philosophy, trans. I. Bernard Cohen and Anne Whitman (University of California Press, 1999).

The Mathematical Principles of Natural Philosophy, trans. Andrew Motte (Benjamin Motte, 1729).

WORKS OF INTEREST

Christianson, Gale E., *In the Presence of the Creator: Isaac Newton and His Times* (Free Press, 1984).

Cohen, I. B., *The Newtonian Revolution* (Cambridge University Press, 1980).

Dobbs, B. J. T. and Jacob, Margaret, *Newton and the Culture of Newtonianism* (Humanities Press, 1995).

Gleick, James, *Isaac Newton* (Pantheon, 2003).

Westfall, Richard S., *Never at Rest: A Biography of Isaac Newton* (Cambridge University Press, 1980).

Rheticus, Georg Joachim

TRANSLATIONS

Three Copernican Treatises, trans. Edward Rosen (Dover, 1959).

WORKS OF INTEREST

Danielson, Dennis, *The First Copernican; Georg Joachim Rheticus and the Rise of the Copernican Revolution* (Walker and Co., 2006).

Riccioli, Giovanni Battista

WORKS OF INTEREST

Graney, Christopher M., *Setting Aside All Authority: Giovanni Battista Riccioli and the Science against Copernicus in the Age of Galileo* (University of Notre Dame Press, 2015).

Wilkins, John

WORKS OF INTEREST

Shapiro, Barbara, *John Wilkins, 1614–1672: An Intellectual Biography* (University of California Press, 1969).

General Overviews

Aït-Touati, Frédérique, *Fictions of the Cosmos: Science and Literature in the Seventeenth Century* (University of Chicago Press, 2011).

Blumenberg, Hans, *The Genesis of the Copernican World*, trans. Robert M. Wallace (MIT Press, 1987).

Cohen, I. Bernard, *The Birth of a New Physics* (W. W. Norton & Co., 1985).

Crowe, Michael J., *Theories of the World from Antiquity to the Copernican Revolution* (Dover, 1990).

Dijksterhuis, E. J., *The Mechanization of the World Picture*, trans. C. Dikshoorn (Oxford University Press, 1961).

Dreyer, J. L. E., *A History of Astronomy from Thales to Kepler*, 2nd edn (Dover, 1953).

Duhem, Pierre, *To Save the Phenomena: An Essay on the Idea of Physical Theory from Plato to Galileo*, trans. Edmund Dolan and Chaninah Maschler (University of Chicago Press, 1969).

Finocchiaro, Maurice A., *The Copernican Revolution and the Galileo Affair: The Blackwell Companion to Science and Christianity* (Wiley-Blackwell, 2012).

Funkenstein, Amos, *Theology and the Scientific Imagination from the Middle Ages to the Seventeenth Century* (Princeton University Press, 1986).

Gingerich, Owen, *The Eye of Heaven: Ptolemy, Copernicus, Kepler* (American Institute of Physics, 1993).

Grant, Edward, *Much Ado about Nothing: Theories of Space and Vacuum from the Middle Ages to the Scientific Revolution* (Cambridge University Press, 1981).

Grant, Edward, *Planets, Orbs, and Spheres: The Medieval Cosmos, 1280–1687* (Cambridge University Press, 1994).

Hallyn, Fernand, *The Poetic Structure of the World: Copernicus and Kepler*, trans. Donald M. Leslie (Zone Books, 1990).

Heilbron, John L., *The Sun in the Church* (Harvard University Press, 1999).

Hirshfeld, Alan, *Parallax: The Race to Measure the Cosmos* (W. H. Freeman, 2001).

Hoskin, Michael, *The History of Astronomy: A Very Short Introduction* (Oxford University Press, 2003).

Howell, Kenneth, *God's Two Books: Copernican Cosmology and Biblical Interpretation in Early Modern Society* (University of Notre Dame Press, 2002).

Kanas, Nick, *Star Maps: History, Artistry, and Cartography* (Springer Praxis, 2007).

Koestler, Arthur, *The Sleepwalkers: A History of Man's Changing Vision of the Cosmos* (Penguin Books, 1990).

Koyré, Alexandre, *From the Closed World to the Infinite Universe* (Johns Hopkins Press, 1957).

Kuhn, Thomas S., *The Copernican Revolution* (Harvard University Press, 1957).

Lattis, James M., *Between Copernicus and Galileo: Christoph Clavius and the Collapse of Ptolemaic Astronomy* (University of Chicago Press, 1994).

Lindberg, David C., *The Beginnings of Western Science* (University of Chicago Press, 2007).

Miller, Daniel Marshall, *Representing Space in the Scientific Revolution* (Cambridge University Press, 2014).

Moss, Jean Dietz, *Novelties in the Heavens: Rhetoric and Science in the Copernican Controversy* (University of Chicago Press, 1993).

Nicolson, Marjorie, *The Breaking of the Circle. Studies in the Effect of the 'New Science' upon Seventeenth Century Poetry* (Northwestern University Press, 1950).

Omodeo, Pietro Daniel, *Copernicus in the Cultural Debates of the Renaissance: Reception, Transformation* (Brill, 2014).

Pedersen, Olaf, *Early Physics and Astronomy: A Historical Introduction* (Cambridge University Press, 1993).

Saliba, George, *A History of Arabic Astronomy: Planetary Theories during the Golden Age of Islam* (NYU Press, 1995).

Taub, Liba Chaia. *Ptolemy's Universe: The Natural Philosophical and Ethical Foundations of Ptolemy's Astronomy* (Open Court, 1993).

Thorndike, Lynn, *The Sphere of Sacrobosco and Its Commentators* (University of Chicago Press, 1949).

Timberlake, Todd and Wallace, Paul, *Finding Our Place in the Solar System: The Scientific Story* (Cambridge University Press, 2019).

Toulmin, Stephen and Goodfield, June, *The Fabric of the Heavens: The Development of Astronomy and Dynamics* (University of Chicago Press, 1999).

Van Helden, Albert, 'The invention of the telescope', *Transactions of the American Philosophical Society* 67 (4), 1977.

Van Helden, Albert, *Measuring the Universe: Cosmic Dimensions from Aristarchus to Halley* (University of Chicago Press, 1985).

Vermij, Rienk, *The Calvinist Copernicans: The Reception of the New Astronomy in the Dutch Republic, 1575–1750* (Royal Netherlands Academy of Arts and Sciences, 2002).

Westman, Robert, *The Copernican Question: Prognostication, Skepticism, and Celestial Order* (University of California Press, 2011).

Index

Page references in *italics* indicate images.

CITY OF GOD

Saint Augustine

'The Heavenly City outshines Rome, beyond comparison. There, instead of victory, is truth; instead of high ranks, holiness'

St Augustine, bishop of Hippo, is one of the central figures in the history of Christianity, and *City of God* is one of his greatest theological works. Written as an eloquent defence of the faith at a time when the Roman Empire was on the brink of collapse, it examines the ancient pagan religions of Rome, the arguments of the Greek philosophers and the revelations of the Bible. Pointing the way forward to citizenship that transcends worldly politics and will last for eternity, *City of God* is one of the most influential documents in the development of Christianity.

Translated with notes by Henry Bettenson
with an introduction by G. R. Evans

ISBN: 978 0 14 044 894 8

CANDIDE, OR OPTIMISM

Voltaire

'You have been to England ... Are they as mad as in France?'

Brought up in the household of a German Baron, Candide is an open-minded young man whose tutor, Pangloss, has instilled in him the belief that 'all is for the best'. But when his love for the Baron's rosy-cheeked daughter is discovered, Candide is cast out to make his own fortune. As he and his various companions roam over the world, an outrageous series of disasters befall them – earthquakes, syphillis, a brush with the Inquisition, murder – sorely testing the young hero's optimism. In *Candide*, Voltaire threw down an audacious challenge to the philosophical views of his time, to create one of the most glorious satires of the eighteenth century.

Translated by Theo Cuffe
with an Introduction by Michael Wood

ISBN: 978 0 14 045 510 6

CRITIQUE OF PURE REASON

Immanuel Kant

'The purpose of this critique of pure speculative reason consists
in the attempt to change the old procedure of metaphysics and to
bring about a complete revolution'

Kant's *Critique of Pure Reason* (1781) is the central text of modern
philosophy. It presents a profound and challenging investigation
into the nature of human reason, its knowledge and its illusions.
Reason, Kant argues, is the seat of certain concepts that precede
experience and make it possible, but we are not therefore entitled
to draw conclusions about the natural world from these concepts.
The *Critique* brings together two opposing schools of philosophy:
rationalism, which grounds all our knowledge in reason, and
empiricism, which traces all our knowledge to experience. Kant's
transcendental idealism indicates a third way that goes beyond
these alternatives.

Translated, edited and with an introduction by Marcus Weigelt
Based on the translation by Max Müller

ISBN: 978 0 14 044 747 7

THE COMPLETE ENGLISH POEMS

John Donne

'The heavens rejoice in motion, why should I Abjure my so much
loved variety?'

No poet has been more wilfully contradictory than John Donne,
whose works forge unforgettable connections between extremes
of passion and mental energy. From satire to tender elegy, from
sacred devotion to lust, he conveys an astonishing range of emotions and poetic moods. Constant in his work, however, is an
intensity of feeling and expression and complexity of argument
that is as evident in religious meditations such as 'Good Friday
1613. Riding Westward' as it is in secular love poems such as 'The
Sun Rising' or 'The Flea'. 'The intricacy and subtlety of his imagination are the length and depth of the furrow made by his passion,'
wrote Yeats, pinpointing the unique genius of a poet who combined
ardour and intellect in equal measure.

Edited by A. J. Smith

ISBN: 978 0 14 042 209 2

PARADISE LOST

John Milton

'Better to reign in Hell, than serve in Heav'n'

In *Paradise Lost* John Milton produced a poem of epic scale, conjuring up a vast, awe-inspiring cosmos and ranging across huge tracts of space and time. And yet, in putting a charismatic Satan and naked Adam and Eve at the centre of this story, he also created an intensely human tragedy on the Fall of Man. Written when Milton was in his fifties – blind; bitterly disappointed by the Restoration and briefly in danger of execution – *Paradise Lost's* apparent ambivalence towards authority has led to immense debate about whether it manages to 'justify the ways of God to men' or exposes the cruelty of Christianity.

Edited with Notes and an Introduction by John Leonard

ISBN: 978 0 14 042 439 3

INFERNO

Dante

'And nothing, where I now arrive, is shining'

Dante's *Inferno* describes his descent into Hell midway through his life with the Roman Virgil as his guide, and is unparalleled in its depiction of the tragedy of sin. It is a work inspired by a profound confidence in human nature, yet also expresses Dante's horror at the way individuals can destroy themselves and each other, creating Hell on Earth. A response to the violent society of thirteenth-century Italy, the *Inferno* reveals the eternal punishment reserved for sins such as greed, self-deception, political double-dealing and treachery. Portraying a huge diversity of characters culminating in an horrific vision of Satan, it broke new ground in the vigour of its language and its storytelling. It has had a particular influence on Modernist writers and their successors throughout the world.

Translated and edited with an introduction, commentary and notes by Robin Kirkpatrick

ISBN: 978 0 14 044 895 5

DISCOURSE ON METHOD AND THE MEDITATIONS

René Descartes

'I think, therefore I am'

Widely regarded as the founder of modern western philosophy, René Descartes (1596–1650) sought to look beyond the established tenets of earlier thinkers and establish a philosophical system more appropriate for the scientific age in which he lived: one based on the power of reason. In the *Discourse*, he details the famous 'Method' by which he strove to develop this system – beginning with the concept of universal doubt and going on to deduce, from the celebrated phrase *cogito ergo sum*, both his own existence and that of God. The *Meditations* is a penetrating exploration of his beliefs, which considers in depth the nature of God, the human soul, and the concept of truth and error. Profoundly influential and original, both works have been the source of outrage and inspiration for generations of readers and philosophers as varied as Leibniz, Hume, Locke and Kant.

Translated with an introduction by F. E. Sutcliffe

ISBN: 978 0 14 044 206 9